RECENT DEVELOPMENTS IN
CONDENSED MATTER PHYSICS

Volume 3 • Impurities, Excitons, Polarons, and Polaritons

Edited by
J. T. DEVREESE
Chairman of the Conference
University of Antwerpen (RUCA and UIA)

L. F. LEMMENS
University of Antwerpen (RUCA)

V. E. VAN DOREN
University of Antwerpen (RUCA)

and

J. VAN ROYEN
University of Antwerpen (UIA)

PLENUM PRESS • NEW YORK AND LONDON

Library of Congress Cataloging in Publication Data

Main entry under title:

Recent developments in condensed matter physics.

". . . papers presented at the first general conference of the Condensed Matter Division of the European Physical Society, held April 9-11, 1980, at the University of Antwerp (RUCA and UIA), Antwerp, Belgium."
  Includes indexes.
  Contents: v. 1. Invited papers — v. 2. Metals, disordered systems, surfaces, and interfaces — v. 3. Impurities, excitons, polarons, and polaritons — [etc.]
  1. Condensed matter — Congresses. I. Devreese, J. T. (Jozef T.) II. European Physical Society. Condensed Matter Division.
QC176.A1R42   530.4'1   80-28067
ISBN 0-306-40648-9   AACR2

Contributed papers presented at the first General Conference of the Condensed Matter Division of the European Physical Society, held April 9-11, 1980, at the University of Antwerp (RUCA and UIA), Antwerp, Belgium

© 1981 Plenum Press, New York
A Division of Plenum Publishing Corporation
233 Spring Street, New York, N. Y. 10013

CONFERENCE CHAIRMAN

J. T. DEVREESE, RUCA & UIA, Antwerpen

LOCAL COMMITTEE

L. F. LEMMENS, RUCA, Antwerpen
V. E. VAN DOREN, RUCA, Antwerpen
J. VAN ROYEN, UIA, Antwerpen

INTERNATIONAL ADVISORY COMMITTEE

M. BALKANSKI, Paris, France
A. ABRIKOSOV, Moscow, USSR
V. M. AGRANOVITCH, Moscow, USSR
P. AVERBUCH, Grenoble, France
G. BENEDEK, Milan, Italy
H. B. CASIMIR, Heeze, The Netherlands
B. R. COLES, London, UK
S. R. DE GROOT, Amsterdam, The Netherlands
A. J. FORTY, Warwick, UK
D. FRÖHLICH, Dortmund, FRG
A. FROVA, Rome, Italy
H. GRIMMEISS, Lund, Sweden
H. HAKEN, Stuttgart, FRG
C. HILSUM, Malvern, UK

W. J. HUISKAMP, Leiden, The Netherlands
G. M. KALVIUS, Munich, FRG
A. R. MACKINTOSH, Copenhagen, Denmark
N. H. MARCH, Oxford, UK
E. MOOSER, Lausanne, Switzerland
N. F. MOTT, Cambridge, UK
K. A. MÜLLER, Zürich, Switzerland
S. NIKITINE, Strasbourg, France
C. J. TODD, Ipswich, UK
M. VOOS, Paris, France
E. P. WOHLFARTH, London, UK
E. WOLF, Stuttgart, FRG
P. WYDER, Nijmegen, The Netherlands
H. R. ZELLER, Baden, Switzerland

INTERNATIONAL PROGRAM COMMITTEE

N. N. BOGOLUBOV, Moscow, USSR
J. BOK, Paris, France
M. CARDONA, Stuttgart, FRG
E. COURTENS, Rüschlikon, Switzerland
S. F. EDWARDS, Cambridge, UK
R. ELLIOTT, Oxford, UK
J. FRIEDEL, Orsay, France
H. FRÖHLICH, Liverpool, UK
F. GARCIA MOLINER, Madrid, Spain
G. HARBEKE, Zürich, Switzerland
H. R. KIRCHMAYR, Vienna, Austria
S. LUNDQVIST, Göteborg, Sweden
S. METHFESSEL, Bochum, FRG

E. MITCHELL, Oxford, UK
F. MUELLER, Nijmegen, The Netherlands
R. PEIERLS, Oxford, UK
H. J. QUEISSER, Stuttgart, FRG
D. SETTE, Rome, Italy
H. THOMAS, Basel, Switzerland
M. TOSI, Trieste, Italy
F. VAN DER MAESEN, Eindhoven, The Netherlands
J. ZAK, Haifa, Israel
A. ZAWADOWSKI, Budapest, Hungary
W. ZAWADZKI, Warsaw, Poland
W. ZINN, Jülich, FRG
A. ZYLBERSZTEJN, Orsay, France

NATIONAL ADVISORY COMMITTEE

S. AMELINCKX, SCK/CEN, Mol
F. CARDON, RUG, Gent
R. EVRARD, ULg, Liège
R. GEVERS, RUCA, Antwerpen
J. P. ISSI, UCL, Louvain-la-Neuve
L. LAUDE, UEM, Mons
A. LUCAS, FUN, Namur

K. H. MICHEL, UIA, Antwerpen
J. NIHOUL, SCK/CEN, Mol
J. PIRENNE, ULg, Liège
D. SCHOEMAKER, UIA, Antwerpen
L. STALS, LUC, Diepenbeek
R. VAN GEEN, VUB, Brussels
L. VAN GERVEN, KUL, Leuven

INTRODUCTION

These volumes contain the invited and contributed talks of the first general Conference of the Condensed Matter Division of the European Physical Society, which took place at the campus of the University of Antwerpen (Universitaire Instelling Antwerpen) from April 9 till 11, 1980.

The invited talks give a broad perspective of the current state in Europe of research in condensed matter physics. New developments and advances in experiments as well as theory are reported for 28 topics. Some of these developments, such as the recent stabilization of mono-atomic hydrogen, with the challenging prospect of Bose condensation, can be considered as major breakthroughs in condensed matter physics.

Of the 65 invited lecturers, 54 have submitted a manuscript. The remaining talks are published as abstracts.

The contents of this first volume consists of 9 plenary papers. Among the topics treated in these papers are:
- electronic structure computations of iron
- the density functional theory
- hydrogen in amorphous Si
- topologically disordered materials
- nuclear antiferromagnetism
- stabilization of mono-atomic hydrogen gas
- covalent and metallic glasses
- nonlinear excitations in ferroelectrics.

The other 56 papers of the first volume are divided among 17 symposia and 12 sessions. The different topics treated are:
- localization and disorder
- metals and alloys
- fluids
- excitons and electron-hole droplets
- surface physics
- dielectric properties of metals
- experimental techniques

- electronic properties of semiconductors
- low-dimensional systems
- defects and impurities
- spin waves and magnetism
- phase changes
- superionic conductors
- dielectric properties
- polarons
- molecular crystals
- superconductivity
- spin glasses
- photo-emission
- polaritons and electron-phonon interaction

A review of these invited talks is given in the Closing Address.

Volumes 2, 3 and 4 contain the contributed papers of the participants to this Conference. All the 170 contributions deal with a wide range of topics including the same subjects reviewed in the invited papers.

The conference itself was organized in collaboration and with the financial support of the University of Antwerpen (Rijksuniversitair Centrum Antwerpen and Universitaire Instelling Antwerpen), the Belgian National Science Foundation and Control Data Corporation, Belgium. Co-sponsors were: Agfa-Gevaert N.V., Bell Telephone Mfg. Co., Coherent, Esso Belgium, I.B.M., Interlaboratoire N.V., I. Komkommer, Labofina, Metallurgie Hoboken-Overpelt, Spectra Physics.

Previous meetings of the Condensed Matter Division of the European Physical Society emphasized topical conferences treating "Metals and Phase Transitions" (Firenze, 1971), "Dielectrics and Phonons" (Budapest, 1974), "Molecular Solids and Electronic Transport" (Leeds, 1977). The promising start of 1971, however, did not evolve into a continuous success, and only 80 participants attended the 1977 meeting.

When the Board of the Condensed Matter Division invited me to act as Chairman for this Conference, several reasons could have been invoked for feeling reluctant about accepting this invitation. It was nevertheless decided to undertake this task, because we were convinced that there exists a genuine need in the European Physical Community for an international forum similar to the successful "March Meeting" of the American Physical Society. This March Meeting is not organized as a meeting for the physical societies of the different states but for all physicists and laboratories individually.

At the early stages of the organization of this Conference, about 800 solid state physicists were members of the Condensed Matter Division. However, as a result of an analysis which was made in Antwerpen, it soon became clear that in Europe not less than 8000 solid state physicists are active in several fields of research.

Therefore as a first step announcements were sent to all these physicists individually. The large response to these announcements, as reflected in the presence of about 600 solid state physicists at this Conference, proves that the need for such a forum indeed is present in the European physics community and that the time is ripe for the organization of an annual conference.

Subsequently, the International Advisory and Program Committees of this Conference were formed with great care. A large number of distinguished physicists were invited to serve as members of these committees. They represented not only the different fields and sections of condensed matter physics but also the different member states of the European Physical Society.

It was also the task of both the International Advisory and Program Committees to guide in establishing the conference program. As a result of a first consultation a large number of topics and speakers were suggested to be incorporated in the program. In order to make a fair and balanced choice, a ballot with all the suggestions was sent out to the Committee members. The result led to a first selection of speakers and topics. This number was gradually supplemented with names and topics resulting from private consultations to a total of 28 topics and 65 invited speakers. The final program consisted of 9 plenary sessions, 11 symposia each with 3 or 4 invited speakers and a maximum of 8 parallel sessions, in which 320 contributed papers were presented. Among the invited and contributed papers, there were several contributions from the U.S.A. Also scientists from Japan and China gave invited and contributed talks.

I should like to thank the invited speakers for their collaboration in preparing the manuscripts of their talks. This volume together with the volumes of the proceedings of the contributed papers will give the scientific community a review of the state of affairs of condensed matter physics in Europe together with several new developments from the U.S.A., Japan and China. It gives a report on the most recent advances in theory as well as experiment. Publication of these proceedings will serve as a guide in the years to come not only to the participants of this conference but to the scientific community at large.

It gives me great pleasure to thank the President of the Universitaire Instelling Antwerpen, Dr. jur. P. Van Remoortere and the Rector, Prof. Dr. R. Clara, for their continuous interest and

support for this Conference and for making available so promptly
and effectively the whole infrastructure of the university conference
center.

Finally, I wish to thank Miss R.M. Vandekerkhove, the administrative secretary of the Conference. My thanks are also due to Miss
H. Evans and Mr. M. De Moor for their administrative and technical
assistance in the organization, and to all those who helped to ensure a smooth operation of the Conference.

J.T. Devreese
Professor of Theoretical Physics
Chairman of the Conference

CONTENTS

1. SEMICONDUCTORS AND SEMIMETALS

3.   EXCITONS AND ELECTRON-HOLE DROPLETS

5. DIELECTRIC PROPERTIES

6. POLARONS AND POLARITONS

TEMPERATURE DEPENDENCE OF p-TYPE DOPING IN BISMUTH

O.P. Hansen
Universitetets Fysiske Laboratorium I
H.C. Ørsted Institutet
2100 Copenhagen Ø, Denmark

and

J. Heremans
Université Catholique de Louvain, Laboratoire PCES,
Place Croix du Sud, B-1348, Louvain-la-Neuve, Belgium

Four tin-doped bismuth samples have been investigated, with hole concentrations at 4.2 K corresponding to three particular positions of the Fermi level with respect to the L-point band structure : in the conduction band, in the middle of the energy gap and in the hole band. The Hall effect and the transverse magnetoresistance are reported for fields up to 6 T and from 4.2 to 300 K.

The motivation for the present investigation originated in the paper by Boxus et al[1] in which electrical resistivity, thermal conductivity and thermopower of sixteen well characterized tin-doped bismuth samples were reported in the temperature range 2-300 K. Tin acts as an acceptor in bismuth and it is accordingly expected that, for example, the thermopower will be positive for sufficiently high dope concentration. This was indeed found by Boxus et al, but at low temperatures only. Around room temperature all samples had a negative thermopower, approaching the one of pure bismuth. This feature provoked us to make Hall effect measurements from liquid helium temperature to room temperature in magnetic fields up to 6 tesla, aiming at a determination of the excess carrier density through the expected saturation of the Hall effect in strong fields. For the sake of completeness we simultaneously measured the magnetoresistance.

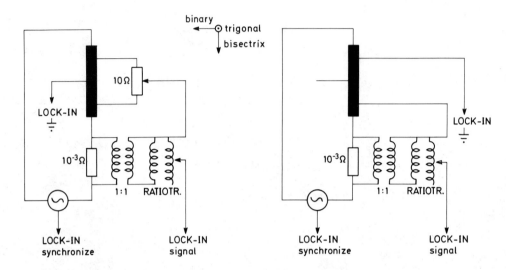

Fig.1: Main characteristics of experimental set up.
        Left: for Hall effect. Right: for magnetoresistance.

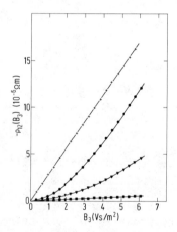

Fig.2: Hall element $\rho 12$ as function of B_3 at temperatures
        25.1 K (+), 191 (⊕), 231 K(▼) and 285 (■).

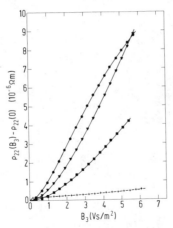

Fig.3 : Resistance element ρ_{22} as function of B_3. Temperatures as in figure 1.

Figures 2 and 3 display the results as function of field at a few selected temperatures for Bi 45'a. The appearance as points arises from a numerical treatment of the continuous recordings, taking averages of deflections for field directions plus and minus. The features seen in figures 2 and 3 are characteristic for all four doped samples : At low temperatures ρ_{12} is linear in the field, but this linear behaviour ceases as temperature is increased. For ρ_{22} the picture looks more complicated, showing crossing, s-shaped curves.

Data for all samples are shown in figures 4 and 5 as function of temperature at the maximum field of 6 tesla. Hereby a simple relation between ρ_{12} and ρ_{22} for the doped samples is revealed. Roughly, ρ_{22} has a maximum when ρ_{12} has dropped to half its low temperature plateau. At temperatures above the maxima, ρ_{22} for the doped samples come close to the pure sample behaviour. The temperatures at which the maxima occur seem to be related to the thermopower measurements of Boxus et al. Thus, above these temperatures the thermopower of the doped samples come close to that of pure bismuth.

The four doped samples to be reported on were spark cut into rectangular rods either directly from the samples of Boxus et al or from the same sample batches. For the history of these (they were grown by Noothoven van Goor at the Philips Research Laboratories, Eindhoven) see reference 1. The main characteristics of the actual samples are summarised in table 1. They were all cut with a bisectrix axis parallel to their long dimension. Further, holes for potential probes were spark cut on the samples. These

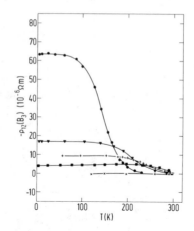

Fig.4 : 6-tesla values of $\rho_{12}(B_3)$ for all samples as function of
        temperature. Sample notations : 67a (●), 45'a (▼), 57'a
        (+), 52a (■), 13-pure (×).

Table 1 : Sample characteristics. Sample numbers for doped samples
          follow those of Boxus et al[1]. $\ell_\perp$ and $\ell_\parallel$: distances
          between Hall probes and resistance probes respectively,
          A : cross sectional area, RRR : residual resistance
          ratio $R_{300}/R_{4.2}$, P-N : excess carrier density calculated
          as P-N = $-B_3/\rho_{12}(B_3)e$ from low temperature plateau of
          $\rho_{12}(B_3)$ at B_3 = 6 tesla, at.% tin : calculated from
          P-N assuming tin to be a monovalent acceptor. The 4.2 K
          Fermi level positions calculated from P-N are : in the
          conduction band (67a), in the middle of the energy gap
          (45'a), and in the hole band (57'a, 52a).

Sample number	$\ell_\perp$	$\ell_\parallel$	A	RRR	P-N at 6T	at.% tin
	mm	mm	mm^2		10^{23}m^{-3}	
67a	2.15	8.01	4.58	4.29	5.89	0.00209
45'a	2.58	10.01	7.81	1.35	22.1	0.00784
57'a	1.92	5.00	3.53		40.2	0.0143
52a	2.22	5.99	4.88	4.10	91.9	0.0326
13-pure	2.15	8.99	4.40	152		

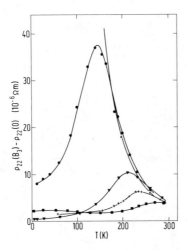

Fig.5 : 6-tesla values of $\rho_{22}(B_3)$ for all samples as function of
temperature. Sample notations as in figure 4.

holes had depth and diameters of respectively 0.10 mm and 0.15 mm.
Care was taken to place them such that the voltages would be
measured along the bisectrix and the binary directions. The
potential probes were copper wires spot welded into the holes, while
current leads were soft soldered to the sample ends.

        The main characteristics of the experimental set up are
shown in figure 1. The sample (shown in dark) was part of a low-
frequency bridge-actually 72.5 Hz was used. It was placed horizon-
tally in the bore of a superconducting magnet giving a vertical
field. The bridge was balanced in zero-field and the magnet
subsequently charged and discharged via a 10 minute time ramp
of the power supply. During this the out-of-balance deflection
on the lock-in detector was recorderd along with the magnet
current on an X-Y recorder. All recordings were made for plus
as well as minus directions of the magnetic field.

REFERENCE

1. J. Boxus, J. Heremans, J.P. Michenaud and J.P. Issi,
   J. Phys. F : Metal Phys. 9, 2387 (1979).

ELECTRON TRANSPORT IN BISMUTH IN NON-QUANTISING

MAGNETIC FIELDS: PSEUDO-PARABOLIC MODEL

O.P. Hansen and I.F.I. Mikhail

Universitetets Fysiske Laboratorium I
H.C. Ørsted Institutet
DK-2100 Copenhagen Ø, Denmark
and
Faculty of Science
A'in Shams University, Cairo, Egypt

The pseudo-parabolic model introduced recently by Heremans and Hansen is analysed further by calculating within this model the effect of strong, non-quantising magnetic fields on electron transport properties. Comparison with experimental data in the range from 20 K to room temperature shows that a variety of complicated field dependences of the thermoelectric power in the intermediate field range, previously unexplained, follow the present analysis. Disagreements in strong fields set in at field values which coincide quite well with the quantum limits.

In a recent paper by Heremans and Hansen[1] a so-called pseudo-parabolic model was introduced to analyse the thermo-electric power of bismuth. The basic features of the model are the Lax non-parabolic dispersion relation[2]

$$\gamma = \varepsilon(1+\varepsilon/\varepsilon_G) = \frac{1}{2}\,\bar{p}\,\bar{\bar{m}}^{-1}\,\bar{p} \tag{1}$$

and the energy dependence of the squared electron-phonon matrix element

$$M^2_{el-ph} \propto \gamma'^{-2} \tag{2}$$

where ε_G is the energy gap, $\bar{p}$ is the momentum $\bar{\bar{m}}^{-1}$ is the inverse mass tensor at the band edge, and the prime on γ means differentiation with respect to energy ε. As the pseudo-parabolic model does not pretend to explain the temperature dependence, nor

the absolute values, of the mobilities, we shall use the
experimental values of these whenever needed. The model does
however presume a specific scattering mechanism, namely the
acoustic intravalley described by a relaxation time which reflects
the density of states and the squared ma trixelement given in (2)

$$\bar{\bar{\tau}} = \bar{\bar{\tau}}_o \left(\frac{\gamma}{kT}\right)^{-1/2} \gamma' \ . \tag{3}$$

We shall leave out of consideration phonon drag and
magnetic quantisation effects. As the measurements by Uher and
Pratt[3] seem to place the region for the dominating phonon drag
below approximately 10K we shall consider this as a lower
temperature limit for the validity of our analysis. The neglect
of magnetic quantisation effects means that the results are
incomplete at field values, for which the spacing between the
Landau levels is larger than the thermal energy.

The transport tensors for a single ellipsoid, cf.(1),
are calculated from the Boltzmann equation in the usual way. The
following notations are used

$$\bar{J} = \bar{\bar{\sigma}}_s(\bar{B})\bar{E} - \bar{\bar{\theta}}_s(\bar{B})\nabla_{\!r}T \tag{4}$$

$$\bar{Q} = T\bar{\bar{\theta}}_s(\bar{B})\bar{E} - \bar{\bar{\xi}}_s(\bar{B})\nabla_{\!r}T \tag{5}$$

where $\bar{J}$, $\bar{Q}$, $\bar{E}$ and $\nabla_{\!r}T$ denote electric current density, heat
current density, electric field and temperature gradient, and $\bar{B}$
denote the magnetic field. The transport tensors are expressed
in the following way

$$\bar{\bar{\sigma}}_s(\bar{B}) = n_s e [\Gamma^{(o)}\bar{\bar{M}}_s + \Gamma^{(1)}\bar{\bar{M}}_s \bar{\bar{B}} \bar{\bar{M}}_s + \Gamma^{(2)} \det(\bar{\bar{M}}_s)\bar{\bar{B}}\bar{\bar{B}}] \tag{6}$$

$$\bar{\bar{\theta}}_s(\bar{B}) = \mp k n_s [\lambda^{(o)}\bar{\bar{M}}_s + \lambda^{(1)}\bar{\bar{M}}_s \bar{\bar{B}} \bar{\bar{M}}_s + \lambda^{(2)} \det(\bar{\bar{M}}_s)\bar{\bar{B}}\bar{\bar{B}}] \tag{7}$$

where Γ's and λ's are energy integrals which depend on the
magnitude, but not on the direction, of the magnetic field,
where $\bar{\bar{M}}$ is the mobility tensor, and where the upper sign is for
electrons, the lower sign for holes.

The total transport tensors are obtained by summing the
contribution from the four groups of carriers, three with
electrons and one with holes, so that

$$\bar{\bar{\sigma}}(\bar{B}) = \Sigma_s \bar{\bar{\sigma}}_s(\bar{B}) \quad \text{and} \quad \bar{\bar{\theta}}(\bar{B}) = \Sigma_s \bar{\bar{\theta}}_s(\bar{B}) \ . \tag{8}$$

The final results can always be expressed in terms of the mobility
tensors of one of the electron ellipsoids $(\bar{\bar{\mu}})$ and of the hole
ellipsoid $(\bar{\bar{\nu}})$. The electron ellipsoid, whose mobility tensor is
usually taken as a reference, is the one with a principal axis

parallel to a binary direction in the common crystallographic frame. Thus, in this frame the two tensors $\bar{\bar{\mu}}$ and $\bar{\bar{\nu}}$ are given according to Hartmann's[4] notations by

$$\bar{\bar{M}}_{electron} = \bar{\bar{\mu}} = \begin{pmatrix} \mu_1 & 0 & 0 \\ 0 & \mu_2 & \mu_4 \\ 0 & \mu_4 & \mu_3 \end{pmatrix} \tag{9}$$

and :

$$\bar{\bar{M}}_{hole} = \bar{\bar{\nu}} = \begin{pmatrix} \nu_1 & 0 & 0 \\ 0 & \nu_1 & 0 \\ 0 & 0 & \nu_3 \end{pmatrix} \tag{10}$$

The densities of electrons and holes are assumed to be equal and are denoted by N :

$$N = 3n_{electron} = n_{hole} \quad . \tag{11}$$

The input data required for the calculations are the Fermi energies, the L-point energy gap, and the mobilities and carrier density of electrons and holes. At low temperatures (T < 80 K) the Fermi energies are assumed to have the values $\varepsilon_F^e = 27.2$ meV and $\varepsilon_F^h = 11.2$ meV. At high temperatures (T > 80 K) the Fermi energies are taken from Heremans and Hansen[1], table 3 (electrons) and from Hansen et al[5], table 2 (holes). They are shown in figure 1 along with the L-point energy gap. For the latter we use the data of Vecchi and Dresselhaus[6].

The mobilities and the carrier density are obtained from least squares minimisation fitting to the experimental values of the weak-field coefficients of either conductivity or resistivity. For T < 80 K we consider the Bethe-Sommerfeld approximation to be sufficiently accurate, and we use the mobility values obtained by Hartman[4] over the range 7.78 to 15.7 K. As has previously been done[1,5] the T^{-2} dependence of these mobilities is extrapolated up to 80 K. As regards ν_3, the uncertainty on this quantity is by far the largest (40 %) and we have therefore used the relation $\nu_3 = \nu_1 m_1^h/m_3^h$, which is equivalent to the assumption of a scalar relaxation time for the holes. For T > 80 K we have made a re-evaluation of the weak-field galvanomagnetic data of Michenaud and Issi[7], using the integral expressions for the Γ's instead of the Bethe-Sommerfeld approximation. The result of the re-evaluation is given in table 1.

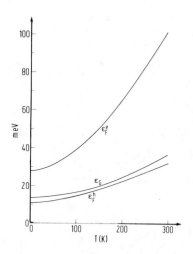

<u>Fig.1:</u> Fermi energies at L- and T-point[1,5] and energy gap at
L-point[6].

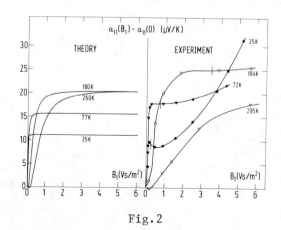

Fig.2

<u>Figs.2-4:</u> (a) Elements of $\bar{\bar{\alpha}}(\bar{B}) - \bar{\bar{\alpha}}(0)$ calculated from equations
(12), (8), (7), (6).   (b) Experimental values from the
following references: Fig.2, ref.8; Fig.3, ref.12;
Fig.4, ref.8.

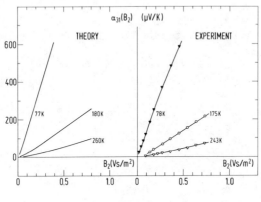

Fig. 3

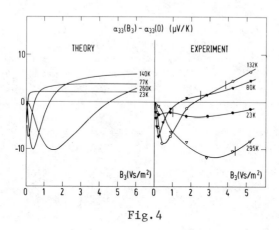

Fig. 4

The thermoelectric power was calculated from the relation

$$\bar{\bar{\alpha}}(\bar{B}) = \bar{\bar{\sigma}}^{-1}(\bar{B})\bar{\bar{\theta}}(\bar{B}) \tag{12}$$

The results are presented in figures 2a to 4a along with their experimental counterparts in figures 2b to 4b. The latter only form part of the results in the references listed. Thus we have chosen for each element and field direction a maximum of four temperatures distributed between 20 K and room temperature.

In most cases the agreement between the experimental and theoretical results is quantitatively good as long as the magnetic field does not exceed the relevant quantum limits, B_L. These limits are marked by vertical bars on the experimental curves. They were calculated from

$$B_L = kT\gamma' m_{\bar{b}}/\hbar e , \tag{13}$$

$$m_{\bar{b}} = (\det \bar{\bar{m}}/\bar{b} \; \bar{\bar{m}} \; \bar{b})^{1/2} , \quad \bar{b} = \bar{B}/|B| \tag{14}$$

taking into account the temperature dependence of the mass elements, of the Fermi energy, and of the energy gap.

As regards the range of fields below the quantum limits we first note that a variety of complicated magnetic field dependences in the intermediate field range, previously un-explained, follow from the present analysis. Second, that the diffusion theory used by Uher and Goldsmid[8] resulted in a marked disagreement at low to intermediate fields. However, the latter theory was based on the assumption of a scalar and magnetic field independent partial thermoelectric power, and, as shown by Cheruvier and Hansen[9], this assumption will be valid only if the relaxation time is independent of energy and the dispersion is parabolic.

Investigations of the Wiedemann-Franz law were performed by Korenblit et al[10] and by Uher and Goldsmid[11] through measurements of the change of thermal conductivity with magnetic field. In both investigations the Lorenz number was found to decrease markedly below 40 K. Here we present the predictions, based on the pseudo-parabolic model, which pertain to the measurement of Uher and Goldsmid[11]. In these measurements the electronic contribution to the thermal conductivity was retrieved form experimental results for $\lambda_{11}(B_2)$ and $\lambda_{11}(B_3)$ in strong magnetic fields, λ denoting the total thermal conductivity.

The electronic contribution λ^E is expressed in terms of the transport tensors in eqs.(4), (5) and (12) as

Table 1 : Re-evaluated average mobility and density values in
units of 10^4 $cm^2V^{-1}s^{-1}$ and 10^{17} cm^{-3} respectively.
Temperature T in kelvin.

T	μ_1	μ_2	μ_3	μ_4	ν_1	N
77	59.1	1.12	32.7	−3.88	9.38	5.24
100	37.8	0.746	19.7	−2.58	6.11	6.60
120	25.2	0.516	13.9	−1.74	4.17	8.31
140	18.0	0.378	10.1	−1.34	3.03	9.85
180	9.96	0.242	5.85	−0.749	1.81	13.8
220	6.51	0.153	3.57	−0.472	1.18	17.7
260	4.33	0.0983	2.17	−0.321	0.767	22.3
300	2.44	0.0876	1.48	−0.223	0.529	30.2

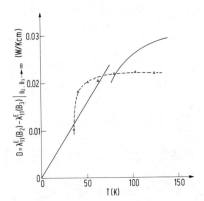

Fig.5 : Experimental values, triangles, from Uher and Goldsmid[11].
Continuous curves are the predictions based on the
pseudo-parabolic model, straight line part represents
the Bethe-Sommerfeld approximation.

$$\bar{\bar{\lambda}}^E(\bar{B}) = \bar{\bar{\xi}}(\bar{B}) - T\bar{\bar{\theta}}(\bar{B})\bar{\bar{\alpha}}(\bar{B}) . \tag{14}$$

We have calculated the difference $\lambda_{11}^E(B_2) - \lambda_{11}^E(B_3)$ in the limit of strong magnetic fields. The result is given in figure 5, where the straight line arises from Bethe-Sommerfeld approximation and the curved part from numerical integration of the transport integrals. The triangles are experimental values[11]. A judgement of the disagreement displayed between theory and experiment is difficult. If significant, it is synonymous to a deviation from the Wiedemann-Franz law.

Acknowledgement - The authors gratefully acknowledge the helpfulness of Dr. C. Uher in supplying numerical experimental data. We are likewise grateful to Professor J.P. Issi for the permission to report the results on $\alpha_{31}(B_2)$ as these were not regularly published before.

REFERENCES

1. J. Heremans and O.P. Hansen, J. Phys. C : Solid St. Phys. 12, 3483 (1979).

2. B. Lax, J.G. Mavroides, H.J. Zeiger and R.J. Keyes, Phys.Rev. Lett. 5, 241 (1960).

3. C. Uher and W.P. Pratt, J. Phys. F : Metal Phys. 8, 1979 (1978).

4. R. Hartman, Phys. Rev. 181, 1070 (1969).

5. O.P. Hansen, E.Cheruvier, J.P. Michenaud and J.P. Issi, J. Phys. C : Solid St. Phys. 11, 1825 (1978).

6. M.P. Vecchi and M.S. Dresselhaus, Phys. Rev. B 10, 771 (1974).

7. J.P. Michenaud and J.P. Issi, J. Phys. C : Solid St. Phys. 5, 3061 (1972).

8. C. Uher and H.J. Goldsmid, Phys. Stat. Solid (b) 63, 163 (1974) and 64, K25 (1974).

9. E. Cheruvier and O.P. Hansen, J. Phys. C : Solid St. Phys. 8, L346 (1975).

10. I. Ya. Korenblit, M.E. Kuznetsov, V.M. Muzhdaba and S.S. Shalyt, Sov. Phys. - JETP 30, 1009 (1970).

11. C. Uher and H.J. Goldsmid, Phys. Stat. Solid (b) 65, 765 (1974).

12. E. Cheruvier, J.P. Michenaud and J.P. Issi, Int. Conf. Physics of Semimetals and Narrow Gap Semiconductors, Nice-Cardiff 1973.

EFFECTS OF ELECTRON-HOLE CORRELATIONS ON RADIATIVE

RECOMBINATION OF DONOR-ACCEPTOR PAIRS IN POLAR SEMICONDUCTORS

E. Kartheuser, A. Babcenco, and R. Evrard

Institut de Physique, Université de Liège

Sart-Tilman, 4000 Liège, Belgium

A variational calculation of the effective-mass electronic states is presented for donor-acceptor (DA) pairs including electron-hole correlations and electron-phonon interactions. The calculations are carried out for unequal electron and hole effective masses. Expressions for the energy levels are obtained in terms of integrals containing trigonometric functions. The variational parameters for the orbital radii and for the electron-hole correlation are determined as a function of the D-A distance. The theory is applied to the zero-phonon spectra of D-A pairs in GaP and in ZnSe and a comparison is made with previous theoretical works. The contribution due to electron-hole correlation increases at small D-A distance and is found to be about 10 meV for GaP.

1. INTRODUCTION

Photoluminescence of donor-acceptor (DA) pairs in semiconductors has been widely investigated both experimentally and theoretically for more than two decades[1]. Since the availability of dye lasers, the new experimental techniques of luminescence excitation spectroscopy (LES)[2] and selective pair luminescence[3] (SPL) has recently stimulated increasing interest in this field. Indeed these sensitive techniques enable a detailed study of the excited states of the impurities involved in the radiative transitions and therefore provide a powerful tool for identification and characterization of impurities in semi-conductors. In addition, those experiments reveal the dependence of the electron-hole recombination energy versus DA pair distance R.

15

   The R-dependence of the zero-phonon (ZP) recombination
spectra has been studied theoretically including the effects of
the departure from spherical symmetry in the electronic charge
distributions at small DA pair distances as well as those of
electron-phonon interaction in polar semiconductors[4].

   The present work is devoted to the study of the effects
of electron-hole correlations on the DA pair recombination energy.
The usual effective mass approximation is used throughout this
work. The theory is applied to polar semiconductors GaP and ZnSe
for which the interaction with the longitudinal optical (LO)
phonons is the dominant electron-phonon interaction. Fröhlich's
formalism[5] is used to describe this interaction which is treated
in the framework of an adiabatic approximation as explained in
a previous paper[4]. Our results are obtained variationally and
therefore give an upper bound to the DA pair energy.

2. CALCULATION OF THE ELECTRON-HOLE CORRELATION EFFECTS

   Within the adiabatic approximation for the particle-
lattice interaction, the effective mass Hamiltonian of the donor-
acceptor complex can be written as[4]

$$H = H_o + V_{ep} \tag{1}$$

with

$$
H_o = -\frac{\hbar^2}{2m_e}\nabla_e^2 - \frac{2}{2m_h}\nabla_h^2
$$

$$
+ \frac{e^2}{\varepsilon_o}\{\frac{1}{|\vec{r}_e - \vec{R}|} + \frac{1}{r_h} - \frac{1}{r_e} - \frac{1}{|\vec{r}_h - \vec{R}|} - \frac{1}{R}\} - \frac{1}{\varepsilon_\infty |\vec{r}_e - \vec{r}_h|} \tag{2}
$$

and

$$
V_{ep} = \sum_{\vec{k}} \frac{|\nabla_k|^2}{\hbar\omega} |\rho_{\vec{k}}|^2
$$

where $\vec{R}$, $\vec{r}_e$ and $\vec{r}_h$ stand for the positions of the acceptor center,
of the electron and of the hole, respectively, the donor impurity
center being taken as origin of the coordinate axes. The first
two terms in H_o represent the kinetic energy of the electron
and hole, respectively, with band masses m_e and m_h. The electron
charge is denoted by e, ε_∞ and ε_o are, respectively, the high-
frequency and static dielectric constants. The quantity ∇_k is
a measure for the strength of the interaction between the electron
(or hole) and the LO-phonons (of frequence ω) as originally

Table I : Variational results for the orbital radii a, α and the correlation parameters β, λ as a function of the D-A pair distance R. The values of m_e, m_h, ε_∞ and ε_o are those appropriate to GaP (see section 2).

R [Å]	a [Å]	α [Å]	β [Å]	λ
12.5	20.09	11.85	12.94	1.215
15	17.08	9.64	8.80	0.719
20	15.77	8.76	7.70	0.692
25	15.50	8.62	8.53	0.753
30	15.47	8.62	2.41	0.855
35	15.52	8.67	11.32	0.889

Table II : Variational results for the D-A pair energy E and different separations R : The values of m_e, m_h, ε_∞ and ε_o are chosen as those appropriate to GaP (see section 2); E_s : Pair energy for spherical orbits; E_{sp} Pair energy in the case of s-p admixture; E_c : Pair energy for spherical orbits with electron-hole correlation; the values between parentheses are those including the effect of electron-phonon interaction.

R [Å]	E_s (meV)	E_{sp} (meV)	E_c (meV)	
12.5	−126.4	−134.6	−138.4	(−140.0)
15	−118.6	−125.0	−130.5	(−133.7)
20	−114.8	−118.1	−126.0	(−131.0)
25	−114.9	−116.4	−124.9	(−131.7)
30	−115.4	−116.1	−124.4	(−131.8)
35	−115.8	−116.1	−123.8	(−132.1)

introduced by Fröhlich. The Fourier transform $\rho_{\vec{k}}$ of the charge distribution is expressed in terms of the electronic part $F(\vec{r}_e, \vec{r}_h)$ of the envelope function by

$$\rho_{\vec{k}} = \int d\vec{r}_e \int d\vec{r}_h |F(\vec{r}_e, \vec{r}_h)|^2 \; (e^{i\vec{k}.\vec{r}_e} - e^{i\vec{k}.\vec{r}_h}) \quad (3)$$

To take the electron-hole correlation effects into account, we use for the electronic part of the wave function :

$$F(\vec{r}_e, \vec{r}_h) = \Psi_o(\vec{r}_e, \vec{r}_h)(1 + \lambda e^{-\frac{|\vec{r}_e - \vec{r}_\beta|}{\beta}}) \quad (4)$$

where $\Psi_o(\vec{r}_e, \vec{r}_h)$ is a product of one-particle hydrogenic 1s functions,

$$\Psi_o(\vec{r}_e, \vec{r}_h) = \frac{1}{\pi(a\alpha)^{3/2}} e^{-r_e/a} e^{-|\vec{r}_h - \vec{R}|/\alpha} \quad (5)$$

The motivation for this choice of the wave function is the following : At short distance, the electron and hole form an exciton whose wave function has the exponential form used in Eq.(4).

Using the trial functions (Eqs.(4) and (5)) the expectation value[6] of the effective Hamiltonian given by Eq.(1) is now minimized with respect to the variational parameters a, α, λ and β[6]. The values of the parameters at the minimum and for different D-A distances R, are given in table I for GaP. Table II shows a comparison between our results for the D-A pair energy and previous calculations for GaP. The values chosen for the different physical parameters are $m_e = 0.365$, $m_h = 0.67$, $\varepsilon_\infty = 9.09$ and $\varepsilon_o = 11.02$. Tables III and IV show similar results for ZnSe with the following values for the parameters : $m_e = 0.16$, $m_h = 0.52$, $\varepsilon_\infty = 6.10$ and $\varepsilon_o = 9.42$.

These results show that allowing s-p mixing in the wave function leads to a lowering of the energy from with 8 meV at R = 12.5 Å for GaP to 0.4 meV at R = 35 Å whereas the contribution of the correlation effects varies from 12 meV to 8 meV in the same range of values of R. Therefore the correlation is more important than s-p mixing for the study of the D-A pairs, contrarily to what is generally assumed.

As the correlation effects are larger for close DA pairs, one expects that they play a far more important role in evaluating the probability of radiative recombination. This is

Table III : Variational results for the orbital radii a, α and the correlation parameters β, λ as a function of the D-A pair distance R. The values of m_e, m_h, ε_∞ and ε_o are those appropriate in ZnSe (see section 2).

R [Å]	a [Å]	α [Å]	β [Å]	λ
16	44.83	13.08	43.38	12.097
18	34.89	11.00	40.01	2.987
20	32.00	10.32	33.41	1.792
25	30.33	9.75	28.61	1.392
30	30.38	9.61	29.84	1.464
35	30.51	9.59	30.27	1.531
40	30.92	9.60	33.01	1.653

Table IV : Variational results for the D-A pairs energy E and different separations R : The values of m_e, m_h, ε_∞ and ε_o are chosen as those appropriate to ZnSe (see section 2); E_s : Pair energy for spherical orbits; E_{sp} : Pair energy in the case of s-p admixture; E_c : Pair energy for spherical orbits with electron—hole correlation; the values between parentheses are those including the effect of electron—phonon interaction.

R [Å]	E_s (meV)	E_{sp} (meV)	E_c (meV)	
16	−113.0	−121.3	−137.2	(−146.9)
18	−108.5	−115.4	−132.4	(−144.4)
20	−106.1	−112.1	−129.7	(−143.1)
25	−103.8	−108.0	−126.4	(−142.1)
30	−103.4	−106.3	−124.7	(−142.1)
35	−103.4	−105.4	−123.4	(−142.3)
40	−103.6	−105.2	−122.3	(−142.4)

confirmed by the calculation of the square of the overlap between
the electron and hole wave functions. Indeed, this quantity
appears as a factor in the probability of recombination. Using
the results of the present work for GaP, one obtains, for instance
at R = 20 Å, a value of 0.36 in the case of uncorrelated spherical
wavefunctions and 0.89 with our wavefunction Eq.(4), which means
an increase of about 2.5 times in the probability of recombination.

REFERENCES

1. P.J. Dean, Progress in Solid State Chemistry (Pergamon Press,
   N.Y., 1973) 8, p.1.

2. P.J. Wiesner, R.A. Street and H.D. Wolf, Physical Review
   Letters 35, 1366 (1975).

3. H. Tews, H. Venghaus and P.J. Dean, Physical Review B 19, 5178
   (1979).

4. E. Kartheuser, R. Evrard and F. Williams, Physical Review B 21,
   648 (1980) and references therein.

5. H. Fröhlich, H. Pelzer and S. Zieman, Philosophical Magazine
   41, 221 (1950).

6. Detailed expressions for the total energy of the D-A pairs
   as a function of the variational parameters a, α, λ, β will
   be published elsewhere.

IONIZED IMPURITY SCATTERING AND DIELECTRIC ENHANCEMENT

OF MOBILITY IN SEMICONDUCTORS AND SEMIMETALS

R. Resta
Laboratoire de Physiqye Appliquée
Ecole Politechnique Fédérale
Lausanne, Switzerland

L. Resca *
Instituto di Fisica
Università di Pisa, Italy

S. Rodriguez
Department of Physics, Purdue University
West Lafayette, Indiana, USA

We generalize the Brooks-Herring theory of ionized-impurity-limited mobility in semiconductors and semimetals, in order to account for dispersive dielectric screening. In a nonzero-gap semiconductor the resulting effect is, contrary to a generalized belief, quite small. In a semimetal the effect of dispersive screening is a strong enhancement of the calculated mobility, in good agreement with the experimental data for α-Sn.

1. INTRODUCTION

The Brooks-Herring (BH)[1] and Dingle[2] theories of ionized impurity scattering in semiconductors are the basic tools for the calculation of collision times and mobilities[3]. Within these theories, the impurity point-charge potential is self-consistently screened by the free-carrier distribution, and the resulting scattering potential assumes in both cases the simple screened Coulombic form.

* NATO Senior Fellow at the Department of Physics, Purdue Yniversity, USA.

$$\phi_D(r) = \exp{(-r/R_D)}/\varepsilon_o r \ , \tag{1}$$

where ε_o is the static dielectric constant of the host and R_D is a characteristic screening length. Atomic units ($\hbar = 1$, $m_e = 1$, $e^2 = 1$) are used throughout this paper. BH use classical statistics, and R_D is simply the Debye-Huckel length, given by

$$R_D^2 = \frac{\varepsilon_o K_B T}{4\pi n} \tag{2a}$$

and where K_B is the Boltzmann constant and n is the concentration of ionized impurities. Dingle[2] has extended the BH theory to quantum statistics; R_D is in this case given in terms of Fermi-Dirac integrals as a function of the carrier effective-mass m, of the temperature and of the impurity concentration. For the completely degenerate case the Dingle screening length assumes the simple form given by

$$R_D^2 = \frac{\varepsilon_o}{4m} \left(\frac{\pi}{3n}\right)^{1/3} \tag{2b}$$

For a normal semiconductor at room temperature the BH expression (2a) is appropriate up to n $\simeq 10^{18}$ cm^{-3}. As a typical example, R_D = 108 a.u. in n-doped Si for a donor concentration of 5.10^{17} cm^{-3}.

In both BH and Dingle theories the dielectric response of the host semiconductor is caracterized by a single parameter : the static dielectric constant ε_o. The mechanism responsible for it is mostly the polarization of valence electrons, and in more recent times much more detailed information has been gathered about such mechanism[4].

Throughout the present paper we work within the model of a homogeneous and isotropic solid, where the dielectric response is dispersive and is characterized by a dielectric function $\varepsilon(k)$. The effects of such a dispersive screening upon the calculated mobilities are discussed in this paper for both semiconductors and semimetals.

2. DIELECTRIC RESPONSE

The dielectric response of an undoped model (homogeneous and isotropic) solid is dispersive, i.e. depends on the momentum of the probe through the relationship

$$\underline{E}(\underline{k}) = \underline{D}(\underline{k})/\varepsilon(k) , \qquad (3)$$

where (k) is by definition the static dielectric function. If we indicate as $\varepsilon^{-1}(r)$ the Fourier antitransform of $1/\varepsilon(k)$, Eq.(3) is equivalent to

$$\underline{E}(\underline{r}) = \int dr' \varepsilon^{-1}(|\underline{r}-\underline{r}'|)\underline{D}(\underline{r}') . \qquad (4)$$

This clearly demonstrates that the screening is non-local in r-space.

The dielectric function $\varepsilon(k)$ has been the object of considerable work. We will use in the following the term "normal semiconductor" to indicate any nonzero-gap semiconductor or insulator, while the term "semimetal" will be used as a synonimous of zero-gap semiconductor, like α-Sn. For a normal semiconductor the basic theories are due to Penn[5] and Resta[6]. These give numerically close results, but the second one is simpler both conceptually and mathematically. In both these theories $\varepsilon(k)$ is regular at low k, and $\varepsilon(0)$ coincides with the static dielectric constant ε . For a semimetal the basic theory is due to Liu and Brust[7], and the dielectric function has therein a low-k behavior exactly intermediate between a metal and a normal semiconductor, i.e. $\varepsilon(k)$ diverges like k^{-1}. The value of $\varepsilon(0)$ is therefore infinite, but a static dielectric constant ε_o can still be defined in the sense that

$$\varepsilon(k) = \varepsilon_o(1 + \lambda k^{-1}) , \qquad (5)$$

this expression being valid only at low k. Numerical values for α-Sn are $\varepsilon_o = 24$; $\lambda = 0.0035$.

We have recently shown that for a doped semiconductor at nonzero temperature the traditional BH and Dingle theories of carrier screening can be straightforwardly generalized to include the dispersive dielectric behavior of the host. Our result[8] is that the doped semiconductor acts as a medium whose effective dielectric function is

$$\varepsilon_{eff}(k) = \varepsilon(k) + \varepsilon_o R_D^{-2} k^{-2} , \qquad (6)$$

where $\varepsilon(k)$ is the dielectric function of the host. The semiclassical reasoning underlying Eq.(6) is valid no matter how $\varepsilon(k)$ has been obtained and can be applied to both normal semiconductors and semimetals.

3. POTENTIAL OF AN IONIZED IMPURITY

Many papers have recently appeared concerning the potential of an ionized impurity in a semiconductor with dispersive screening[9],[10]. These papers have in common an unsound starting point : the picture of a Penn (or Resta) model semiconductor as a "semiconductor with spatially variable dielectric constant". As discussed in Section 2, the screening is "spatially invariant" but nonlocal. Besides that, the first papers of the series[9] suffered from serious mathematical errors[10]. Different and contradictory potential forms have therefor been proposed.

We have very recently solved[8] the same physical problem following a quite different path, and using a much simpler and physically transparent scheme. Our screened potential is in k-space

$$\phi(k) = \frac{4\pi}{k^2 \varepsilon_{eff}(k)} \quad , \tag{7}$$

where $\varepsilon_{eff}(k)$ is the same as in Eq.(6). The Fourier antitransform of Eq.(7) has been evaluated[8], and found indistinguishable from $\phi_D(r)$, Eq.(1), at r larger than one bond - length in Si for any realistic concentrations. This result was expected on a physical ground, since one bond-length is the typical length appearing in valence screening[5],[6], and this always much smaller than R_D.

After our approach has been known, the author of Ref.9 and others have been able to manipulate their theories in order to rather closely reproduce the correct results[10]. However these theories remain based on the unjustified initial assumptions discussed above and result unnecessarily complicated and involved.

4. CALCULATED MOBILITIES IN NORMAL SEMICONDUCTORS

The collision times are obtained through the usual angular integration[1-3], and with the use of the first Born approximation. The calculation has been performed[11] for normal semiconductors with use of Eqs.(6) and (7). The effect of dispersive screening is in this case a *quite small reduction* of the calculated mobility with respect to the traditional theories, of the order of a few percent for any concentration value. This appears to be physically sound, because of the well known fact that ionized impurity scattering in semiconductors is dominated by low-momentum transfer processes. In a normal semiconductor, from Eq.(6), the effect of the dispersive dielectric fuction $\varepsilon(k)$ is easily recognized as small at the important values of $k (k \lesssim 1/R_D)$, or equivalently the impurity potential, as discussed above, is not influenced by dispersive screening at large $r (r \gtrsim R_D)$.

Many other authors[12] have recently attempted to evaluate the importance of this same physical effect on the calculated mobilities in normal semiconductors. They obtained the puzzling and unphysical result of a *quite large change in mobility,* either a reduction or an enhancement. The reasons for that is the use of incorrect impurity potentials, from the traditional one, Eq.(1), at large r (small k). This "dilemma" has been finally solved by us[8,11].

5. CALCULATED MOBILITIES IN SEMIMETALS

In a semimetal the dispersive dielectric function $\varepsilon(k)$ at low k, from Eq.(5), is singular and therefore quite different from ε_o. Strong differences in calculated mobilities are thus expected, as originally guessed by Liu and Brust[7].

We apply the theory described above, with no change, to the case of a semimetal. Due to the dielectric singularity the potential of an ionized impurity is quite different from the traditional one, Eq.(1), even at larger r. In $\underline{k}$-space the potential is

$$\phi(k) = \frac{4\pi}{\varepsilon_o (k^2 + \lambda k + R_D^{-2})} \tag{10}$$

and the low-k Born approximation cross section is strongly different with respect to the one calculated with a constant ε_o. As a result, the calculated mobilities turn out to be *strongly enhanced* with respect to the standard Dingle ones.

We report in figure 1 our calculation, for n-doped α-Sn at 4.2 K, and converted to the experimental units. The carrier gas is completely degenerate at concentrations of 10^{14} cm^{-3} or above. In that range the expression (2b) for R_D holds. Some experimental findings are also shown[13].

Tosatti and Liu have previously published a calculation[14] of this same effect, obtained through a modified RPA scheme. Their calculation is quite complicate and many approximations are performed; the results are also shown in figure 1. It comes out that the very simple scheme outlined in this paper gives results in satisfying agreement. More recently Bailyn and Liu[15] have published a more general theory of the dielectric response

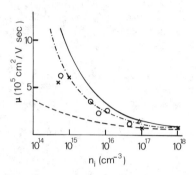

<u>Fig.1</u> : Plot of the calculated mobility in α-Sn. Solid line :
this work; dash-dotted line : Ref.14; dashed line :
Dingle theory.
The experimental data at 4.2 K are also reported, Ref.13.
Triangle : Tufte and Ewald; circles : Hinkley and Ewald;
crosses : Lavine and Ewald.

of a doped semimetal. Their final formulas are much involved and
no numerical result is shown. However, we argue that for the
static ($\omega = 0$) response their result should not be too far from
Eqs.(5) and (6).

The main adventage of the scheme presented here, besides
its much greater simplicity, is the fact that one same theory goes
with no change for both normal semiconductors and semimetals.

Acknowledgements   -   Work supported by National Swiss Foundation,
by CNR/GNSM Unit in Pisa and by NSR/MRL Program No. DMR 77-23798.

REFERENCES

1. H. Brooks, Phys. Rev. <u>83</u>, 879 (1951); C. Herring (unpublished).

2. R.B. Dingle, Philos. Mag. <u>46</u>, 831 (1955); R. Mansfield, Proc.
Phys. Soc. B <u>69</u>, 76 (1956).

3. H. Brooks, Advances in Electronics and Electron Physics,
edited by L. Marton (Academic, New York, 1955), Vol.7, p.85;
K. Seeger, Semiconductor Physics (Springer-Verlag, Wien 1973),
Chap.6.

4. A. Baldereschi, R. Car and E. Tosatti, Solid State Commun. <u>32</u>,
757 (1979); A. Baldereschi at the 1980 EPS/CMD Conference
(unpublished).

5. D.R. Penn, Phys. Rev. 128, 2093 (1962); G. Srinivasan, ibid. 178, 1244 (1969).

6. R. Resta, Phys. Rev. B 16, 2717 (1977).

7. L. Liu and D. Brust, Phys. Rev. 173, 777 (1968).

8. R. Resta, Phys. Rev. B 19, 3022 (1979).

9. P. Csavinszky, Phys. Rev. B 14, 1649 (1976); P. Csavinszky, Int. J. Quantum Chem. 13, 221 (1978); P. Csavinszky, Phys. Rev. B 20, 4372 (1979).

10. J.R. Meyer, Phys. Rev. B 20, 1762 (1979); P. Csavinszky and R.A. Morrow, ibid. 21, in press.

11. R. Resta and L. Resca, Phys. Rev. B 20, 3254 (1979).

12. M.A. Paesler, Phys. Rev. B 17, 2059 (1978); L.M. Richardson and L.M. Scarfone, ibid. 18, 5892 (1978); M.A. Paesler, D.E. Theodorou and H.J. Queisser, ibid. 18, 5895 (1978); L.M. Richardson and L.M. Scarfone, ibid. 19, 925 (1979); D.E. Theodorou and H.J. Queisser, ibid. 19, 2092 (1979); L.M. Richardson and L. Scarfone, ibid. 19, 5139 (1979).

13. O.N. Tufte and A.W. Ewald, Phys. Rev. 122, 1431 (1961); E.D. Hinkley and A.W. Ewald, ibid. 134, A 1261 (1964); C.F. Lavine and A.W. Ewald, J. Phys. Chem. Solids 32, 1121 (1971).

14. L. Liu and E. Tosatti, Phys. Rev. B 2, 1926 (1970).

15. M. Bailyn and L. Liu, Phys. Rev. B 10, 759 (1974).

INFLUENCE OF A MAGNETIC FIELD ON HOPPING CONDUCTION

IN n-TYPE INDIUM PHOSPHIDE

G. Biskupski and H. Dubois

Laboratoire de Spectroscopie, associé au C.N.R.S.
U.E.R. de Physique - Université de Lille I
59655 Villeneuve d'Asco Cédex

The hopping regime in the low temperature conduction is studied as a function of the magnetic field for two n-type InP crystals with impurity concentration $N_D - N_A$ equal to $6.6 \ 10^{15} \ cm^{-3}$ and $1.2 \ 10^{16} \ cm^{-3}$ respectively. These two samples are located on the insulation side of the metal-nonmetal transition and present in the lowest temperature region, the ε_3 (hopping) conduction process, i.e. $\rho_3 = \rho_{03} \exp\left(\frac{\varepsilon_3}{kT}\right)$. The variation of the preexponential factor ρ_{03} is investigated for magnetic fields up to 7.5 Tesla. The values of ρ_{03} are very well described by the theoretical expressions derived from a percolation model for the different magnetic field regions. In magnetic field below 4.75 Tesla, the preexponential factor is proportional to $\exp\left(\frac{t a_0}{\lambda^4 N_D}\right)$, where λ is the characteristic magnetic length, a_0 the Bohr radius, and t have values of 0.048 and 0.066 for the two samples respectively while the theoretical value is $t = 0.04$. In the high magnetic field region above 4.75 Tesla, ρ_{03} is proportional to $\exp[q(\lambda^2 a_B N_D)^{-1/2}]$, where a_B is the magnetic field dependent Bohr radius, N_D the donor concentration. The experimental values of q are respectively 0.75 and 0.91 while the theory predicts $q = 0.98$. The discrepancies between the experimental and theoretical values of t and q are discussed.

29

1. INTRODUCTION

Electrical resistivity of single crystals of n type Indium Phosphide has been studied for temperatures between 100 K and 4.2 K in magnetic fields up to 7.5 T [1]. The impurity concentrations of the two samples labelled InP 105 and InP 202 are respectively $n_1 = 6.6 \ 10^{15} \ \text{cm}^{-3}$ and $n_2 = 1.6 \ 10^{16} \ \text{cm}^{-3}$, the donors concentrations are $N_{D1} = 10.18 \ 10^{15} \ \text{cm}^{-3}$ and $N_{D2} = 17.10 \ 10^{15}$ cm^{-3}, the compensations are $K_1 = 0.35$ and $K_2 = 0.3$. According to the MOTT criterion these samples are on the insulating side of the metal non metal (MNM) transition. At low temperature, impurity conduction is observed and the different conduction processes with activation energies $(\varepsilon_1, \ \varepsilon_2, \ \varepsilon_3)$ are observed in different temperatures ranges.

In this work attention is drawn to the hopping conduction $(\varepsilon_3$ process) for which the resistivity can be written

$$\rho_3 = \rho_{30} \ \exp \ (\varepsilon_3/KT) \tag{1}$$

The activation energy ε_3 is observed to be quite constant when the magnetic field is varied, and the preexponential coefficient (ρ_{30}) dependence on magnetic field is reported and compared with the results of percolation theory.

Since Mikoshiba's work [2] on weak field magneto resistance, the development of percolation theory has brought a better understanding of the hopping conduction mechanisms [3,4]. Shklovskii [5] has given a synthesis of the percolation theory results for the hopping conduction in lightly doped semiconductors, and has derived expressions for ρ_{30} in the different magnetic field ranges. The numerical coefficients appearing in the pre-exponential factor have been refined with numerical calculations accomplished by the same authors [6-8] for the low and strong magnetic field regions.

A comparison of theory with experimental results is now available for Ga As [9] and n type InP [10] in the low magnetic field range, but few results are available for strong magnetic fields [9].

2. WEAK MAGNETIC FIELD RANGE B < 4.5 T

The values of ρ_{03} for each sample were derived from resistivity versus inverse temperature curves obtained for different magnetic fields up to 8.5 T [1] and plotted versus B(T) (fig.1). To analyse these results, the values of ρ_3 were plotted versus the square magnetic field for different temperatures (fig. 2,3). In the low magnetic field region, the values of ρ_3 cannot be fitted with a straight line. This is caused by the existence of a negative magnetoresistance which saturates when the magnetic field increases. When this negative magnetoresistance is sub-tracted (fig.2,3) the data are quite well fitted with a straight line.

Shlovskii [5] has modified the Mikoshiba theory [2] for weak magnetic fields; when the magnetic field length

$\lambda = (\frac{c\hbar}{eB})^{1/2}$ is greater than $a/(N_D a^3)^{1/6}$ he gave the expression:

$$\rho_{30}(B) = \rho_{30}(o) \exp (\frac{ta_0 B^2 e^2}{N_D c^2\hbar^2})$$ (2)

with t = 0.06 in a first work [5] and t = 0.04 in a later calculation [8]. For InP a_0 = 77 Å [10, 12] or 84 Å [11]. With this second value of a_0, a least squares fit of the data gives t_1 = 0.048 for InP 105 and t_2 = 0.066 for InP 202. Emel'yanenko [10] found t = 0.04 for a pure sample and t = 0.06 for a sample wit $N_D \geqslant 10^{16}$ cm^{-3}; and explained these different values with the onset of a MNM transition in the second case. This argument cannot hold in our case since the two samples are on the insulating side of the MNM transition. Kahlert [9] found also t = 0.06 for Ga As samples on the insulating side of the transition. It must be noted that our values of t were derived from data extended to B = 4.5 T, while Emel'yanenko data are given for B $\leqslant$ 2.8 T.

3. STRONG MAGNETIC FIELD RANGE B > 4.75 T

When $\lambda < a_0/(N_D a_0^3)^{1/6}$, the binding energy of the donors increases with B [1], and the electron orbitals change their shape from spheres to double paraboloids. In this case Shklovskii [5-8] has given the following equation :

$$\rho_{30}(B) = \rho_o \exp \{q\ (\lambda^2 a_B N_D)^{-1/2}\}$$ (3)

where a_B is the Bohr radius derived from the field dependent binding energy of donors by the expression:

$$a_B = \hbar/(2m^* \varepsilon_1(B))^{1/2}$$ (4)

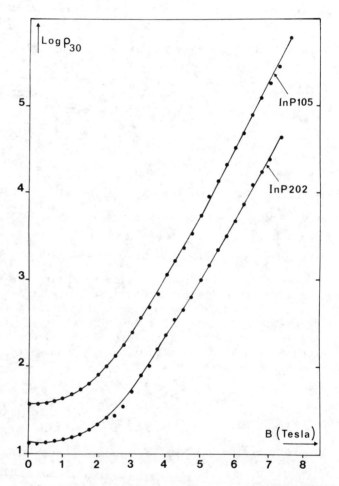

<u>Fig.1</u> : Logarithm of the preexponential coefficient ρ_{03} (Ω cm)
versus magnetic field in T for two samples InP 105 and
InP 202.

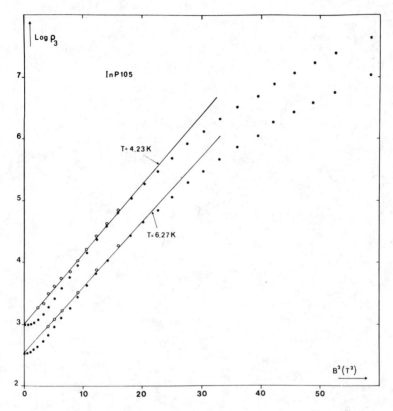

Fig.2 : Logarithm of the resistivity ρ_3 (Ω cm) versus the square
magnetic field at two temperatures for sample InP 105.

● Experimental values, o values of ρ_3 when the negative
magnetoresistance has been substracted.

So the exponential coefficient in (3) may be written

$$q \left\{ \left(\frac{eB}{ch^2 N_D} \right)^{1/2} (2m^* \varepsilon_1)^{1/4} \right\} \tag{5}$$

ε_1 has been derived from the resistivity versus inverse
temperature curves [1] and $\rho_{30}(B)$ has been plotted versus $\varepsilon_1^{1/4} B^{1/2}$
(fig.4). Equation (3) is quite well verified in each sample for
$B > 4.75$ T (fig.4). For the two samples $\lambda = (a_o N_D^{-1/3})^{1/2}$ when
$B = 1.7$ T and $B = 2.02$ T respectively. Hence it follows that the

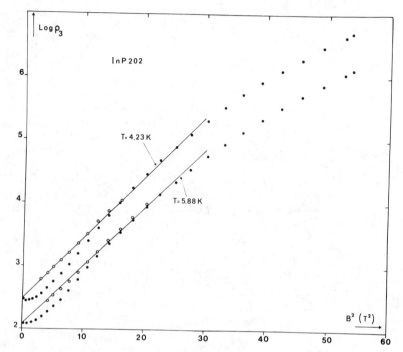

Fig.3 : Logarithm of the resistivity ρ_3 (Ω cm) versus the square magnetic field at two temperatures for sample InP 202.

● Experimental values, o values of ρ_3 when the negative magnetoresistance has been susbstracted.

Shklovskii criterion becomes

$$\lambda < (0.6 \ (a_o \ N_D^{-1/3})^{1/2} \ .$$

A least squares fit of the results of figure 4 above 4,75 T hields q_1 = 0.746 and q_2 = 0.914 respectively while Shklovskii values are q = 0.9 [5] and q = 0.98 [7] . It should be noted the good qualitative agreement between experimental data and theory. A higher values of q_1 may be obtained with a higher value of N_{D1}; this would yield to a higher value of t_1 getting closer to 0.06 which the first value derived by Shklovskii.

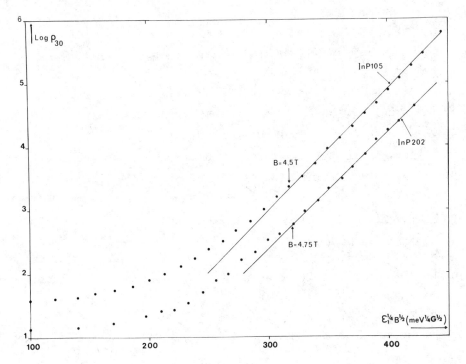

Fig.4 : Logarithm of the preexponential coefficient ρ_{30} versus $\varepsilon_1^{1/4} B^{1/2}$ for the two samples, ε_1 (meV), B(Gauss).

4. CONCLUSION

The transverse magnetoresistance of two n-type InP crystals has been investigated in the hopping regime of the impurity conduction for several values of the magnetic field.

Account has been taken of the negative magnetoresistance in the low magnetic field region. The preexponential coefficients are in good agreement with theoretical expressions derived from the percolation theory. Nevertheless the numerical coefficient in the exponential describing the magnetic field dependance of the resistivity are in better agreement with the first values derived by Shklovskii than with the values derived in his last work.

REFERENCES

1. Biskupski G., Dubois H., Laborde O., Zotos X., Phil. Mag.
   (1980) to be published.

2. Mikoshiba N., Phys. Rev. 127, 1962 (1962).

3. Shklovskii B. I., Efros A. L., Sov. Phys. JETP 33, 468 (1971).

4. Pollak M., J. non cryst. solids, 11, 1 (1972).

5. Shklovskii B. I., Sov. Phys. Semicond., 6, 1053 (1973).

6. Shklovskii B. I., Sov. Phys. JETP, 34, 1084 (1972).

7. Skal A. S., Shklovskii B. I., Sov. Phys. Semicond., 7, 1058 (1974)

8. Shklovskii B. I., Efros A. L., Sov. Phys. Usp., 18, 845 (1975).

9. Kahlert H., Landwehr G., Schlachetzki A., Salow H., Z. Physik
   B 24, 361 (1976).

10. Emel'yanenko, O. V., Masagutov, K. G., Nasledov D. N., Timchenko
    I. N., Sov. Phys. Semicond., 9, 330 (1975).

11. Blood P., Orton J. W., J. Phys. C, 7, 893 (1974).

12. Hilsum C., Fray S., Smith C., Solid st. commun., 7, 1057 (1969).

ELECTRICAL PROPERTIES OF MoTe$_{2-x}$ SINGLE CRYSTALS DOPED WITH BROMINE

A. Bonnet, P. Saïd, and A. Conan

Laboratoire de Physique des Matériaux et Composants
de l'Electronique, Institut de Physique
Université de Nantes - 2, rue de la Houssinière
44072 Nantes Cedex (France)

Thermoelectric power and d.c. electrical conductivity
measurements are performed on MoTe$_{2-x}$ single crystals in a wide
temperature range (77 - 770 K).
A new experimental technique for the electrical conductivity
and T.E.P. measurements has been used : an automatic data
acquisition system is controlled by a H.P. 9815 A calculator
connected to a H.P. interface bus.
The samples were prepared from the stoichiometric crystals by
tellurium depletion and annealed at 780 K. The experimental results
are analysed on the basis of impurity conduction. It is conjectured
that three processes contribute to the total conductivity :
hopping conduction between impurity sites, scattering by polar
modes and, in the intrinsic domain, scattering by impurities.

In a recent paper[1], electrical conductivity and thermoelectric power (T.E.P.) measurements have been performed on MoTe$_2$
single crystals. The samples were prepared by the vapour transport
method with bromine as transport reagent and it has been shown
that the experimental results can be interpreted in terms of
impurity conduction : at low temperature hopping-conduction
between impurity sites, in the intermediate range of temperature
scattering by optical phonons and in the intrinsic region
scattering of the electrons of the conduction band by impurities,
leading to a T^3 dependence of the preexponential term of the
conductivity. The interaction between electrons and modes
carrying an electric polarization of the crystal can be explained
by the presence of lacunar sites. So, for a best knowledge of the
room temperature behaviour of these compounds it was necessary to

37

study the departure from stoichiometry : in this paper, transport properties of non-stoichiometric MeTe$_{2-x}$ crystals have been measured. They were prepared from the stoichiometric MoTe$_2$[1,2] crystals by tellurium depletion and annealed at 550°C during two or three days. The experimental results, performed in a large temperature range (77-770 K), along a direction perpendicular to the c-axis, are analysed using the same model as in[1]. The crystals, cut in a rectangular shaped and bound on a thin mica sheet, were cast with a high temperature epoxy resin. Electrical connections were made by mean of two copper and two constantan wires, wrapped round the mica sheet so as to minimize the thermal losses. The T.E.P. was measured using a method described previously[3] : the sample is placed between two heaters. The first one is regulated at a given temperature, the second one acts so as to controll the difference of temperature between the two ends of the sample : the heating is adjusted so as to obtain a linear time-dependance of the increasing temperature. The thermal emfs between the two copper and the constantan wires are measured by a Keithley digital voltmeter and the T.E.P. is deduced. For electrical conductivity measurements, the two constantan wires acted as current leads while the two copper wires were used to measure the potential difference in the crystal. The intensity of current through the sample, obtained from a constant current supply, was very low (less than 100 nA) because of the non-ohmic behaviour of the sample[4].

All the experiment is calculator controlled : an automatic data acquisition system is controlled by a H.P. 9815 A calculator connected to the H.P. interface bus which is merely a set of sixteen wires to which all devices on that bus are connected. Eight of these wires serve to carry the data messages back and forth over the bus. To maintain order, only one device at a time can place information on these data lines, and that device is known as the active talker. Any or all of the other devices on the bus may sense the information on these data lines and act on that information. The digital voltmeter can be either a talker or a listener. It is made a listener so that it can be programmed for the correct voltage range and told when to take a reading. It is then made a talker so that it can put the results of that reading on the data bus. Temperature difference measurements ΔT between the two ends of the sample — and the corresponding thermal emfs — are stored in data — storage registers. When the temperature gradient is small, the variations of the thermoelectric emf of the sample ΔE are linear with the temperature. The T.E.P. of the sample against copper is computed from ΔT and ΔE measurements by using a least square method, the coefficient of determination giving an indication about the validity of the measurements. Measurement storage allows to make and store voltage measurements at rates as high as three measurements per second. So, each stored value is a mean between ten measurements, and the error on the last digit is lowered. The storage of a great number of measurements and the use of statistical functions for the determination of the slope corresponding to $\Delta E = f(\Delta T)$ variations, combine speed precision and repeatibility and lead to the obtainment of more significant results. The value of the thermoelectric emf can be related to the temperature by a third degree equation, the coefficients of which have been fitted on computer in the ranges between 77-300 K and 300-800 K.

Measured values of the conductivity σ (Ω^{-1} cm^{-1}) on MoTe$_2$, MoTe$_{1,985}$, MoTe$_{1,97}$, MoTe$_{1,95}$ over the temperature range 77-770 K are given in figure 1. The results are plotted as Ln σ versus $10^3/T$. The general behaviour of the four compounds is about the same.

There can be three processes leading to conduction in these crystals and their relative contributions change markedly in different temperature ranges : at low temperature, we can see that the conductivity is thermally activated. The low value of the activation energy shows that conduction can occur by thermally assisted tunneling between impurity-sites. At higher temperature the conductivity is an increasing function of the reciprocal temperature. This result can be explained if the charge - carriers are scattered by optical modes. The same carriers being scattered in two ways (impurity atoms and optical lattice vibrations), the electrical resistivity due to each component may be added together directly to give the total resistivity. At still higher temperature carriers are excited in the extended states of the valence or conduction band. In the intrinsic domain, the plot of the conductivity as a function of the temperature shows a T-dependence, characteristic of scattering by ionized impurities. The resultant electrical conductivity is obtained by adding the conductivities of each group supposed acting independtly.

The T.E.P. variations of three MoTe$_{2-x}$ non-stoichiometric compounds as a function of the reciprocal temperature are shown in figure 2.

As for MoTe$_2$, the graphs show a N-type behaviour for the two first compounds while the MoTe$_{1,95}$ crystal is P-type in the high temperature limit. As for the conductivity, the same three processes govern the mechanism leading to the total T.E.P. At low temperatures, the experimental results show a linear T-dependance of the T.E.P. At higher temperatures, the conduction is governed by the interaction with the optical modes, leading also to a linear T-dependence of the T.E.P. At still higher temperatures, the T.E.P. is given by $S = k(\Delta E/kT + 4)/q$ where ΔE,

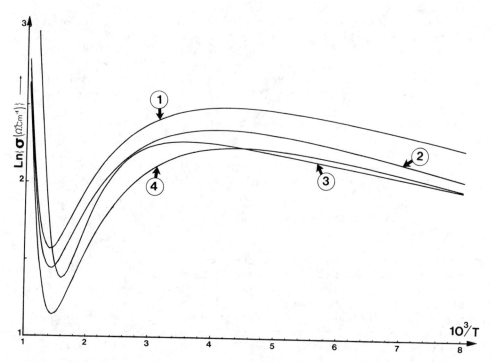

Fig.1 : Theoretical results for the electrical conductivity
of MoTe$_{2-x}$
(1) MoTe$_2$, (2) MoTe$_{1,985}$, (3) MoTe$_{1,97}$, (4) MoTe$_{1,95}$.

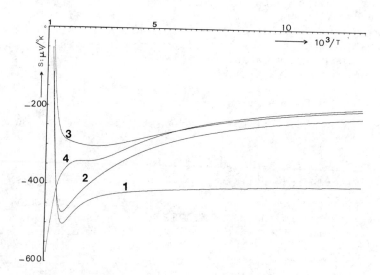

Fig.2 : Theoretical results for T.E.P. of MoTe$_{2-x}$
(1) MoTe$_2$, (2) MoTe$_{1,985}$, (3) MoTe$_{1,97}$, (4) MoTe$_{1,95}$.

the band width between the Fermi level and the valence or
conduction band is a linear function of T over the temperature
range measured : $\Delta E = \Delta E_o - \lambda T$.

For the three first single crystals, we can observe the
following features. The numerical value $\gamma \simeq 1,28 \ 10^{-3}$ eV K^{-1}
deduced from the T.E.P. measurements is very similar to that found
by several authors on different chalcogenides by optical and T.E.P.
measurements[5,6]. The results obtained in the low temperature limit
show that the mean distance between impurity sites R_0 remains
nearly constant with a precision of 2 % ($\varepsilon \simeq 8,4$ meV) and
consequently the number of impurities ($R_0 \simeq 24$ Å). Moreover,
the thermoelectric power measurements show a linear T-dependance
with a constant which decreases in absolute value showing that
the scattering mechanisms become more and more elastic and we
can reasonably think that for MoTe$_{1,96}$[7], this constant must vanish
(metallic behaviour). The main feature is observed at higher
temperatures, in the intrinsic domain, where the type of
conductivity changes and becomes P-type. The Fermi level deduced
from the slope of the T.E.P. and conductivitiy graphs is about
the same for the three first compounds ($\sim 0,495$ eV below the
conduction band) while for the fourth one, it is about $0,515$ eV
above the valence band. So, the total gap is $1,01$ eV which is the
value found by optical measurements ($1,03$ eV)[1].

These results show that the transport properties of
such materials depend on the preparation of the sample, the
vapour-phase method leading to heavily doped semi-conductors.
It is interesting to notice that all the processes which take
part in the conduction mechanisms are governed by the presence
of bromine and lacunar sites. For non-stoichiometric MoTe$_{2-x}$
single crystals, the vacancies act as trapping centers
for the impurity electrons. So, when the departure from
stoichiometry is sufficiently high, the conduction occurs by
holes excited in the valence band (vacancies acting as acceptor
levels) leading to a P-type conduction.

REFERENCES

1. A. Conan, D. Delaunay, A. Bonnet, A.G. Moustafa and
   M. Spiesser, Physica Status Solidi (b) 94 (1979).

2. F. Levy, Il nuovo cimento, 38B, 3, 359 (1976).

3. A. Conan, G. Goureaux and M. Zoater, Journal de Physique
   Appliquée, 6, 383 (1971).

4. A. Conan, M. Zoaeter and G. Goureaux, Physica Status Solidi (a)
   38, 505 (1976).

5. B.T. Kolomiets and B.V. Pavlov, Soviet Physics Semiconductors, 1, 350 (1967).

6. C.H. Hurst and E.A. Davis, Journal of Non-Crystalline Solids, 16, 343 (1974).

7. A. Conan, G. Goureaux, M. Zoaeter, Journal of Physical and Chemical Solids 36, 315 (1975).

ULTRASONIC QUANTUM OSCILLATIONS IN ANTIMONY IN THE

INTERMEDIATE REGION

H. Çelik and T. Alper

Hacettepe University, Department of Physics

Beytepe - Ankara, Turkey

The Fermi Surface of antimony has been studied with oscillatory variation of longitudinal ultrasonic wave attenuation in the frequency range of 10-270 MHz at 4.2 K. Investigation of the angular dependence is as follows : q//x, H in y-z plane; q//z, H in x-y plane; q//x, H in x-z plane; q//x, H in x-y plane. The results have been discussed by using the ellipsoidal model of the Fermi Surface of antimony. The ellipsoidal parameters have been compared with the parameters obtained by ultrasonic and other techniques.
So far, studies on ultrasonic quantum oscillations have concentrated on either the dHvA region or on the giant quantum oscillation region. In this study the intermediate region (13 to 23 kG in Sb) has been investigated in details. Attempt has been made to compare the cyclotron masses deduced by the angular dependence of period, line shape and amplitude analysis of the oscillations.

1. INTRODUCTION

Ultrasonic studies on Antimony accumulated on either low or high magnetic field regions giving de Haas van Alphen (dHvA) type[1,2] or Giant Quantum Oscillations (GQO)[3-6] respectively. There is no detailed study in the intermediate region except for that of Matsumoto and Mase[6] who studied dHvA to GQO transitions. In our work, we have put emphasis on the transition region (H = 13 - 23 kG). In addition complementary measurements were taken on Geometric Resonance (GR).

The Fermi Surface of antimony consist of ellipsoidal packets. In the first Brillouin zone there are three electron and six hole packets. The ellipsoidal parameters can give information on effective masses. Throughout the analysis we used α_{ij} the components of the inverse mass tensor. The packets were named with the conventional nomenclature.

In the dHvA region the period (P) of the quantum oscillations is[7]

$$P = H_n^{-1} - H_{n-1}^{-1} = e\hbar/m_c^* cE_f = 2\pi e\hbar/cA_{ext} . \qquad (1)$$

In this equation m_c^* is the effective cyclotron mass and A_{ext} is the external cross sectional area of the Fermi Surface perpendicular to the magnetic field. The period gives information on α_{ij} via A_{ext}. In general the period may be written in the form

$$P = (e\hbar/m_o cE_f)(C_1 Cos^2\theta + C_2 Sin^2\theta + C_3 Sin\theta Cos\theta)^{1/2} \qquad (2)$$

where C_i's represent combinations of the ellipsoidal parameters and θ is angle between H and a reference direction[1].

2. EXPERIMENTAL METHODS

To measure the attenuation of the ultrasonic waves in antimony, we used the single ended pulse-echo technique. Ultrasonic waves were launched using 10 mm diameter quartz transducer with nominal fundamental frequency of 10 MHz. The impedance matching was achieved by stup tuners and/or coaxial cable delay lines. The pulse-echo system consisted of a Matex 9000 Ultrasonic Attenuation Comprator and a Matec 2470 Automatic Attenuation Recorder. The system was capable of detecting as low as 0.01 dB attenuation changes. Most of the measurements were taken from two echoes. At high frequencies where the attenuation was very high only one echo was visible and in this case changes were monitered from the height variation of that echo. 30 and 50 MHz ultrasonic frequencies were found to be quite satisfactory, except that in GR much higher frequencies were used.

All measurements were taken at liquid helium temperature. A spatial jig was designed to align the sample with respect to the magnetic field inside the metal dewar. The estimated accuracy of the alignment was better than 0.5 degrees. In angle dependent measurements a 15 inch electromagnet was rotated through its vertical axis.

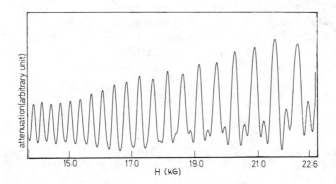

Fig.1 : q // z, H // x-y,   θ = 30° from x-axis, f = 30 MHz.

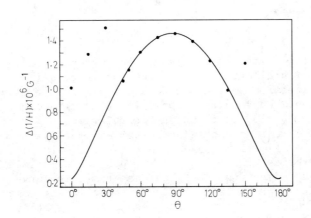

Fig.2 : CASE I: q // x, H // y-z, θ from z-axis.

The sample used was prepared from a zone refined ingot
purchased from Material Research Co. The quoted purity was
99.999 %. The residual resistance ratio was measured to be 510.
The crystallographic axis were determined with back reflection
Laue technique. As also confirmed by cleavage, the accuracy of
the prepared planes, with respect to the desired crystallographic
direction, were much better than 0.5 degree. The tolerance on the
flatness and parallelism of the surface required by the high
frequency ultrasonic technique, was achieved by a spark erosion
wire cutter and a planer.

3. RESULTS AND DISCUSSION

To reveal the complete picture of the Fermi Surface,
measurements were repeated in four different cases. In the course
of measurements the magnetic field was swept in five minutes and
the attenuation variation was recorded. A typical recorder output
is presented in figure 1. In most cases the attenuation change
did not exceed 0.2 dB. At higher fields oscillations behave as
peaks rather than having a sinusoidal shape which is characteristic
of the GQO. This might indicate a transition from dHvA to GQO
region. Observed spin splitting at higher fields also confirms this
transition. The period of the oscillations was found from the slope
of the peak number versus 1/H plot. To obtain information on the
angular dependent period, a similar procedure was repeated with
different magnetic field directions. The studied four cases are
as below.

CASE I : $q \, // \, x$ and H in y-z plane.

In this notation x, y and z indicates binary, bisectrix
and trigonal crystallographic axis respectively, q is the
propagation direction of the sound wave and H is the magnetic
field direction. The same notation will be used throughout. The
angle dependent period is shown in figure 2. The experimental
points on the solid curve originate from an a-ellipsoid. Few points
above the solid line might be due to hole ellipsoid contributions.
The best theoretical curve (Eqs.2) through the experimental point
was obtained by adjusting the C_i parameters. This curve has a
maximum at 87.5 degrees indicates that the tilt angle measured
from the trigonal plane is 2.5 degrees. From the estimated C_i's
the ellipsoidal parameters can be obtained.

CASE II : $q \, // \, x$ and H in x-z plane

This configuration was derived from CASE I by tilting
the sample 30 degrees along the z-axis. As also shown in figure 3

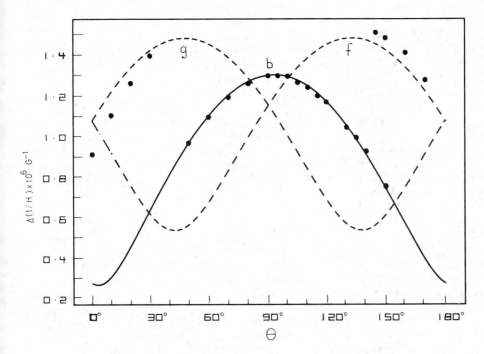

Fig.3 : CASE II, q // x, H // x-z, θ from z-axis.

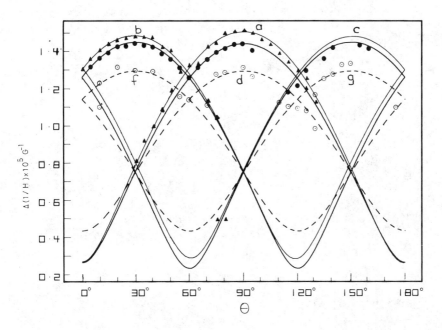

Fig.4 : CASE III and CASE IV, θ from x-axis.

the angle dependent period gives information on b, f and g-
ellipsoids. Since there are not sufficient experimental points
belonging to f and g-packets, the dotted line in this figure
was drawn using Mori and Mase's[3] parameters.

CASE III : q // z and H in x-y plane.

        The angle dependent period is shown in figure 4. As
also indicated in this figure the contributions come from a, b
and c electron ellipsoids (filled circles) and f, d and g-hole
ellipsoids (dotted circles). The periods belonging to hole ellipsoids
were separated from the best frequencies on the oscillations. The
broken curve was drawn with the parameters obtained by Mori and
Mase[3].

CASE IV : q // x and H in x-y plane.

        This and the previous case (CASE III) probe the same
ellipsoids. The only difference is the angle between q and H
In CASE III sound waves propagated perpendicular to the magnetic
field. In CASE IV this angle varies by changing the magnetic field
direction. As shown in figure 4 (triangles) the period is greater
than the period obtained from the previous case. This might imply
GQO. In order to obtain information on the period originating from
this phenomenon, A_{ext} in Eqs.(1) must be replaced with the cross
sectional area of the Fermi Surface perpendicular to the magnetic
field at $k_z = k_o$ ($k_o = m^*V/\hbar Cos\phi - q/2Cos\phi$, where V is the velocity
of sound and ϕ is the angle between H and q)[7]. Since cross
sectional area is always smaller than the extremal area, the
period resulting from the GQO is greater than the period of
dHvA type oscillations.

GR : q // x , H in y-z plane.

        In this investigation GR experiments were performed
to deduce accurately some of the parameters. A typical oscillation
is shown in figure 5. The oscillations start at around 50 G and
the amplitude of the oscillations increase with increasing
magnetic field. The attenuation change is around 20 dB. The
period analysis allows us to obtain independently α_{33} with an
accuracy of better than 7 %. This will permit to solve the
ellipsoidal parameter.

Line Shape and the Amplitude Analysis :

        Since the slope of the $ln\alpha(H)H^{1/2}$ versus 1/H curve
may be taken as $-2\pi^2kTcm_c^*/e\hbar$, the amplitude analysis may be used
to obtain the cyclotron masses[6] without the use of Fermi energy[6].

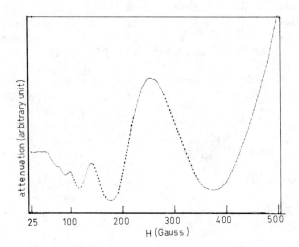

<u>Fig.5</u> : GR; q // x, H // y-z, θ = 70° from z-axis; f = 210 MHz.

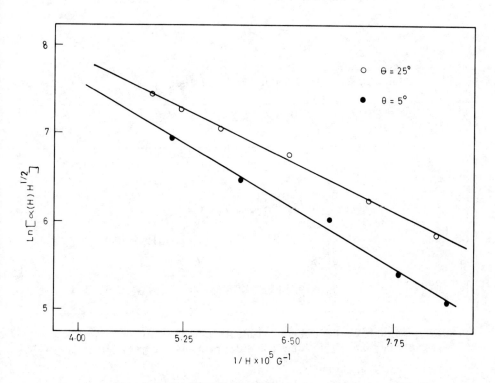

<u>Fig.6</u> : Amplitude Analysis for CASE IV.

An attempt was made to investigate the amplitude and the line
shape behaviour for CASE IV where the oscillations result from
a single source. A typical plot for the two θ values are shown
in figure 6. From the slope, the effective masses were calculated
to be 0.091 m_o and 0.083 m_o for 5 and 25 degrees respectively,
whereas the values obtained from the period analysis are 0.090 m_o
(for 5 deg.) and 0.083 m_o (for 25 deg.); the agreement is good.

In the GQO region the half width of the oscillation is[7]
3.53 kTPH$_n$ cm$^*_c$/eh. The effective cyclotron masses from period and
half width analysis are within 15 %.

One of the main objects of this study is to find the
ellipsoidal parameters. In table 1 the parameters of the electron
packets, Fermi energy, tilt angle and the carrier densities are
presented for our sample as well as for others.

The cyclotron masses calculated with Eqs.2 from the
measured period, along the major crystallographic axis are
presented in table 2. In this table the values other than ours
were taken from table II of ref. (11). From the two tables it
may be concluded that our results fall into a general scheme.

Table 1 : Comparison of α Tensor Elements, Fermi Energy E_f,
Density of Carriers n_e and Tilt Angle ψ.

Exp. Method	α_{11}	α_{22}	α_{33}	α_{23}	E_f	ψ	n_e
GM and SdH[8]*	6.0	0.7	17.3	1.0	13		3.8
CR[9]*,**	10.7	0.94	11.5	−0.83		4°	4.2
US[2]*	9.4	0.43	14.5	0.62	13		4.2
GM[10]					15.2	3°	4.2
US (GQO)[3]	10.3	0.36	14.9	−0.54	15.2	2°8'	5.5
Present work	10.5	0.66	13.9	−0.62	15.2	2.5°	4.2

GM Galvanomagnetic; US Ultrasonics; SdH Shubnikov de Haas;
E_f Fermi energy in 10^{-14} erg ; n_e Carrier density in 10^{19} cm^{-3}.
* Originally attributed to hole
** In the original work effective masses were given which are
here transformed into α_{ij}.

Table 2 : Comparison of cyclotron masses m_c^*/m_o along the crystallographic directions

Ref.No.	HOLES			ELECTRONS		
	x	y	z	x	y	z
9	.215	.075		.315		
	.080	.12	.103	.097	.315	.315
12		.082			.084	
13			.114			
8	.31	.069	.090	.303		.48
				.11		
11		.080	.105			
present work	.093	.081	.105	.098	.083	.128

3. CONCLUSION

On theoretical basis, GQO may be observed when the direction of the sound wave is not perpendicular to the magnetic field. In our experiment this condition is provided in CASE II and IV. In practice, the exact right angle condition can not be fullfilled and at high fields GQO may be observed. This effect could be seen in CASE III and IV where the results probe the same ellipsoids. Due to the angle relation between q and H the CASE IV results give slightly larger periods. This is seen in figure 4. A greater period separation could be expected for higher fields.

Although the cross sectional area at $k_z = k_o$ is smaller than the extremal area, the difference may be neglected for the ellipsoid like packets, GQO and dHvA periods one very close together. As shown in table 1 the parameters obtained from both and the present work are within experimental error; the transition from the dHvA to GQO region could not be observed on period analysis. Peak-line shape and the spin splitting of the oscillations of high magnetic fields (figure 1) imply the intermediate region.

REFERENCES

1. J.B. Ketterson, Physical Review, 129, 18 (1963).

2. L. Eriksson, O. Beckman, S. Hörnfeldt, Journal of the Physics and Chemistry of Solids, 25, 1339 (1964).

3. H. Mori and S. Mase, Journal of the Physical Society of Japan 31, 738 (1971).

4. Y. Suido, Y. Yosida, S. Mase, Journal of the Physical Society of Japan, 39, 109 (1975).

5. Y. Yosida, S. Mase, Y. Suido, Journal of the Physical Society of Japan, 39, 1661 (1975).

6. Y. Matsumoto, S. Mase, Journal of the Physical Society of Japan, 38, 1328 (1975).

7. Y. Shapira, Physical Acoustics, part 5. Edited by W.P. Mason (New York; Academic Press) pp.1-58 (1968).

8. G.N. Rao, N.H. Zebouni, C.G. Grenier, J.M. Reynolds, Physical Review, 133, A 141 (1964).

9. W.R. Datars and J. Vanderkooy, IBM Journal of Research and Development, 8, 247 (1964).

10. Ö. Öktü, G.A. Saunders, Proceedings of the Physical Society, 91, 156 (1967).

11. Y. Ishizawa, Journal of the Physical Society of Japan, 25, 150 (1968).

12. L.R. Windmiller, Physical Review, 149, 472 (1966).

13. N.B. Brandt, N.Ya. Minina, Kang Chu Chuen, Soviet Physics - JETP, 24, 73 (1967).

EVIDENCE OF THE ELECTRON VELOCITY RUNAWAY IN

POLAR SEMICONDUCTORS

J.P. Leburton

Siemens AG, Research Laboratories

Otto-Hahn-Ring 6, D-8000 München 83, W.-Germany

A new approach of treating the Boltzmann equation for electron-polar optical phonon (P.O.P.) interaction scattering is described. It is based on electron wave vector direction conservation for the high electron energy P.O.P. scattering, what gives a simple form to the collision integral. Afterwards, information on the distribution function is derived from a Fokker-Planck-like equation. It is shown that the tail of the distribution function exhibits a minimum before saturating up to infinity. The possibility of this velocity runaway is then discussed.

In polar semi-conductors for high electron energy, the Boltzmann-equation (B.E.) can be derived into a relatively tractable form allowing to get information on the tail of the carrier distribution function (D.F.)[1]. In this framework, runaway is of interest because for high electron processes e.g. for impact ionization and Gunn effect, the calculation of the D.F. must take this phenomenon into account in the determination of the collision frequencies.

Several authors[2-4] studied this problem, neglecting the electron-electron interaction. Their analysis starts with the D.F. limited to two terms of the Legendre polynomial expansion. Although they obtain a confirmation of the velocity runaway in polar semi-conductors, we think that this method is not appropriate to a situation where the D.F. is strongly "extruded" in the electric field direction[5].

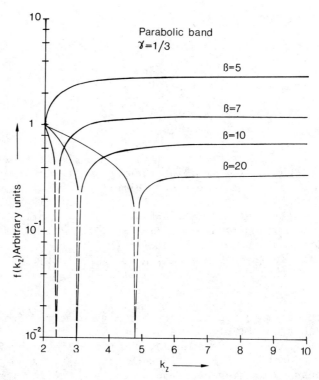

<u>Fig.1</u> : Profile of the D.F. in the electric field direction for different values of the parameter β. The unit of k_z is $(2m\omega/\hbar)^{1/2}$. The D.F. has been normalized to one for $k_z = 2$.

Our approach is based on the "quasi" conservation of the electron wave vector direction during the collision process of highly energetic carriers with the polar optical phonons (P.O.P.)[6]. This is a consequence of the q^{-1} (q = phonon wave vector)-dependence of the matrix element of the electron-P.O.P. interaction[7-8] which yields a rather simple form for the transition probability between $\vec{k}$ and $\vec{k}'$, respectively before and after the P.O.P. scattering[1].

$$W(\vec{k}-\vec{k}') \sim \delta(\cos \Theta - 1) \, \delta(\varepsilon - \varepsilon' \pm \omega) \qquad (1)$$

Θ is the angle between $\vec{k}$ and $\vec{k}'$, and the second δ-function expresses the energy conservation for the absorption and the emission of phonons with a frequency ω. Then, for $\hbar^2 k^2/2m \geqslant 4\hbar\omega$ where m is the effective mass, a Fokker-Planck-like equation is derived from the B.E. neglecting the second order terms, this equation is :

$$\frac{\partial}{\partial k_z} f(\vec{k}) = \frac{\beta}{k} [(\text{sh}^{-1}(\hbar k^2/2m\omega)^{1/2} - \cosh^{-1}(\hbar k^2/2m\omega)^{1/2}) f(\vec{k})$$

$$- \frac{2m\omega}{\hbar} \frac{(\text{sh}^{-1}(\hbar k^2/2m\omega)^{1/2} - \gamma\cosh^{-1}(\hbar k^2/2m\omega)}{2(1-\gamma)k} \frac{\partial}{\partial k} f(\vec{k})] \qquad (2)$$

for a parabolic band. Here, k_z is the wave vector component in the electric field direction, k is the absolute value of $\vec{k}$, $\beta = 2\alpha\omega(2m\hbar\omega)^{1/2}/eE$, α is the electron-phonon coupling constant, E is the electric field and $\gamma = \exp(-\hbar\omega/k_B T)$ where T is the temperature and k_B the Boltzmann constant. In polar coordinates, omitting the φ-dependence of the D.F. (azimuthal symmetry), we obtain :

$$[\frac{2m\omega\beta}{(1-\gamma)\hbar} \frac{(\text{sh}^{-1}(\hbar k^2/2m\omega)^{1/2} - \gamma\cosh^{-1}(\hbar k^2/2m\omega)^{1/2}}{k^2} - \cos \theta] \frac{\partial}{\partial k} f(k,\theta)$$

$$+ \frac{\sin \theta}{k} \frac{\partial}{\partial \theta} f(k,\theta)$$

$$+ \frac{\beta}{k} (\text{sh}^{-1}(\hbar k^2/2m\omega)^{1/2} - \gamma\cosh^{-1}(\hbar k^2/2m\omega)^{1/2}) f(k,\theta) \qquad (3)$$

if θ is the polar angle of k_z. Setting $\theta = 0$, the D.F. profile along k_z can directly be deduced. It is seen that the D.F. exhibits a minimum going to zero when the first term between brackets of eq.(3) vanishes. In figure 1 we observe effectively a minimum after which the D.F. saturates. The minimum point is displaced towards the high k_z-values with β, as expected. Beyond

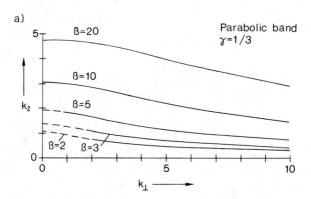

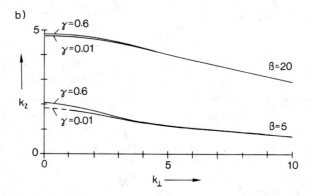

Fig.2 : (a) Curves of the D.F. minima for different values of β. k_z and $k_\perp = (k_x^2 + k_y^2)^{1/2}$ are expressed in $(2m\omega/\hbar)^{1/2}$ units. The dashed parts of the curves are for $\hbar^2 k^2/2m \leqslant 4\hbar\omega$ where the model fails.
(b) Variation of the minimum curve for different values of γ with β = 20 and β = 5. A γ-variation of more than 60 times shows only a slight shift towards the high values of k_z.

this point, the phonon scattering is not able to dissipate the
carrier energy gained from the electric field and the D.F. takes
progressively a constant value.

More generally, from eq.(3), we obtain the dependence
of the D.F. minimum on $\vec{k}$. This is given by both relations
$\partial f/\partial \theta = 0$ and $\partial f/\partial k = 0$[9]. The expression of $f(k,\theta)$ is not needed
and the minimum curve can be deduced from the coefficient of
$\partial f/\partial k$, setting this equal to zero[1]. From figure 2a it is seen
that the runaway is not only an energy function but spread for
very high values of $k_\perp$. The slight decrease of the curves with k
is due to the "radial" form of the electron-P.O.P. scattering
which, for large $k_\perp$ cannot balance the electron momentum gain
from the vertical electric field. The minimum front decreases
with E and for $E = 2\alpha\omega(2m\hbar\omega)^{1/2}/10$ the runaway occurs already
at $\hbar^2 k_z^2/2m \simeq 10\ \hbar\omega$. Using a method developed by Levinson[3], Aas
and Bløtekjaer have found a runaway critical energy[4]

$$\varepsilon_c = \sqrt{3}\ \hbar\omega\ \frac{(1 + \gamma)E_o}{(1 - \gamma)E}\tag{4}$$

With $E_o = \alpha\omega(2m\hbar\omega)^{1/2}$, ε_c is $\sqrt{3}\ 10\ \hbar\omega$ for $\gamma = 1/3$, which has the
same order of magnitude as our results. The relation (4) gives a
rather sensitive variation of this critical energy with the
temperature ($\gamma \to 1$, $\varepsilon_c \to \infty$). However Figure 2b shows that, in
reality the effect of the temperature is quite small and that,
for high electron energy, the phonon emission is the only
important mechanism of energy and momentum relaxation. Indeed,
in this case, that part of the emission transition probability
which is proportional to the phonon population, N_g, balances the
absorption transition probability in the total collision rate.
Therefore, the contribution of these temperature dependent
factors is negligible.

Practically, in GaAs for example, if we neglect the
non parabolicity of the conduction band, the velocity runaway
takes place at $E = 2.3$ kV/cm, below the threshold field
($E_c \sim 3$ kV/cm) of the Gunn effect[10]. Obviously, a more rigourous
analysis of the process must include the non parabolicity of the
band, and, in this case, an important effect on the transport
properties at high electron energy, is expected[1].

Acknowledgements - The author would like to express his thanks
to Dr. G. Dorda for reading and commenting on the manuscript.

REFERENCES

1. J.P. Leburton, to be published.

2. H. Fröhlich, Proceedings of Royal Society of London 188A
   532 (1947).

3. I.B. Levinson, Soviet Physics-Solid state 7, 1098 (1965).

4. E.J. Aas and K. BlØtekjaer, Journal of Physics and Chemistry
   of Solids 35, 1053 (1974).

5. D. Kranzer, H. Hillbrand, H. Pötzl and O. Zimmert, Acta
   Physica Autriaca 35, 110 (1972).

6. W.P. Dumke, Physical Review 167, 783 (1968).

7. This point is of crucial importance. So, the optical
   deformation potential matrix element is either constant or
   proportional to q^{11-13}, giving for high energy a relaxation
   time proportional to $\varepsilon^{-1/2}$ and $\varepsilon^{-3/2}$ respectively. Keeping in
   mind a Fröhlich's argument[2] - roughly speaking, the rate of
   energy transfer from the field to an electron is on an average
   over many collision proportional to the relaxation time,
   whereas the rate of energy transfer from electron to the
   lattice vibrations is inversely proportional to it" - even
   for fast electrons the relaxation time decreased with the
   electron energy and the carriers can always dissipate energy
   into the lattice. Therefore, in covalent semi-conductors like
   Si or Ge, such a runaway effect can not be expected.

8. H. Fröhlich, Philosophical Magazine Suppl. 3, 325 (1954).

9. M. Abromovitz and I.A. Segun, Handbook of Mathematical
   Functions, DOVER publ. Inc. New York (1968).

10. E.M. Conwell and O.M. Vassel, Physical Review 166, 717 (1968).

11. E.M. Conwell, Solid State Physics, Suppl. 9, F. Seitz, D.
    Turnball, H. Ehrenreich editors Academic Press Inc. New York
    (1967).

12. W.A. Harrison, Physical Review 104, 1281 (1956).

13. D.K. Ferry, Surface Sciences 57, 218 (1976).

LUMINESCENCE AND RESONANT RAMAN SCATTERING IN

Mn$_x$Cd$_{1-x}$Te MIXED CRYSTALS

M. Picquart, E. Amzallag, and M. Balkanski

Laboratoire de Physique des Solides, associé au CNRS
Université Pierre et Marie Curie
4 Place Jussieu - 75230 Paris Cedex 05, France

The photoluminescence of Mn$_x$Cd$_{1-x}$Te is studied for various Mn contents $0.05 \leqslant x \leqslant 0.5$ at 2 K. The free exciton energy is observed to depend linearly on x. A broad band located at 16100 cm^{-1} is attributed to electronic transitions within Mn^{2+} ions.
Resonant Raman scattering measurements performed for x = 0.2, 0.3, 0.4 at 77 K, using a tunable dye laser, show resonance effects of the LO$_1$ (CdTe-like), LO$_2$ (MnTe-like), LO$_1$ + LO$_2$ phonon modes, on the scattered photon.

1. INTRODUCTION

Mn$_x$Cd$_{1-x}$Te alloys are being studied extensively as large gap semiconductor and magnetic materials. Their strong magneto-optical effects[1,2] have been explained by means of exchange interaction between the localized magnetic moments of Mn^{2+} ions and exciton or band electron states. Previous reflectivity[3], infrared absorption and Raman scattering measurements[4] have allowed to investigate the lattice vibrations of these mixed crystals. In particular, when the Mn concentration is varied from x = 0 to x = 0.6, the frequencies of the long-wavelength optical phonons are observed to follow a "two-mode" behaviour, i.e. the normal modes related to each one of the constituents can be separately observed.

Photo-luminescence measurements have been reported on these materials[5] which were restricted to Mn contents up to x = 0.3. This paper presents luminescence data extended up to x = 0.5 as well as resonant Raman scattering results, obtained on samples with Mn concentrations of 0.2; 0.3; 0.4, using a cw tunable dye laser.

2. EXPERIMENTAL

The $Mn_xCd_{1-x}Te$ single crystals studied come from the Institute of Physics of Polish Academy of Sciences (Warsaw) where they were grown by means of a modified Bridgman technique. Their composition was determined by chemical analysis and their zinc-blende structure was confirmed by an X-ray examination.

Luminescence experiments were performed on samples with Mn contents varying from x = 0.05 to x = 0.5, in a liquid helium cryostat pumped to 2 K, using the 4880 Å exciting line of an Ar^+ laser.

Resonant Raman scattering was measured at 77 K, for three concentrations x = 0.2, 0.3, 0.4. Three different dye lasers, pumped with a 12 W Ar^+ laser were used : Rhodamine B for x = 0.2, Rhodamine B and 6G for x = 0.3, and Coumarin 7 for x = 0.4. The light power on the sample was about 5 to 10 mW depending on the dye. A premonochromator was set up in order to eliminate the dye luminescence. The instrumental resolution was approximately 20 Å. All the measurements were made in a quasi-back-scattering geometry, the scattered light was dispersed by a Coderg PHO double monochromator and detected by a cooled S20 photomultiplier.

3. RESULTS AND DISCUSSION

The luminescence spectra obtained for an excitation energy of 2.54 eV (above bandgap) are gathered on fig.1. The highest energy line is attributed to free exciton luminescence. For x = 0.05 its energy is 1.67 eV, value which was obtained previously[5] using the same technique. When Mn content is varied from 0.05 to 0.5, the free exciton energy is found to depend linearly on composition according to the following equation

$$E_x = 1.595 + x1.56 \text{ (eV)}$$

This variation is comparable to that proposed by Gaj et al[6] from exciton reflection measurements, except that their data correspond to 77 K, instead of 2K here. On the other hand, this line is

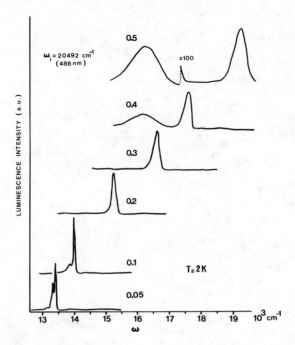

<u>Fig.1</u> : Luminescence spectra of $Mn_x Cd_{1-x} Te$ for different Mn
        concentrations at 2 K with the exciting line 488 nm.

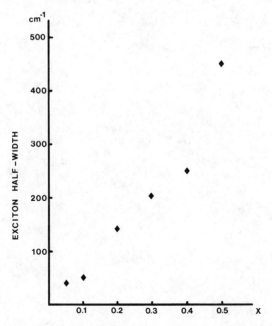

<u>Fig.2</u> : Luminescence exciton half-width in $Mn_x Cd_{1-x} Te$ versus Mn
        concentration at 2 K.

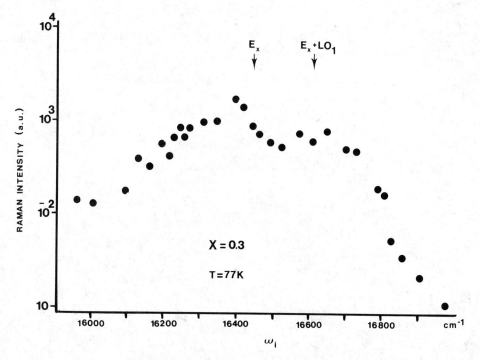

Fig.3 : Raman intensity of $Mn_{0.3}Cd_{0.7}Te$ versus incident light
frequency at 77 K, for the Stokes components.
a) LO_1 phonon (165 cm^{-1})

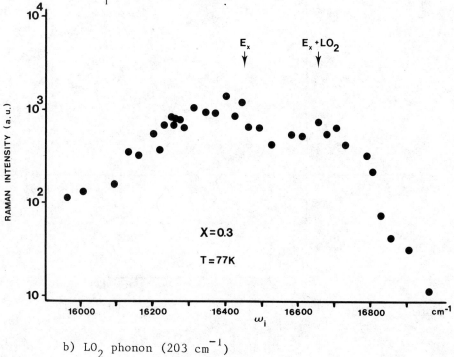

b) LO_2 phonon (203 cm^{-1})

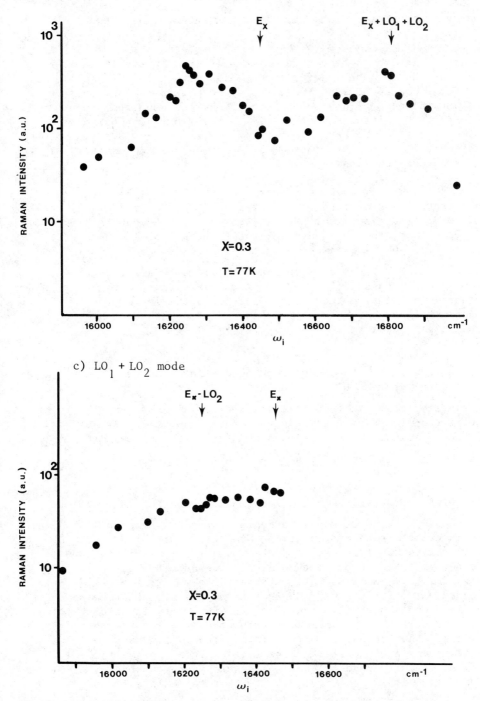

c) $LO_1 + LO_2$ mode

Fig.4 : Raman intensity of $Mn_{0.3}Cd_{0.7}Te$ versus the incident light
frequency at 77 K for the anti-Stokes LO_2 phonon.

observed to broaden with increasing Mn contents, as shown on
fig.2.

At the low energy side of the free exciton line, a small
peak is clearly observed for the samples with x = 0.05 and
x = 0.1, located at 1.66 and 1.725 eV respectively. This peak
was attributed by Planel et al[5] to an exciton bound to a neutral
acceptor. The binding energy is found to be here approximately
10 meV for x = 0.05 and 15 meV for x = 0.1. For x $\geq$ 0.2, the bound
exciton is not resolved, but it probably causes the asymetry of
the free exciton line.

The luminescence spectra for x = 0.4 and x = 0.5 are
characterised by a broad band centered at $\sim$16100 cm^{-1} (2 eV) whose
half-width increases from 800 cm^{-1} for x = 0.4 to 950 cm^{-1} for
x = 0.5. Such a broad band below the free exciton line was also
observed on photoluminescence spectra of $Mn_xCd_{1-x}Se$ by Giriat and
Stankiewicz[7]. This band is most probably due to electronic
transitions within Mn^{2+} ions. Koi and Gaj[8] suggested these
transitions might be responsible for the deviation of the
fundamental absorption edge in these mixed crystals from the
usual exponential shape, when x $\geq$ 0.4.

The resonant Raman scattering (RRS) spectra of $Mn_xCd_{1-x}Te$
crystals show that the more pronounced effects appear on the
lines involving the LO_1, LO_2 and $LO_1 + LO_2$ phonon modes. It should
be recalled that, in these alloys, the optical phonons at Γ point
are of two-mode type[4]. For example, for x = 0.3, the frequencies
of the modes LO_1 (CdTe-like) and LO_2 (cubic MnTe-like) are 165
and 203 cm^{-1} respectively, at liquid nitrogen temperature.

An investigation of the intensities of the Stoke
components LO_1, LO_2, $LO_1 + LO_2$, carried out for x = 0.2, 0.3 and
0.4 showed that these modes resonate on the scattered photon, i.e.
the relevant resonance conditions are those for which the
scattered photon frequency is equal to the free exciton transition
frequency. This can be seen on figure 3, where the intensities of
LO_1, LO_2, $LO_1 + LO_2$ lines are represented versus the exciting light
frequency ω_i, for x = 0.3. Whether these modes resonate also on
the incident photon is not well established yet. One reason for
this could be that we did not correct the data for the absorption.
Besides, other resonance effects are probably existing for
exciting energies lower than the free exciton energy (fig.3c).
They might correspond to Raman processes where the energy of a
transition within Mn^{2+} ions, i.e. with intermediate states on
the localized Mn 3d states. Measurements on a sample with x = 0.4,
which are in progress, may clear up this point. As can be seen in

figure 1, for x = 0.4, the free exciton energy is well above the
luminescence band.

At last, figure 4 shows the anti-Stokes of the LO_2 mode
to resonate also on the scattered photon.

All these luminescence and RRS measurements are to be
considered as preliminary results. Selective photo-excitation
experiments may provide more data for an elaborate analysis of
the influence of Mn localized states on the light scattering in
these mixed crystals.

Acknowledgement — Thanks are due to Prof. R.R. Galazka for
providing the samples.

REFERENCES

1. A.V. Komarov, S.M. Ryabchenko, O.V. Terletskii, I.I. Zheru,
   R.D. Ivanchouk, Journ. of Exp. and Theor. Phys. 46, 318 (1977).

2. J.A. Gaj, J. Ginter, R.R. Galazka, Phys. Stat. Sol. b 89, 655
   (1978).

3. W. Gebicki, W. Nazarewicz, Phys. Stat. Sol. b 86, K135 (1978).

4. W. Gebicki, E. Amzallag, M. Picquart, Ch. Julien, M. Le
   Postollec, Intern. Meeting on Magn. Semicond. CNRS, Montpellier,
   France, Sept. 1979 (to be published in Journal de Physique).

5. R. Planel, J. Gaj, C. Benoit à la Guillaume, op.cit. in
   ref.4.

6. J.A. Gaj, R.R. Galazka, M. Nawrocki, Solid Stat. Comm. 25,
   193 (1978).

7. W. Giriat, J. Stankiewicz (private communication).

8. N.T. Khoi, J.A. Gaj, Phys. Stat. Sol. b 83, K133 (1977).

MEASUREMENT OF LUTTINGER PARAMETERS OF THE VALENCE BAND

OF ZnSe BY RESONANT BRILLOUIN SCATTERING

B. Sermage
Centre National d'Etudes des Télécommunications
196 rue de Paris, 92220 Bagneux (France)

G. Fishman
Groupe de Physique des Solides de l'Ecole Normale
Supérieure, Université Paris VII
4 Place Jussieu, 75005 Paris (France)

We report the experimental observation of heavy and light excitons
by resonant Brillouin scattering of exciton polaritons in ZnSe
in the (100), (110) and (111) directions. Using the three branch
polariton model, which is needed in the case of a degenerate
valence band, we fit the experimental Brillouin curves and obtain
the values of the masses of the two kinds of excitons. The
Luttinger parameters of the valence band are then deduced.

It is now well established since the pioneering work of
Brenig, Zeiher and Birman[1] that the Brillouin Scattering is very
efficient when the excitons are involved in the resonance with
acoustic phonons. The first experimental evidence of this effect
was performed by Ulbrich and Weisbuch in GaAs[2] and afterwards
several authors made the same kind of experiment in CdS[3] and CdSe[4].
As suggested in Ref.1 and experimentally checked[2] the Resonant
Brillouin scattering on excitons (R.B.S.) is a particularly
adequate tool to plot the energy dispersion curve of excitonic
polariton[5,6]. The case of GaAs, which has a zinc-blende structure
and whose valence band has a Γ_8 symmetry, is more complex than
the case of wurtzite structure. According to [7] the problem of
exciton dispersion curve has no exact solution; Kane[8] succeeded
to solve this problem using a perturbation method to second order.
For the direct gap semiconductors which are more particularly

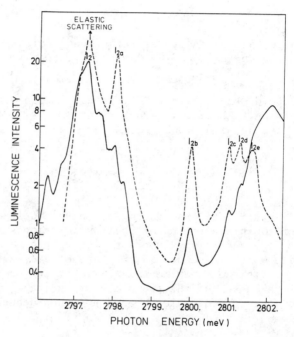

<u>Fig.1</u> : The full line is the luminescence spectrum (excitation
         energy ∿3 eV). The dotted line is the excitation spectrum
         of I_2.

concerned in this paper the calculation of Ref.8 shows that there are two kinds of excitons (heavy and light) with can be simply characterized by their effective masses M_h and M_l with M_h larger than M_l. However the R.B.S. experiments reported on GaAs[2] or CdTe[9] do not clearly show the existence of the two exciton branches. Anyway, the interaction between a photon on the one hand and the heavy and light exciton on the other hand needs a more complete treatment than the usual one used when a single exciton branch exists. For this purpose the knowledge of the wavefunction of excitons is necessary to be able to calculate the coupling with the light. From Ref.8, this leads to a three branch polariton dispersion curve, the intermediate one corresponding to light exciton. Indeed, in Ref.10, the author was dealing only with a wave vector parallel to <100> direction and strictly speaking these results cannot be used just as they are. However, a more precise theory leads to complicated calculation and will be given elsewhere.

If one wishes to observe heavy and light excitons it is necessary to select the "best candidate", i.e. to use a cubic semiconductor with the largest exchange interaction possible. Because linear k term is too large in copper bromide[11], the most suitable compound is zinc selenide. Moreover, it is necessary to have a sample which is oriented in the <100> direction; otherwise, the identification of different lines is too uncertain due to scattering by T.A. phonons as well as L.A. phonons[9]. On the contrary in the <100> direction the scattering by T.A. phonon is forbidden in backscattering[12] (but not in other experimental configuration[13]). The first experimental evidence of the existence of heavy and light exciton was recently reported in ZnSe[14] for the <100> direction and the purpose of this paper is to show that further and indeed necessary information can be obtained from experiments performed in other directions.

The samples were oriented close the <100>, <110> and <111> direction and the main uncertainty in this field proceeds certainly from the angle between incident and scattered light for obvious experimental reasons.

The comparison of spectra corresponding to different directions has allowed us to notice that the splitting between the Rayleigh line and some other lines does not depend on the direction. This rules out the possibility for these lines to belong to the R.B.S. spectra because the phonon velocity depends on the direction of wave vector. A preliminary study of the I_2 complex is necessary to understand the interpretation in details, so we start with this point. Figure 1 presents two spectra. The first one is the luminescence spectrum of ZnSe from

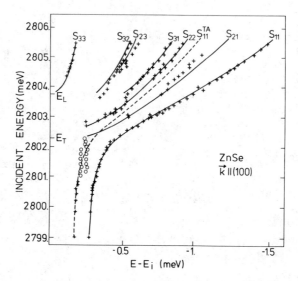

Fig.2 : Experiment : Circles and crossed lines display respectively
        scattering or excited states on I$_2$ (see text), and
        resonant Brillouin Scattering on excitonic polaritons.
        Theory the full (dashed) line displays scattering by
        L.A. (T.A.) phonon.

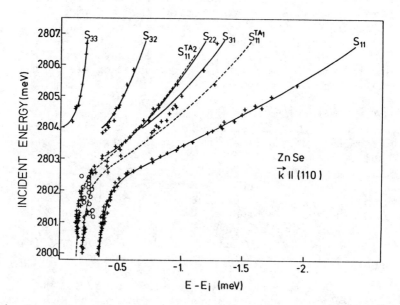

Fig.3 : The various lines have the same meaning as in fig.2.

an excitation energy near 3 eV. This spectrum must be compared
with the excitation spectrum. It must be noticed that the
excitation energy is in this case 2797 meV which is equal to I_2
energy : thus the lines we call I_{2a}, I_{2b}, I_{2c}, I_{2d} and I_{2e} are
identified as excited states of I_2. Only further studies in this
area could enable to know what kind of excited state are involved
in this spectrum[15]. Now a detailed analysis between the three
kinds of spectra gives the following results (Fig.2). Some lines
(marked as 0 in Fig.2 and Fig.3) have a shift (versus the Rayleigh
frequency) which is equal to the splitting between the excited
states of I_2. The order of magnitude of this shift is very near
to the shift of possible scatterings due to T.A. phonon which are
not strictly forbidden due to experimental reason explicited above
and also close to the shift of possible scattering of an intermediate branch in a slightly different interpretation[14] : the
spectra in several directions remove all ambiguity. A resonance
on I_2 was already observed in $CdSe$[4]; the lines reported here have
an intensity which is maximum for an energy equal to that of I_{2e}.
Although everything is not clear it is likely that these peaks
correspond to a resonance on these excited states. Other peaks
are due to T.A. phonon scattering : the intensities are weaker
than the intensities due to L.A. phonon scattering by about two
order of magnitudes.

The measurement of the index (using a prism of 13°) gives
$n = 2.86 + 0.05$ at low temperature ($\sim$ 10 K) at a wavelength equal
to 4462 Å (2778 meV). The Brillouin shifts for T.A. and L.A.
phonons are measured so that the sound velocity is deduced.
Fig.2 presents the set of experimental results for Stokes peaks.
The indexes of lowest, intermediate and upper most branches are
respectively 1, 2 and 3. S_{ij} is a peak which corresponds to a
transition from a branch i to branch j in backscattering. So S_{11},
S_{22} and S_{33} correspond to intrabranch scatterings. Such peaks as
S_{21}, S_{31} ... correspond to interbranch scatterings. All the above
peaks are due to L.A. phonon scattering. On the contrary, the
peaks S_{11}^{TA} are due to T.A. phonon scattering which become visible
near the excitonic resonance and disappear when the I_2 excited
peaks begin to play a part. We can note that in the short energy
range where they are visible, the peaks due to T.A. phonon "follow"
the S_{11} peaks contrarily to the peak originating from I_2.

We can remark that, below the energy E_L there are four
peaks for a given energy : it is not possible to explain these
peaks by a two branch polariton model even qualitatively. As
usual the fit matches first with the intrabranches transition :

it is excellent and even a little astonishing, considering that
Kane's theory is a perturbation theory. The fit is then tempted
for interbranch transitions. The fit is not as good as for the
intrabranch transitions. As a matter of fact, this problem is
not peculiar to the three branch polariton model and exists
even in simple model : see for example the Fig.2b of Ref.2.
Up to now, this point has not been cleared up.

From Fig.2 it is possible to deduce the following
parameters : the longitudinal transverse splitting E_{LT}; the
exchange energy Δ and the two translational masses M_1 and M_h
which describe the behaviour of light and heavy exciton in k-space.
Here it can be useful to note that our new interpretation of peaks
I_α does not qualitatively change the contents of Ref.14 but alters
quantitatively the parameters in a significant way. Another cause
of change is relative to the exchange energy Δ only. We have taken
Δ equal to the splitting δ which is the difference, at $\vec{K} = 0$,
between the energy of excitons $|1 + 1\rangle$ and $|2M\rangle$ and not equal to
$\delta + 1/3\ E_{LT}$ like in Ref.14 (See a discussion in Ref.16). Using
the values of sound velocity we have measured, we obtain the
following parameters E_{LT} = 1.45 $\pm$ 0.05 meV, $\Delta = \delta$ = -0.1 $\pm$ 0.2 meV.
M_1 = 0.38, M_h = 1.11. (It must be noticed that $\delta + 1/3\ E_{LT}$ = 0.5
meV, which is the value with which the exchange energy of Ref.14
must be compared).

Another point calls for a comment. The equation which
gives the polariton dispersion curve depends on two oscillator
strengths $4\pi\beta_1(K)$ and $4\pi\beta_2(K)$ where the index 1 refers to heavy
excitons (at least for large enough wave vector) and the index 2
refers to light excitons. In general case, these two oscillator
strengths are K-dependent and only their sum $4\pi\beta_o$ is constant
(as in CuBr where the linear k term is predominant[11]). The
agreement is such that the theory can be considered as quite
convenient. Nevertheless due to the smallness of δ (which is a
more relevant parameter than the exchange energy Δ), the wave
function of each exciton-branch is very close to be "heavy" or
"light" exciton, almost without mixing. In these conditions the
coupling between excitons and photons which is proportional to
the square of the transversal part of the wave function is
constant and $4\pi\beta_1 \sim 3/4\ 4\pi\beta_o$ and $4\pi\beta_2 \sim 1/4\ 4\pi\beta_o$: this case is
precisely the so-called "very particular case" considered in
Ref.14 and the oscillator strength of each exciton branch does
not depend on K in first approximation.

The two masses M_h and M_l depend on three Luttinger parameters so that the measurement of the translational masses in another direction than <100> is needed. The experimental results in <110> direction are given in Fig.3. Of course, the values of the background dielectric constant, the longitudinal transverse splitting and the exchange energy are the same as in the <100> direction. However the energy of longitudinal exciton is slightly different from the one used in <100> direction because the forbidden gap is not exactly the same for two different samples (see a discussion in Ref.17). Fig.3 is interpreted in a similar way to Fig.2. In this direction, the scattering by the fastest T.A. phonon is allowed even in backscattering configuration and indeed because of the slight disorientation between incident and scattered light, the two kinds of scattering can be observed. Theoretically, the representative curves of $S_{11}^{TA}2$ (the lowest one) and S_{22} are confused. It is obvious that from Fig.3 alone, it is not possible to ascribe most of the lines (except for S_{11} and S_{33}) without ambiguity; in the <110> direction, inside the above model we obtain M_h = 1.95 and M_l = 0.37. The light mass, whose inverse is the sum of two terms is not very sensitive to the directions contrarily to the heavy mass whose inverse is the difference of two same terms.

From all these results and within Kane's theory framework we conclude that γ_1 = 3.9 $\pm$ 0.8, γ_2 = 1.1 $\pm$ 0.2, γ_3 = 1.5 $\pm$ 0.3. The agreement between the points of the R.B.S. spectrum in the <111> direction and their computed position with the above parameters is in reasonable agreement; this will be discussed in more details elsewhere.

REFERENCES

1. Brenig W., Zeiher R. and Birman J.L., Physical Review, B6, 4617 (1972).

2. Ulbrich R.G. and Weisbuch C., Physical Review Letters 38, 865 (1977).

3. Winterling G. and Koteles E.S., Solid State Communications, 23, 95 (1977).

4. Hermann C. and Yu P.Y., Solid State Communications, 28, 313 (1978).

5. Pekar S.I., Soviet Physics, JETP 6, 785 (1958).

6. Hopfield J.J., Physical Review 112, 1555 (1958).

7. Dresselhaus G., Journal of Physics and Chemistry of Solids $\underline{1}$, 14 (1956).

8. Kane E.O., Physical Review, B$\underline{11}$, 3850 (1975).

9. Ulbrich R.G. and Weisbuch C., Festkörper Probleme (Advances in Solid State Physics) Volume XVIII, Page 217, J. Treusch (ed.), Vieweg, Braunschweig (1978).

10. Fishman G., Solid State Communications, $\underline{27}$, 1097 (1978).

11. Suga S., Cho K. and Bettini M., Physical Review B$\underline{13}$, 943 (1976).

12. Meijer H. and Polder D., Physica, $\underline{19}$, 255 (1953).

13. Vacher R. and Boyer L., Physical Review, $\underline{6}$, 639 (1972).

14. Sermage B. and Fishman G., Physical Review Letters, $\underline{43}$, 1043 (1979).

15. Romestain R. and Magnea N., Solid State Communications, $\underline{32}$, 1201 (1979).

16. Bonneville R. and Fishman G., This conference.

17. Sermage B. and Voos M., Physical Review, B$\underline{15}$, 3985 (1977).

THE POTENTIAL WELL OF Mn^{2+} IN $SrCl_2$

F. Hess and H.W. den Hartog

Solid State Physics Laboratory of the
University of Groningen

1 Melkweg, 9718 EP Groningen, The Netherlands

The shape of the potential well of Mn^{2+} as an impurity in $SrCl_2$ is calculated for low temperatures by means of a polarizable-point-ion model. The computed well is very flat in the <111>, <100> and <110> directions, in agreement with experimental findings. A major improvement on previous work by van Winsum et al. is the complete absence of a polarization catastrophe in the potential energy calculations.

1. INTRODUCTION

$SrCl_2$ is an ionic crystal with the fluorite structure. Experimental work[1,2] on the electric-field effect in EPR of $SrCl_2:Mn^{2+}$, as well as previous theoretical investigations[3] of this system, indicate that the potential well of Mn^{2+} is very flat. This may well be due to the guest ion's being smaller than the host ion (Sr^{2+}) it replaces and to the high polarizability of the surrounding Cl^- ions.

This paper deals with static theoretical calculations of the Mn^{2+} potential well in the <111>, <100> and <110> directions. Basically we used the same model and methods as did van Winsum et al.[3]. However, some modifications, both in the theoretical model and in the numerical methods, were introduced, which yield a drastic improvement on the previous results. The reader is referred to van Winsum et al.[3], and the references contained therein, for the sources of our input data, except where other references are explicitly given.

2. MODEL

        We use a polarizable-point-ion (PPI) model. Hence, the
ions are supposed to interact electrostatically as point charges
and induced point dipoles. In addition, between each ion pair a
short-range repulsion acts, which has a potential of the familiar
Born type :

$$\phi = b.\exp(-r/\rho)$$

where r is the distance between the ions, and b and ρ are parameters characteristic for the ion pair. As the positive ions in
the fluorite structure are relatively far removed from each other,
this short-range repulsion is omitted for positive ion pairs. The
very small van der Waals term for the Cl^--Cl^- interaction used in
ref.3 has been dropped as it hardly affects the results (A much
larger van der Waals term might, however, be appropriate).

        Our theoretical crystal consists of three regions.
Region I contains N_1 ions, including the impurity. Each of these
ions is allowed to relax from its lattice position. Region II
surrounds region I and contains N_2-N_1 ions which are fixed at
their lattice sites. All of the N_2 ions are polarizable. Finally,
region III, which surrounds region (I + II), is considered as a
continuum, having the shape of a large slab with a low-frequency
dielectric constant ε_0. Across this slab an external electric
field can be applied of E_{ext} Volts/Å. Region (I + II) is chosen
as a sphere centered around the impurity's lattice site. It can
be shown[4] that the external electric field gives rise to a
homogeneous field in region (I + II) of magnitude :

$$E_{loc} = E_{ext}.(\varepsilon_0 + 2)/3$$

        In region 1 the positive and negative host ions have
electronic polarizabilities α_e^{++} and α_e^- respectively. In region II
we use additional polarizabilities to simulate the (generally
very slight) displacement of the ions, which otherwise would not
be taken into account : α_d^{++} and α_e^- respectively. These we
determined as follows. For $SrCl_2$ at low temperatures[5] : $\varepsilon_0 = 6.94$
and $\varepsilon_\infty = 2.72$ (high-frequency dielectric constant). Defining :

$$\alpha_e = \alpha_e^{++} + 2\alpha_e^-, \quad \alpha_d = \alpha_d^{++} + 2\alpha_d^-$$

we have, because of the Clausius-Mossotti relation :

$$\frac{\varepsilon_0 - 1}{\varepsilon_0 + 2} \cdot \frac{\varepsilon_\infty + 2}{\varepsilon_\infty - 1} = \frac{\alpha_e + \alpha_d}{\alpha_e}$$

From the values for ε_0 and ε_∞, and the electronic polarizabilities we find : α_d = 6.173. By putting (rather arbitrarily) : α_d^{++} = 1/2 α_d, α_d^- = 1/4 α_d, we obtain the values for the total polarizabilities in region II as listed in Table I. The effect of adding α_d (which was not done by van Winsum et al.[3]) is to slightly lower the slopes of the Mn^{2+} potential well.

The most important modification of the model concerns the polarization energy. This is given by[6] :

$$E_p = -1/2 \sum_i \alpha_i (\vec{E}_{ci} + \vec{E}_{di}) \cdot \vec{E}_{ci}$$

where the summation extends over all ions in region (I + II), α_i is the polarizability of ion i, $\vec{E}_{ci}$ is the electric field at ion i due to point charges and $\vec{E}_{di}$ the electric field at ion i due to induced point dipoles. In order to obtain self-consistent values for the vectors $\vec{E}_{di}$ they must be calculated by an iterative method. These dipolar contributions to the electric field were omitted by van Winsum et al.[3]. Their results were seriously affected by an instability, called the polarization catastrophe. This instability is sometimes held against the PPI model[7,8]. However, it turns out that by taking the dipolar field correctly into account, the polarization catastrophe does not occur at all. Van Winsum et al.[3] attempted to prevent this catastrophe by not allowing the ions to approach each other beyond a certain minimum distance. Although our model contains a similar device (minimum distance for an ion pair : 80 % of the sum of the ionic radii), in none of the computed cases its use was required.

More details about the model can be found in references 3, 9 and 10.

Table I : Input parameters[3]

Lattice constant : r_0 = 6.984 Å					
Repulsion parameters			Polarizabilities ($Å^3$) :		
ion pair	b(eV)	ρ(Å)	ion	region I	region II
Mn^{2+}-Cl^-	9197	0.2664	Mn^{2+}	0.912	--
Sr^{2+}-Cl^-	190812	0.2257	Sr^{2+}	1.550	4.637
Cl^--Cl^-	1227	0.3214	Cl^-	2.974	4.517

3. CALCULATIONS

The shape of the potential well of Mn^{2+} in $SrCl_2$ was calculated by putting the impurity at a certain position and then minimizing the potential energy of region (I + II) as a function of the positions of the $N_1 - 1$ host ions in region I. We took $N_1 = 21$, $N_2 = 421$. Region I comprises the ions at distances $< \frac{1}{2} \sqrt{2} \, r_0$ from the impurity's lattice site, and region (I + II) the ions at distances $\leq 2r_0$, where r_0 is the lattice constant. Figure 1 indicates the ions in region I.

The number of independent parameters needed to determine the unknown positions of the 20 relaxable host ions would be 60 in the most general case. It is 10 for a displacement of the impurity in the <100> direction (C_{4v} symmetry), 13 for the <110> direction (C_{3v} symmetry) and 17 for the <110> direction (C_{2v} symmetry). We used Powell's minimization algorithm[11], which requires no derivatives. The computer programs were written in Pascal and run on a CDC Cyber 74-16. Passing once through the iterative loop for the calculation of the dipolar field vectors $\vec{E}_{di}$ took about 33 seconds CP time (this time is proportional to N_2 squared). One evaluation of the potential energy, excluding the above loop, took about 3 seconds CP time. By suitably intertwining the minimization steps and the dipolar field iterations an efficient numerical method was obtained. The computation of one point of the Mn^{2+} potential well at an accuracy of about 10^{-3} eV took typically between 400 and 1000 seconds CP time.

4. RESULTS

Some numerical results are shown in figures 2 and 3. All computations are precise to within about $\pm$ 0.001 eV. In figure 2 the potential well of Mn^{2+} in the <111> direction is compared to the potential well of Sr^{2+} (in pure $SrCl_2$) in the same direction. The relative flatness of the Mn^{2+} well is striking! Figure 3 shows the Mn^{2+} potential well in respectively the <111>, <100> and <110> directions (drawn curves). In all these directions the well is very flat. The dramatic effect of omitting the dipolar contributions to the electric field is illustrated by the dashed curves.

When the Mn^{2+} sits at its lattice site (the center of the well) n the eight surrounding Cl^- ions are shifted approximately 0.21 Å radially inward from their lattice sites and the ions in region I maintain the O_h symmetry in the <100> and <110> cases, as expected, but not quite in the <111> case. We take this as an indication of the input parameters not being correct.

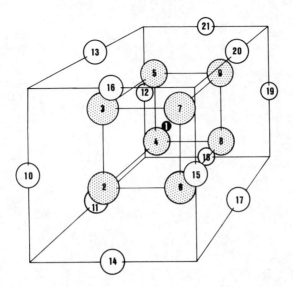

Fig.1 : Lattice positions of the relaxable ions in region I.
Ion 1 : Mn^{2+}, ions 2...9 : Cl$^-$, ions 10...21 : Sr^{2+}.

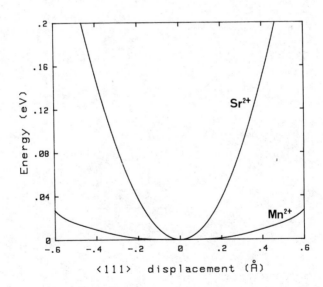

Fig.2 : Potential wells of Sr^{2+} and Mn^{2+} in SrCl$_2$ in the <111>
direction.

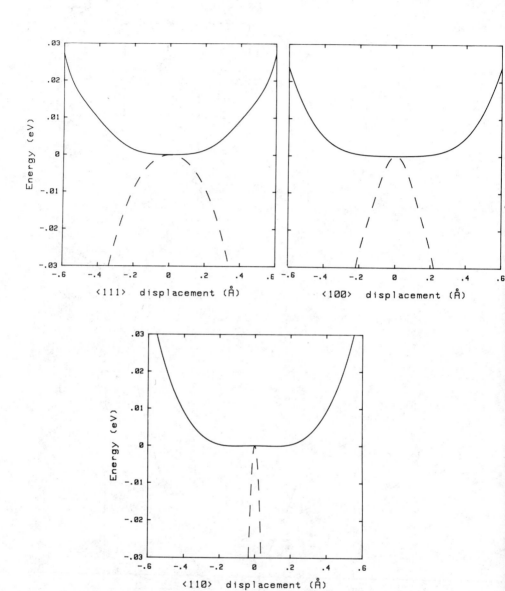

Fig.3 : Drawn curves : potential well of Mn^{2+} in $SrCl_2$ in (from left to right) the <111>, <100> and <110> directions. Dashed curves : corresponding potential wells if the dipolar contributions to the electric field are omitted.

The potential well in each of the three directions was also computed with an externally applied electric field of 0.005 V/Å parallel to the direction considered. In all three cases this resulted in potential energy curves which can be obtained from the ones without external field by subtracting a slope of approximately 0.008 eV/Å in the field direction.

How are the numerical results affected by changes in the input parameters? Some preliminary calculations indicate that if the Mn^{2+}-Cl^- repulsion is increased, or the Cl^--Cl^- repulsion decreased, the Mn^{2+} energy well becomes somewhat less flat. In the opposite cases the well tends to obtain shallow off-center minima.

5. DISCUSSION

The flatness of the Mn^{2+} potential well in $SrCl_2$ is in excellent qualitative agreement with experimental findings. However, whereas the experiments indicate that the Mn^{2+} can be shifted by an applied electric field in the <111> direction and not easily in other directions[1,2], we did not find a significant difference between the <111>, <100> and <110> directions. This may be due to incorrect input data.

We realize that a PPI model has limitations and shortcomings, which stem from its simple nature. In particular, the interactions between ions at close distances are hard to describe in a simple manner. The more so where an impurity is involved. A step further would be to take deformations of the slightly overlapping electron clouds into account, as is done e.g. in the deformation-dipole model[12]. However, in view of its success in predicting the positions of ions near impurities in ionic crystals (see e.g. Bijvank[10]), the PPI model appears to be a quite useful approximation. At present considerable uncertainties exist regarding the short-range potential input data (This would be just as true for any other model). Further work on this topic might well result in more accurate quantitative predictions.

Acknowledgements - We thank Dr. E.J. Bijvank and Prof. A.J. Dekker for efficient suggestions and discussions. This work forms part of the research program of the "Stichting voor Fundamenteel Onderzoek der Materie" (FOM) and is financially supported by the Netherlands Organization for the Advancement of Pure Research (ZWO).

REFERENCES

1. K.E. Roelfsema and H.W. den Hartog : Physical Review B $\underline{13}$, 7, p.2723-2728 (1976).

2. K.E. Roelfsema and H.W. den Hartog : Journal of Magnetic Resonance $\underline{29}$, 2, p.255-273 (1978).

3. J.A. van Winsum, H.W. den Hartog and T. Lee : Physical Review B $\underline{18}$, 1, p.178-184 (1978).

4. H. Fröhlich : "Theory of Dielectrics" (Oxford, 1950).

5. W. Hayes (ed.) : "Crystals with the Fluorite Structure" (Oxford, 1974).

6. C.J.F. Böttcher : "Theory of Electric Polarization" (Amsterdam, 1973).

7. A.B. Lidiard and M.J. Norgett in : "Computational Solid State Physics", eds. F. Herman, N.W. Dalton and T.R. Koehler (New York, 1972), p.385-412.

8. C.R.A. Catlow, K.M. Diller, M.J. Norgett, J. Corish, B.M.C. Parker and P.W.M. Jacobs : Physical Review B $\underline{18}$, 6, p.2739-2749 (1978).

9. W.D. Wilson, R.D. Hatcher, G.D. Dienes and R. Smoluchowski : Physical Review $\underline{161}$, 3, p.888-896 (1967).

10. E.J. Bijvank : "The Crystal Field Probe Gd^{3+} in a Systematic Series of Dipolar Defects" (Thesis, Groningen, 1979).

11. D.M. Himmelblau : "Applied Non-Linear Programming" (New York, 1972).

12. J.R. Hardy and A.M. Karo : "The Lattice Dynamics and Statics of Alkali Halide Crystals" (New York, 1979).

MINORITY CARRIER TRAPS IN ELECTRON IRRADIATED n-TYPE GERMANIUM

F. Poulin and J.C. Bourgoin

Groupe de Physique des Solides de
l'Ecole Normale Supérieure, Université Paris 7
Tour 23, 2 Place Jussieu
75221 Paris Cedex 05, France

Minority carrier traps created by electron irradiation at room temperature in slightly doped n-type germanium have been studied using transient capacitance spectroscopy. Levels at 0.16 (H_1), 0.30 (H_2), 0.37 (H_3) and 0.52 (H_4) eV from the valence band are found. The introduction rates of these traps, from which the threshold energy for their creation is deduced, have been determined. The traps H_2, H_3 and H_4 are found to be associated with a threshold of 30 ± 10 eV and the trap H_1 with a threshold of 45 ± 5 eV. It is suggested that H_1 corresponds to a level associated with the divacancy.

INTRODUCTION

N-type germanium with a free carrier concentration of 10^{13} cm^{-3} has been irradiated at room temperature with electrons of energy ranging from 0.6 to 2.9 MeV. The electron doses used, from 10^{15} to $2 \cdot 10^{16}$ cm^2, are chosen in such a way that the concentration of the electrically active defects introduced by the irradiation remains approximatively constant when the electron energy varies. In a previous paper[1] we have given the energy levels associated with the four electron traps we observed and their introduction rates (i.e. their concentration produced by one incident electron) versus electron energy. Two traps (E_1 and E_2) are found to correspond to a threshold energy of 20 ± 5 eV which, we concluded, is the threshold energy for atomic displacement Td. It was also suggested that the two other traps (E_4 and E_5), because they are in equal concentration and because

83

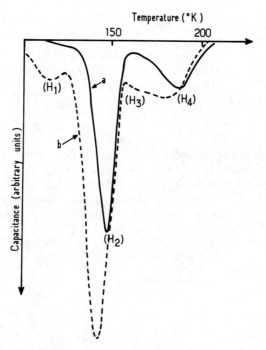

<u>Fig.1</u> : D.L.T.S. spectra (rate window 55 ms) due to hole traps in
n-type germanium irradiated with 2×10^{16} cm^{-2} electrons
of 0.6 MeV (a) and with 10^{15} cm^{-2} electrons of 1.6 MeV (b).

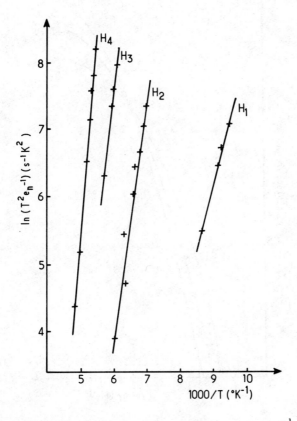

Fig.2 : Variation of the emission rate l_n with T^{-1} for the four traps observed.

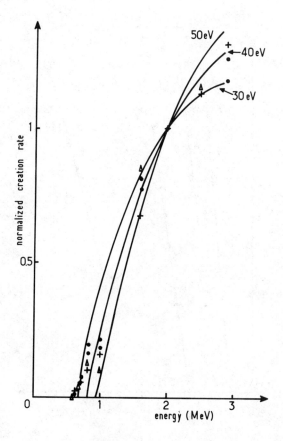

Fig.3 : Variation of the normalized creation rates for the H_2 (+),
H_3 (Δ) and H_4 (o) traps versus electron energy.

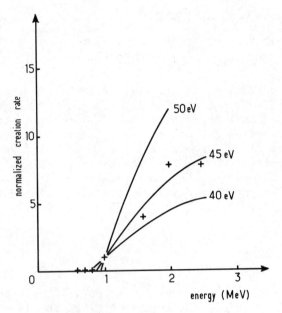

Fig.4 : Variation of the normalized creation rate for the H_1 trap
        versus electron energy.

they correspond to a threshold energy approximatively equal to
2 Td, are associated with the divacancy.

        The aim of this communication is to report the results
of a similar study performed on hole traps. The technique used,
a transient capacitance technique (D.L.T.S.), is described in
ref.1. The hole traps are observed by injecting holes in a junction
(with the help of a pulse which biaises the junction in forward
direction).

RESULTS

        Typical D.L.T.S. spectra obtained are given in figure 1.
These spectra are deduced by difference between the experimental
spectra without and with injection. At low energy the irradiation
creates only two traps (labelled H_2 and H_4) (Fig. 1a). At high
energy two new traps (H_1 and H_3) are present (Fig. 1b). The
signatures of these traps, i.e. the variation of the logarithm
of the emission rate l_n versus the inverse of the temperature T,
are given in figure 2. The slopes of these curves provide the
following activation energies : 0.16 $\pm$ 0.02 eV for H_1, 0.30 $\pm$ 0.03

eV for H_2, 0.37 ± 0.06 eV for H_3 and 0.52 ± 0.1 eV for H_4. The larger uncertainty on H_3 and H_4 comes from the fact that these traps are not well resolved.

The variations of the relative concentrations of the traps with the energy of irradiation are given in figures 3 and 4. We have not been able to measure the absolute concentrations because this requires the use of a sophisticated pulse generator which we do not have. Consequently the concentrations are normalized to 1 for an arbitrary energy and the experimental data are compared to normalized theoretical curves[2] calculated for various values of the threshold energy. The data are consistent with a threshold of 30 ± 10 eV for H_2, H_3 and H_4 and of 45 ± 5 eV for H_1.

CONCLUSION

The values of the threshold energies are consistent with those reported in ref.1. The trap H_1 exhibit the same threshold than the electron traps E_4 and E_5 and consequently we suggest that it is also associated with the divacancy. Three different charge states for the divacancy in the forbidden gap seems a reasonable picture for the divacancy since this is the case in silicon. However, it is possible that, because the sum of the activation energies of the H_1 and of the E_5 is practically equal to the width of the gap, these two traps are associated with the same level. The threshold energy (30 ± 10 eV) found for the H_2, H_3 and H_4 traps is slightly larger than the threshold (20 ± 5 eV) found for the E_1 and E_2 traps, probably because of the difficulty to extract the amplitudes of the different traps from a spectrum. The threshold energies for these H and E traps can only be the same, i.e. the H_2, H_3 and H_4 traps also correspond to defects due to the association of vacancies, or interstitials, with impurities.

REFERENCES

1. F. Poulin and J.C. Bourgoin, Revue Phys. Appl. 15, 15 (1980).

2. J.C. Bourgoin, P. Ludeau and B. Massarani, Revue Phys. Appl. 11, 2791 (1976).

X-RAY STUDIES OF POINT DEFECTS IN RARE GAS SOLIDS

R. Balzer and E.J. Giersberg

Institut für Angewandte Physik der
Technischen Hochschule Darmstadt

6100 Darmstadt, Germany

Point defects in solid Argon have been created by X-irradiation at 10 K. Lattice parameter change can be used as a measure of defect concentration. From dose curves the defect formation yield was determined. The absorbed energy per Frenkel pair was found to be $E_{FP} = (20 \pm 5)$ eV. The volume of spontaneous recombination was determined to be $v = 60$ atomic volumes in Argon after irradiation at 10 K. From isochronal and isothermal annealing measurements at least 7 different annealing stages can be separated. The first 5 stages with annealing temperatures up to 23 K are due to single defect recombination. The activation energy for uncorrelated single defect recombination was found to be $w = 55$ meV.

Point defects in rare gas solids can be created by X-rays very efficiently[1]. From combined lattice parameter change and diffuse X-ray scattering measurements defect properties as volume change and mechanical anisotropy of the interstitial defect have been determined[2]. In Argon the formation volume of an interstitial atom has been measured to be $\Omega_{FP} = 0.9 + 0.2$ atomic volumes, where in the analysis use has been made of the theoretical result that the formation volume of a vacancy is negligible in rare gas solids ($\Omega_v = -0.02$ atomic volumes)[3-5].

Lattice parameter change measurements on solid Argon have been performed after irradiation with different X-ray doses and during isothermal and isochronal annealing. In our measurements the lattice parameter change has been obtained from the shift of the (600)-Bragg-reflection using CuK_α-radiation.

Fig.1 shows typical dose curves for X-irradiated Argon
crystals. The defect induced lattice parameter change can be used
as a measure of the defect concentration. Lattice parameter changes
of 10^{-3} and more can be created within one hour of irradiation.
From the dose curves it is possible to get information about the
defect formation mechanism and about defect reactions as spontaneous
recombination and clustering.

Competing defect formation mechanisms are Knock-on
processes by photo- or Compton-electrons, defect production by
the recoil energy during generation of photo-electrons, double
ionisation of neighbouring atoms, repulsion of the atoms of an
excited molecule after relaxation to the ground state (excimer
mechanism).

Knock-on processes are effectively only above a
threshold energy, which can be calculated to be about 5-10 keV
for the photoelectrons in order to transmit the defect formation

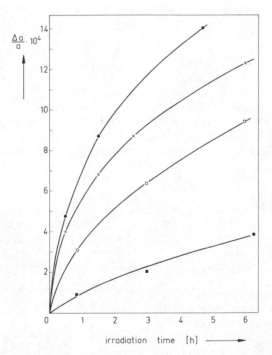

Fig.1 : Relative lattice parameter change of solid Argon versus
        irradiation time, irradiated at 10 K by the unfiltered
        spectrum of a tungsten anode, operated at 50 kV/12 mA (□),
        80 kV/12 mA (x) and 120 kV/12 mA (o). For comparison, a
        dose curve 50 kV/22 mA from a copper anode is included (■).

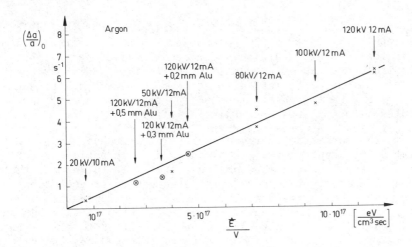

Fig.2 : $(\frac{\dot{\overline{\Delta a}}}{s})_o = \lim_{t \to o} \frac{d(\frac{\Delta a}{a})}{dt}$ as a measure of the defect formation
rate versus absorbed energy per volume and time for
different X-irradiation spectra as indicated.

energy of about 200 meV to the Argon atom. The defect production
rate for knock-on processes therefore depends on the energy of
the photons of the X-ray spectrum.

        It turns out however, as it is shown in fig.2, that the
initial slope of all dose curves depends only on the total
intensity absorbed in the crystal. The initial slopes of different
dose curves are plotted against the total intensity absorbed in the
crystal. The absorbed energy has been calculated and measured by
separate experiments. The X-ray spectrum has been varied by using
tube-voltages between 20 and 120 kV and by suppressing the soft
part of the X-ray spectrum by using Aluminium filters. From the
linear dependence it followes that only ionisation processes can
be responsible for the defect formation in solid rare gases. A
careful discussion, which will be published elsewhere, rules out
all proposed defect production mechanisms but the excimer mechanism.

        The defect formation yield turns out to be $E_{FP} = (20 \pm 5)$
eV of absorbed energy per Frenkel pair. Ionisation or excitation in
rare gas crystals leads to the formation of localized excitons[6].
These localized excitons can be understood as relaxed excited
molecules with strongly reduced internuclear separation (Excimer).
After relaxation to the lowest excited state a radiative transition
to the ground state takes place. Due to the repulsive energy the
atoms are separated and can be displaced in opposite directions.

Thus it is possible to produce two Frenkel pairs per ionisation,
if the repulsive energy is large enough. Since the nearest neighbours in fcc-crystals are always directed in [110]-directions,
the displacements take place in [110]-directions, which are known
as the directions of easiest displacement in fcc-crystals. Rare
gas dimers are well investigated and the repulsive energy can be
calculated without large uncertainties[6,7,8].

        The formation energy of a Frenkel pair has been
calculated for Krypton[4] and Xenon[5] and can be approximated for
Argon using the law of corresponding states[9]. It turnes out that
the repulsive energy in Argon is large enough to produce two
displacements per ionisation. That means, every ionisation leads
to the formation of two Frenkel pairs. In Krypton defects can only
be produced if the repulsive energy is transmitted to one of the
two atoms of the molecule predominantely. This reduces the defect
production rate considerably.

        This tendency is affirmed by the experiment : 150 times
more energy is necessary to form a Frenkel pair in Krypton
compared to Argon. In Xenon finally this mechanism should be
ineffective. Experiments on Xenon have been started and there
will be an answer to this question very soon.

        If the measured lattice parameter change in plotted
against the absorbed energy, all dose curves reduce to one curve

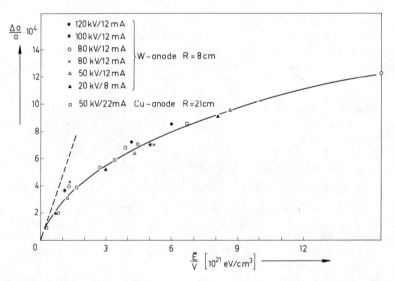

Fig.3 : Relative lattice parameter change of solid Argon versus
        absorbed energy per volume for different X-irradiation
        spectra.

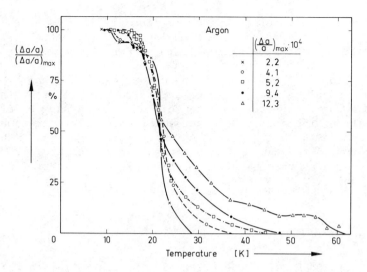

<u>Fig.4</u> : Isochronal annealing curves for different defect
concentrations. Relative lattice parameter change as
measure of defect concentration versus annealing
temperature.

(fig.3). With the assumption that the interstitial defect can
react during displacement with vacancies already present in the
crystal and recombine (σ_V) or react with other interstitial
defects to be stabilized (σ_I) the defect formation rate is found
to be

$$\frac{dc}{d\Phi} = \sigma_D \left\{ 1 - \frac{\sigma_v}{\sigma_v + \sigma_I} (1 - e^{-(\sigma_v + \sigma_I)cr_p}) - vc \right\} \qquad (1)$$

with :

σ_D : the initial defect formation rate,

v : the volume of spontaneous recombination,

r_p : the mean length of the displacement sequence in the crystal
with negligible defect concentration,

c : defect concentration.

Using equation 1 the volume of spontaneous recombination
for defects in solid Argon after irradiation at 10 K has been
determined to be v = 60 atomic volumes and $r_p \approx 60$ lattice spacings
was found.

Figure 4 shows isochronal annealing curves after X-
irradiation at 10 K for different defect concentrations. In

crystals with small defect concentrations all defects anneal
below 30 K, whereas in samples with large initial defect
concentrations annealing stages up to 60 K have been measured.

From isothermal annealing measurements at least 7
different annealing stages can be separated. The first 5 stages
with annealing temperatures up to 23 K are due to single defect
recombination. The activation energy for uncorrelated recombination
of single defects in Argon in the temperature rage 20-23 K has
been determined to be w = 55 meV. Annealing stages above 23 K can
be interpreted as recombination of clustered interstitial defects
with vacancies.

REFERENCES

1. R. Balzer and E.J.Giersberg, Physica Status Solidi (a) 57,
   K141 (1980).

2. R. Balzer, O. Kroggel, and H. Spalt : to be published in
   Journal of Physics C.

3. H. Kanzaki : Journal of Physics and Chemistry of Solids 2, 24
   and 107 (1957).

4. R.M.J. Cotterill and M. Doyama : Physics Letters 27A, 35
   (1967).

5. M. Doyama and R.M.J. Cotterill : Physical Review B1, 832
   (1970).

6. I.Y. Fugol : Advances in Physics 27, 1 (1978).

7. Topics in Applied Physics : Vol.30 Excimer Lasers.

8. J.A. Barker and A. Pompe : Australien Journal of Chemistry 21,
   1683 (1948).

9. J. de Boer : Physica 14, 139 (1948).

SYSTEMATIC INVESTIGATIONS OF THE CRYSTAL FIELD OF Gd^{3+}–M^+

CENTERS IN ALKALINE EARTH FLUORIDE CRYSTALS

E.J. Bijvank and H.W. den Hartog

Solid State Physics Laboratory

1 Melkweg, 9718 EP Groningen, The Netherlands

Local charge compensation centers have been observed in EPR in alkaline earth fouoride crystals which had been doped simultaneously with GdF_3 and MF (M = Li, Na, K, Rb and Ag). These centers are orthorhombic (C_{2v}) and the principal x, y and z-directions are aligned along the crystallographic [001], [1$\bar{1}$0] and [110] directions, respectively. These charge compensation complexes have been studied systematically with EPR, electric field effect (EFE) in EPR and ENDOR. Employing a computer program we have calculated the ionic positions in the neighborhood of the complex by minimizing the potential energy of the system.

From the combined studies we conclude that : the electrostatic model applied to calculate B_2^0 and B_2^2 using the ionic positions as evaluated from the potential energy minimization gives results which are in pair agreement with the experimental observations. This also applies to the theoretical values for the EFE!

I. INTRODUCTION

        In this paper we want to assess the capabilities of the polarizable point charge model in calculating the crystal field parameters B_2^0 and B_2^2 for Gd^{3+} in typical ionic materials such as the alkaline earth halides. Another model that has been used to understand the zero field splitting parameters mentioned above is the superposition model introduced by Newman et al. [1,2]. This model however was found to be unable to describe the typical properties observed for a systematic series of complexes of the

type $Gd^{3+}-M^+$ in the alkaline earth fluorides [3]. An important
reason for this is probably that in ionic materials we are dealing
with long range interactions, whereas in the superpositions model
only short range interactions are supposed to exist.

For our calculations we have considered both the shell
and polarizable point charge model and preferred to apply the latter
one because in our method we did not observe any sign of a
"polarization catastrophe", i.e. a very large negative potential
energy when neighboring ions approach each other too closely. We
have minimized the potential energy of the impurith system
together with a number of neighboring ions with respect to their
positions including effects associated with Coulomb, polarization
and repulsion interactions. This method provides us with the
equilibrium positions of 44 ions.

Using the results for the ionic positions associated
with minimum potential energy we are able to evaluate the crystal
field parameters of Gd^{3+} on the basis of some of the coupling
schemes that we have investigated in our recent papers [3-5]. It
is important to note here that apart from the conventional
contributions to the crystal field parameters B_2^0 and B_2^2 which
have been described by Hutchison et al. [6] and Wybourne [7] there
is another significant contribution to B_2^0. This contribution comes
from the odd terms in the crystal potential and gives rise to an
extra crystal field splitting in second order in accordance with
the theory given by Kiel et al. [8,9] and Parrot [10]. This
effect has been studied extensively for $Gd^{3+}-M^+$ complexes in CaF_2
by Lefferts et al. [5] and therefore it has been taken into
account in the treatment given in this paper.

It will be shown that for orthorhombic defect centers
of the type $Gd^{3+}-M^+$ in GaF_2 (see Fig.1) the crystal field parameter B_2^0 can be calculated from the known properties of the GaF_2
lattice, the charges, polarizabilities and the repulsive potentials
associated with the ions under consideration. We emphasize that
the results that will be given here have been obtained without
fitting any of the involved parameters!

II. DESCRIPTION OF THE CALCULATION METHOD

We have assumed that the alkaline earth fluoride lattice
consists of polarizable point charges; the charges of Ca^{2+} and F^-
are $+2e$ and $-e$, respectively. The polarizability of these ions

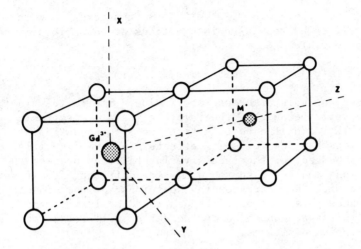

Fig.1 : Schematic representation of a Gd^{3+}-M$^+$ complex in CaF$_2$. The system of principal axes have been indicated by x, y and z.

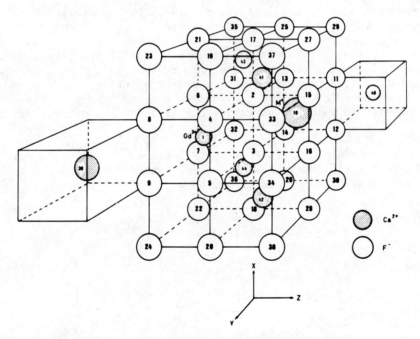

Fig.2 : Three dimensional view of region I containing 44 ions which are allowed to relax towards new equilibrium positions.

are $\alpha(Ca^{2+}) = 0.98$ Å and $\alpha(F^-) = 0.76$ Å in accordance with
Franklin [11]. The repulsive interactions are described in terms
of Born potentials

$$\phi_{ij}(r_i - r_j) = B_{ij} \exp(-r_i - r_j / \rho_{ij}) ; \tag{1}$$

the van der Waals interactions are included implicitly in the
repulsion terms because of the relatively small electronic
polarizabilities of the host lattice ions. Introduction of
impurity ions has the following effects : a. the impurities carry
effective charges, b. the repulsive interactions between the
impurity ions and their surroundings are different from that
associated with the corresponding host ions, c. the effective
charges cause polarization of the surrounding lattice and d.
the ions in the neighborhood of the defect will relax towards
new equilibrium positions.

In order to tackle the problem in a suitable way we
define two regions in the crystal under consideration :
- In region I (the immediate vicinity of the defect) the ions
  are allowed to relax; in the present calculations the total
  number of ions in region I is 44.
- In region I and II the ions are polarized by the charges and
  induced dipoles present in region I and II.

The numbers of ions in regions I and II are 44 and 246,
respectively. The contributions of the rest of the crystal has been
taken into account by expanding the potential energy associated
with long range Coulomb interactions in the neighborhood of a
lattice site in a Taylor series as has been done by van Winsum
et al. [12].

By adding the various contributions in a suitable way
we obtain for the energy deviations as compared to the perfect
lattice energy a Coulomb contribution :

$$\Delta E_s = \sum_{i \in I} \{ \sum_{\substack{j \in I \\ j > i}} \frac{e_i' e_j'}{|\vec{r}_i' - \vec{r}_j'|} - \sum_{\substack{j \in I \\ j \neq i}} \frac{e_i' e_j'}{|\vec{r}_i' - \vec{r}_j|} \} +$$

$$+ \sum_{i \in I} e_i' V(\vec{r}_i' - \vec{r}_i) +$$

$$+ \sum_{\substack{i \in I \\ i \neq Gd}} \frac{e_i'}{|\vec{r}_i' - \vec{r}_{Gd}|} - \sum_{\substack{i \in I \\ i \neq M}} \frac{e_i'}{|\vec{r}_i' - \vec{r}_M|} +$$

$$+ \sum_{i \in I} \sum_{j \in R} \left\{ \frac{e_i' e_j}{|\vec{r}_i - \vec{r}_j|} - \frac{e_i e_j}{|\vec{r}_i - \vec{r}_j|} \right\} +$$

$$+ \sum_{\substack{i \in I \\ i \neq j}} \sum_{j \in I} \frac{e_i' e_j}{|\vec{r}_i' - \vec{r}_j'|} - \sum_{\substack{i \in I \\ j \neq i}} \sum_{j \in I} \frac{e_i e_j}{|\vec{r}_i - \vec{r}_j|} \quad ; \qquad (2)$$

a polarization contribution

$$\Delta E_p = -\frac{1}{2} \sum_{i \in C} \alpha_i' (\vec{E}^{(i)'}_{\text{point-charges}} + \vec{E}^{(i)'}_{\text{induced dipoles}}) \cdot$$

$$\cdot \vec{E}^{(i)'}_{\text{point charges}} \qquad (3)$$

and a repulsive contribution

$$\Delta E_r = \sum_{\substack{j > i \\ j, i \in C}} \left\{ \phi_{ij}' (\vec{r}_i' - \vec{r}_j') - \phi_{ij} (\vec{r}_i - \vec{r}_j) \right\} \qquad (4)$$

In equation (2) to (4) the primes refer to actual charges, polarizabilities, positions and repulsive potentials; the unprimed quantities are associated with the pure lattice; i.e. the charge for the Gd³⁺ ion is +3e. The input parameters for the calculations have been given in Table I.

It is obvious that the number of ions in region I and II have to be limited because : i. region I contains the relaxing ions and the total potentials energy is minimized with regard to the positions of these ions and ii. the polarization energy has to be calculated in an iterative way in order to obtain self-consistent results!

III. RESULTS OF MINIMIZATION PROGRAM

It can be seen easily from Fig.2 that taking 44 ions in region I leads to 37 displacement parameters for an orthorhombic Gd³⁺-M⁺ center in CaF₂.

An interesting result of the potential energy calculations is that the energy of the Gd³⁺-Li⁺ defect in CaF₂ is 1.6 eV lower than the Gd³⁺-Cs⁺, indicating that it is easier to produce defects of the former type, which is in agreement with our observations. Infact it was hard to produce sufficient numbers of Gd³⁺-Cs⁺ complexes to be able to carry out EPR experiments.

Table I : Input parameters as used in the calculations

Lattice constant [Å] : 5.436			
ion	polarizability [Å^3]	ion	polarizability [Å^3]
F$^-$	0.76	Na$^+$	0.255
Ca^{2+}	0.98	K$^+$	1.201
Gd^{3+}	1.04	Rb$^+$	1.797
Li$^+$	0.029	Cs$^+$	3.137
ion i – ion j	B_{ij} [× 10^{-10} erg]		ρ [Å]
F$^-$ – F$^-$	7.33		0.282
Ca^{2+} – F$^-$	27.46		0.282
Ca^{2+} – Ca^{2+}	102.96		0.282
Gd^{3+} – F$^-$	60.00		0.282
Li$^+$ – F$^-$	2.40		0.291
Na$^+$ – F$^-$	6.41		0.296
K$^+$ – F$^-$	12.44		0.315
Rb$^+$ – F$^-$	15.10		0.323
Cs$^+$ – F$^-$	28.06		0.323

In Fig.3 we give a survey of the displacements of the nearest 24 ions for a series of Gd^{3+}-M$^+$ complexes in CaF$_2$. It should be noted that the displacements from the lattice sites have been enlarged by a factor of three.

IV. CRYSTAL FIELD PARAMETERS

From the calculated positions of the ions we have evaluated the important crystal field parameters B_2^0 and B_2^2 and also the odd crystal field parameter which can be determined by means of electric field effect experiments in EPR [5]. The theory necessary to understand the connection between the electrostatic crystal field potential parameters c_2^0 and c_2^2 and the magnetic splitting parameters B_2^0 and B_2^2 has been given elsewhere [4,5].

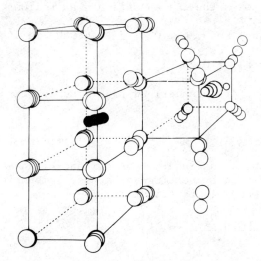

Fig.3 : Three dimensional view of the Gd^{3+}-M$^+$ complexes
(M = Li, Na, K, Rb and Cs). The results of the
displacement parameters can be taken from this picture.
The displacements have been enlarged by a factor of three.

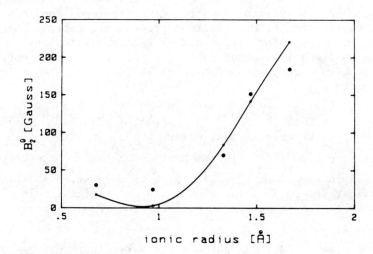

Fig.4 : Calculated and experimental parameters B$_2$ as a function of
the M$^+$ radius. The experimental values are connected with
each other by the drawn line.

The relations between the c's and B's are as follows :

$$B_2^0 (\text{Gauss}) = 10.20 \times 10^{-18} \; c_2^0 + 4.980 \times 10^{-19} \; (c_1^0)^2$$

$$B_2^2 (\text{Gauss}) = 10.20 \times 10^{-18} \; c_2^2 \; ,$$

(5)

where the c parameters are derived from a series expansion as given by :

$$V(\vec{r}) = c_0^0 + c_1 P_1^0 + c_2^0 P_2^0 + c_2^2 P_2^2 + \ldots \ldots .$$

(6)

P_ℓ^m's are functions as defined by Abragam and Bleaney [13] and the c's can be calculated by setting-up lattice sums which can be shown to be for monopole contributions :

$$c_1^0 = \frac{e}{4\pi\varepsilon_0} \; \Sigma \; \frac{Q_i Z_i}{R_i^3}$$

$$c_2^0 = \frac{e}{16\pi\varepsilon_0} \; \Sigma \; \frac{Q_i (3Z_i^2 - R_i^2)}{R_i^5}$$

$$c_2^2 = \frac{3e}{16\pi\varepsilon_0} \; \Sigma_i \; \frac{Q_i (X_i^2 - Y_i^2)}{R_i^5}$$

c_1^0 is given in units V/m whereas c_2^0 and c_2^2 are in V/m^2. Similar lattice sums can be written down for contributions from induced dipoles. Similar to the minimization process described in section 2 we have made the polarization effect self-consistent by employing an iteration method including the nearest 400 ions. A more detailed description of this method will be given in a later paper.

The results obtained for B_2^0, B_2^2 and c_1^0 have been given (together with the experimental observations) in Fig.'s 4, 5 and 6. It can be seen from these figures that for B_2^0 and c_1^0 the calculated results are in excellent agreement with experiment. Also the trends for these parameters as a function of the M^+-radius are in agreement with the experimental results. The results obtained for B_2^2 deviate significantly from the experimental observations, probablt region I and II do not contain sufficient numbers of ions. This is also suggested by the large scatter of the calculated values of B_2^2, which is in contrast with experiment showing a definite gradual increase with increasing M^+-radius.

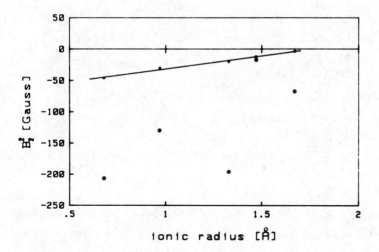

Fig.5 : Calculated and experimental parameters B_2^2 as a function of the M$^+$ radius. The experimental values are fitted by means of a straight line.

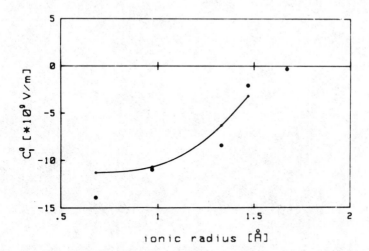

Fig.6 : Calculated and experimental c_1 as a function of the M radius. The experimental points are connected with the drawn line.

REFERENCES

[1] D.J. Newman and W. Urban, Adv. Pyys. 24, 793 (1975).

[2] A. Edgar and D.J. Newman, J. Phys. C 8, 4023 (1975).

[3] E.J. Bijvank, H.W. den Hartog and J. Andriessen, Phys. Rev.
     B 16, 1008 (1977).

[4] E.J. Bijvank and H.W. den Hartog, Phys. Rev. B 12, 4646 (1975).

[5] A.N. Lefferts, E.J. Bijvank, and H.W. den Hartog, Phys. Rev.
     B 17, 4214 (1978).

[6] C.A. Hutchison, B.R. Judd and D.F.D. Pope, Proc. Phys. Soc.
     B 70, 514 (1957).

[7] B.G. Wybourne, Phys. Rev. 148, 317 (1966).

[8] A. Kiel, Phys. Rev. 148, 247 (1966).

[9] A. Kiel and W.B. Mims, Phys. Rev. 153, 378 (1967).

[10] R. Parrot, Phys. Rev. B 9, 4660 (1974).

[11] A.D. Franklin, J. Phys. Chem. Solids 29, 823 (1968).

[12] J.A. van Winsum, H.W. den Hartog, and T. Lee, Phys. Rev. B 18,
     178 (1978).

[13] A. Abragam and B. Bleaney, Electron Paramagnetic Resonance
     of Transition Ions (Clarendon, Oxford, 1970).

THE EFFECT OF UNIAXIAL PRESSURE ON THE EPR SPECTRUM OF $SrCl_2:Mn^{2+}$

H.W. den Hartog and K. Post

Solid State Physics Laboratory

1 Melkweg, 9718 EP Groningen, The Netherlands

In order to obtain more information about the remarkable impurity system $SrCl_2:Mn^{2+}$ we have, in addition to the electric field effect (EFE) experiments – carried out in our laboratory recently – performed uniaxial pressure investigations at various temperatures between 4-300 K. An important conclusion drawn from the EFE experiments was that the application of a moderately strong external electric field leads to relatively large displacements of the small Mn^{2+} impurity in $SrCl_2$ if the direction of the electric field is chosen along [11$\bar{1}$].

We have attempted to explain the observed effects, which show up as an extra crystal field splitting of second degree, using the bulk elastic properties of the $SrCl_2$ crystal. From the elastic constants c_{11}, c_{12} and c_{44} one can straightforwardly calculate the relative positions of the ions under stress conditions. Subsequently, we have calculated the classical formulae given by Roelfsema and den Hartog. The results obtained from these calculations yield values which are far too small as compared to the results gathered from the uniaxial pressure experiments along [111] and [110].

It is concluded that the experimental results can be explained only if it is assumed that the local elastic properties in the neighborhood of the Mn^{2+} impurity deviate appreciably from the bulk values.

1. INTRODUCTION

In earlier papers [1,2] we have reported about the unusual electric field effect of the system $SrCl_2:Mn^{2+}$ observed in EPR between 4 and 300 K. This electric field effect is due to large displacements of the crystal Mn^{2+} impurity. The interpretation has been supported by the results of theoretical calculations performed by van Winsum et al. [3] and Hess et al. [4] indicating that the potential well seen by the impurity in $SrCl_2$ is very flat.

In this paper we shall present the results of additional experiments on this interesting impurity system. We have performed uniaxial stress experiments in EPR in the temperature range 4-300 K. From the electric field effect results it is clear that the observations are rather anisotropic and therefore we have also studied the effects along the important axes [100] , [110] , [111] employing a set-up as published by Szumowski et al. [5] .

We have observed significant variations of the stress effects along the axes investigated. Applying stress along [100] only minimal effects could be observed, whereas for the axes [110] and [111] large effects have been detected. The explanation for the observed phenomena is not the displacement of the Mn ion, but we have to consider relatively large displacements of the neighboring Cl ions (see Fig.1) due to the applied stress. These displacements can not be explained by the bulk elastic properties of $SrCl_2$ as determined by Lauer et al. [6] .

2. EXPERIMENTAL RESULTS

In Fig.2 we show the EPR spectrum of $SrCl_2:Mn^{2+}$ as measured at 50 K using a LHe flow cryostat. In this experiment (upper spectrum) no stress has been applied. The spectrum clearly shows the hyperfine splitting due to the nuclear spin (5/2) of manganese. A very small crystal field splitting combined with the second order hyperfine splitting causes each of the hyperfine lines to split into five S_z lines. The nature of this spectrum has been discussed in detail in our previous papers [1,2] . The effect of an externally applied stress of 500 bars along [111] has also been given in Fig.2 (lower trace). Our experimental set-up is such that the stress is applied along the vertical axis while the magnetic field is in the horizontal plane.

It can be seen from Fig.2 that in contrast with the upper trace the lower spectrum has central lines in each group of fine lines which are equal in intensity. In addition these groups now consist of five lines each; of course the different groups partly overlap. By studying the behavior of these groups

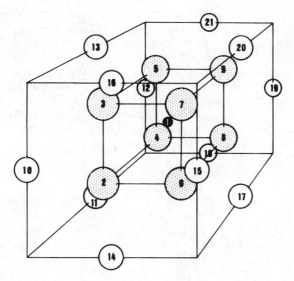

Fig.1 : Schematic representation of Mn^{2+} in SrCl$_2$. Only the nearest and next nearest neighbors of the impurity have been indicated. 2-9 are Cl$^-$-ions, 10-21 are Sr^{2+}-ions.

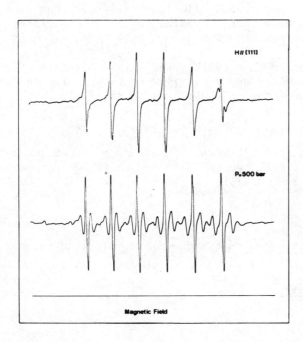

Fig.2 : EPR spectra of SrCl$_2$:Mn^{2+} at 50 K with and without uniaxial stress along [111].

as a function of the strength of the stress we obtain results as shown in Fig.3. Like the electric field effect the uniaxial stress effect increases linearly with the applied pressure. A feature of interest is that the slopes of the straight lines fitting the data points differ by about a factor of two, indicating that the stress induced shifts of the EPR lines are due to an extra crystal field contribution of degree two. In addition we have observed that the effect, if the stress is applied along [111], is axialmy symmetric with respect to this axis. Now we can write the extra term in the spin Hamiltonian induced by the stress along [111] as

$$H_{stress} = B_2^0(stress)O_2^0 \qquad\qquad (1)$$

Similar experiments as for [111] have been carried out for [110] and [100]. For [100] the stress induced splittings are small as compared with those for [111]. If the stress is applied along [110] the observed splittings are of approximately the same magnitude as for [111]. In Fig.4 we show the EPR spectrum of $SrCl_2:Mn^{2+}$ with and without stress measured with H_0 along [110]. The corresponding spectra have been taken at 70 K. Rotation of the sample about the stress axis reveals a strong angular dependence of the uniaxial stress effect; the effect for H_0 along the [100] axis in the $(1\bar{1}0)$ plane is approximately zero. A rotational diagram of the uniaxial stress effect for $P/[1\bar{1}0]$ has been given in Fig.5. Because also for stress $\parallel [1\bar{1}0]$ the effect is linear with the magnitude of the stress the extra crystal field giving rise to the line shifts are of second degree.

Keeping in mind that the uniaxial stress effect does not show axial symmetry about the $[1\bar{1}0]$ axis we write the stress induced terms of the spin Hamiltonian as

$$H_{stress} = B_2^0(stress)O_2^0 + B_2^2(stress)O_2^2 . \qquad (2)$$

From the rotational diagram together with the above mentioned experimental details we are able to give the stress induced crystal field parameters in Table I.

Table I : Stress induced crystal field parameters

Stress direction	B_2^0 (G/kBar)	B_2^2 (G/kBar)
[100]	<0.2	0
[110]	+12.3	+12.3
[111]	+13.7	0

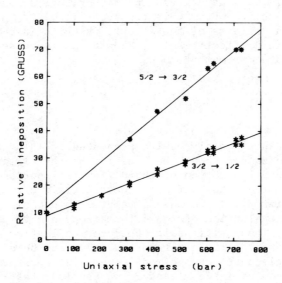

Fig.3 : The effect of uniaxial stress along [111] on the fine
structure of the group of EPR lines at highest magnetic
fields (see Fig.2).

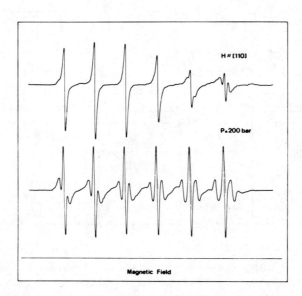

Fig.4 : EPR spectra of SrCl$_2$:Mn^{2+} with $\vec{H}$// [110] at 70 K with and
without uniaxial pressure along [1$\bar{1}$0] .

In Fig.6 we have given the temperature dependence of the stress induced shifts. From this figure it is clear that at about 60 K there is a decrease of the effect, which is in contrast with the electric field effect results, showing an increase at this temperature and a maximum at 70 K.

3. DISCUSSION

The experimental results presented in this paper allow us to draw conclusions about the behavior of the Cl ions neighboring the Mn^{2+}. Because the motion of the Cl ion is important the stress results deviate from those of the electric field effect experiments published earlier [1,2]. We have attempted to explain the observations by means of an evaluation of the deformations due to externally applied stress. The positions in the deformed lattice are used to calculate the crystal field parameters; this can be done on the basis of a simple point charge model (see also Roelfsema et al. [1,2]). From a comparison of Tables I and II we conclude that the bulk elastic properties are not suited to explain the observations. We have to assume that c_{44} for the Cl ions neighboring the Mn^{2+} ions is reduced. This can also be concluded from the potential energy calculations carried out by Hess et al. [4]. In Table II we have therefore employed a reduction factor of 10 for c_{44}.

Table II : Theoretical stress induced parameters

Stress direction	B_2^0(G/kBar)		B_2^2(G/kBar)	
	bulk	$c_{44}/10$	bulk	$c_{44}/10$
[100]	0.4	0.8	0	0
[110]	-0.3	-7.0	-0.7	-8.0
[111]	-0.8	-11.1	0	0

At low temperatures the Mn^{2+} ion, which moves in a rather flat potential well [3,4], remains in the central part of the well. When the temperature is increased to about 70 K the Mn^{2+} ion is spread over a larger space leading to a large electric field effect. For the surrounding Cl ions this leads to a larger repulsion giving rise to smaller displacements of the Cl ions under the influence of an externally applied uniaxial stress.

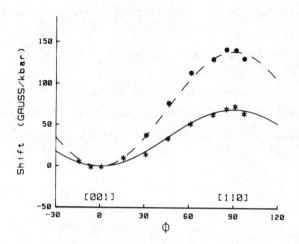

Fig.5 : Rotational diagram of the stress effect with $\vec{P} /\!/$ [110]
and H$_0$ in (1$\bar{1}$0) as measured at 70 K.

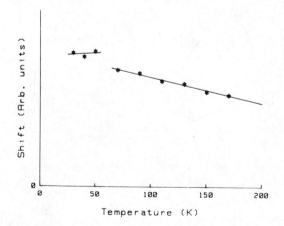

Fig.6 : Temperature dependence of the stress effect as measured
for $\vec{P} /\!/$ [111] .

REFERENCES

[1] K.E. Roelfsema and H.W. den Hartog, Phys. Rev. B 13, 2723 (1976).

[2] K.E. Roelfsema and H. W. Den Hartog, J. Magn. Res. 29, 255 (1978).

[3] J.E. van Winsum, H.W. den Hartog, and T. Lee, Phys. Rev. B 18, 178 (1978).

[4] F. Hess and H.W. den Hartog, to be published.

[5] J. Szumowski and K. Falkowski, Rev. Sci. Instr. 47, 252 (1976).

[6] H.V. Lauer, K.A. Solberg, D. H. Kühner, and W.E. Bron, Phys. Lett. 35, 219 (1971).

DISLOCATION MOTION INVESTIGATED BY MEANS OF

NUCLEAR SPIN RELAXATION MEASUREMENTS

W.H.M. Alsem*, J.T.M. De Hosson*, H. Tamler[+],
H.J. Hackelöer[+], and O. Kaner[+]

* Department of Applied Physics
  Materials Science Centre
  Rijksuniversiteit Groningen
  Nijenborgh 18, 9747 AG Groningen, The Netherlands

[+] Institute of Physics, University of Dortmund,
  Postfach 500500, 46 Dortmund 50, Western Germany

The technique of pulsed nuclear magnetic resonance is used to study the movement of dislocations in deforming alkalihalide single crystals of different crystallographic orientation. The spin-lattice relaxation time $T_{1\rho}$ depends strongly on the mean free path L covered between dislocation jumps. Since the work-hardening behaviour of solids is related to the distance L, it is possible by this experiment to obtain detailed information about the microscopic plasticity of crystals.

1. INTRODUCTION

The structure of dislocations and the dislocation density are objects worthy of study. Generally, the existence of dislocations permits solids to be plastically deformed with ease, a circumstance upon which our modern technology is so dependent. Actually, the ease with which solids can be deformed was the first crucial experimental observation that lead to the discovery of the dislocation. Strengthening is brought about by obstructing the movement of dislocations.

In the past, great attention was paid to the strengthening and work-hardening mechanisms of alkali halides. The most widely quoted mechanical properties were those determined by simple tensile tests. However, the current understanding of work-hardening of alkali halides is still rather unsatisfactory since different models have been scrutinized for the same plastic deformation stage. It is well known that when deforming an alkali halide crystal plastically by glide, the stress needed to cause further deformation, the flow stress, increases with strain. This is strain hardening and in suitable conditions ionic crystals show a shear stress (τ)-shear strain (a) curve with three stages of strain hardening. The extent of a particular stage and the value of the hardening rate $\theta = d\tau/da$ will depend on factors such as temperature of testing, strain rate, impurity content and crystal orientation.

The present paper will outline the experimental technique of pulsed nuclear magnetic resonance for studying the deformation mechanism of ultra pure sodiumchloride single crystals at room temperature (imp conc $\leq$ 10 ppm). The purpose of this investigation is twofold. First, to provide information about the orientation dependence of work-hardening. The results applying four different deformation axes are presented : <100>, <110>, <9 13 10>, <111>. Secondly, to investigate the self-consistency of the work-hardening theory, NMR theory of moving dislocations and its compatibility with mechanical experiments. In particular, the mean free path L of mobile dislocations derived from pulsed nuclear magnetic resonance experiments are compared with the values predicted by work-hardening theory.

In that connection, it is to be understood that the physical model of dislocation motion applied here, assumes a thermally activated jerky motion of mobile dislocations between strong obstacles viz. forest dislocation and impurities. The plastic shear strain rate $\dot{a}$ is then given by $\dot{a} = b\rho_m L/\tau_j$, where ρ_m is the mobile dislocation density, τ_j represents the mean time between two jumps of dislocations and b is the magnitude of the Burgers vector. The spin-lattice relaxation rate in the rotating frame $R_D{}^\rho$ is given by

$$R_D{}^\rho = \left(\frac{1}{T_{1\rho}}\right)_{dyn} = C\,\frac{\dot{a}}{bL} \,, \tag{1}$$

where C contains the local fields ($H_L{}^2$) and the mean quadrupolar distortion (A_Q) caused by the presence of dislocations :

$$C = \frac{A_Q}{H_1{}^2 + H_L{}^2} \,, \tag{2}$$

$T_{1\rho}$ is the spin-lattice relaxation time. For a comprehensive survey of $R_D{}^\rho$ reference should be made to [1].

2. EXPERIMENTAL TECHNIQUE

Although the principle of the experiments is simple, the technical feasibility is not. A single crystal is plastically deformed by a tensile machine of which the crosshead moves forward a driving rod with a constant velocity, while at the same time a dynamic nuclear relaxation rate measurement is made (see fig.1). Therefore a r.f. coil is mounted coaxially with the specimen and is connected with a Bruker SXP 4-100 spectrometer, which supplies and receives the r.f. pulses for the NMR pulse experiment. The frame in which the driving rod moves, is placed in the gap of a 1.4 T Bruker BE 25 magnet. The spectrometer is triggered by the electronic control of the hydraulic Zonic deformation equipment. Subsequently, a spin-locking experiment is performed during plastic deformation. The pulse sequence used [1] consisting of a $\pi/2$ pulse and a locking pulse of strength H_1 yields a value of R^{ρ}_{total}. Immediately before and after the plastic deformation a static relaxation rate R^{ρ}_{static} is measured. The dynamic enhancement of R^{ρ}_{D} caused by jumping dislocations is obtained by

$$R_D{}^{\rho} = R^{\rho}{}_{total} - R^{\rho}{}_{static} \cdot \qquad (3)$$

During the experiments the stress on the crystal and the total elongation is recorded as a function of time, from which stress-strain curves could be determined.

3. RESULTS AND DISCUSSION

Because the components of the stress in the glide direction and in the glide plane depend on the direction of the compression axis, the stress-strain (ρ vs ε) curves have to be reduced to shear stress - shear strain (τ vs a) curves according to $\tau = \phi\upsilon$ and a $= \varepsilon/\phi$, in which ϕ is the Schmid factor. The τ vs a curves, resolved on (110) planes for [100] and [110] compression axes and resolved on [100] planes for [111] and [9 13 10] deformation axes are depicted in fig.2 using the Schmid factors of the active glide planes as have been listed in table I.

In figure 3 the accompanying results of the spin-locking experiments have been indicated, where $R_D{}^{\rho}$ has been plotted as a function of the shear strain rate $\dot{a}$ ion {110} (fig.3a) and {100} (fig.3b). The linear relation between $R_D{}^{\rho}$ and $\dot{a}$ is expressed in equation 1. From this proportion and with the aid of experimentally $(H_L{}^2)$ and theoretically (A_Q) determined values for C the mean free path L for dislocations is obtained (see table II).

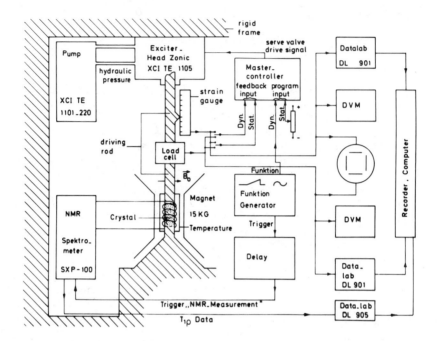

<u>Fig.1</u> : Scheme of the experimental set up.

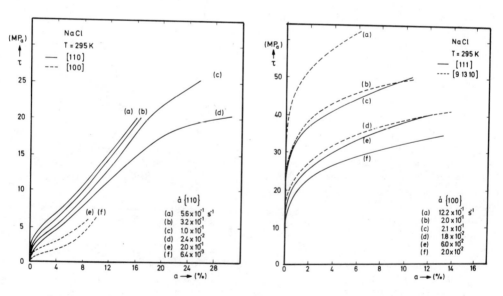

<u>Fig.2</u> : Shear stress τ as a function of the shear strain a for
different oriented NaCl single crystals, for shear on
{110} (left) and for shear on {100}(right).

Table I : The Schmid factor δ for <110> slip direction

Deformation axis	{100}	{110}	{111}
<100>	0.	0.5	0.41
<110>	0.35	0.25	0.41
<9 13 10>	0.5	0.13	0.36
<111>	0.47	0.	0.27

Table II : The mean free path L calculated from $R_D{}^\rho$ vs à
(units 10^{-6} m)

Deformation axis	slip on	{100}	{110}
[100]		–	0.57
[110]		0.28	0.35
[9 13 10]		0.22	0.55
[111]		0.24	–

From figure 2 it may be concluded that there exists an orientation dependence of work-hardening. Work-hardening is much higher applying <110> deformation axis than for compression along the <100> axis. In addition, the work-hardening along <111> axis is somewhat smaller than along the <9 13 10> compression axis. However, the predominant slip systems in NaCl-type crystals are four <110>{1$\bar{1}$0} which are all equally stressed for [100] and [110] deformation axis. Nevertheless, the [110] compression axis hardens stronger than the [100] because there exists a finite shear stress on {100} planes applying <110> compression which is absent using <100> deformation. Extra hardening could be caused by activation of slip on these secondary {100} slip planes and the inference with primary {110} slip planes. Matucha [2] concluded from electron-microscopic slipline observations that in <110> and <100> oriented NaCl crystals, deformation takes place by single slip on one of the {110} planes right from the onset of macroscopical plastic flow (stage-I). In stage-II where a higher constant work-hardening exists than in stage-I, about 1 % of slip occurs on {100} planes in <110>-oriented specimens.

The dependence of the work-hardening rate on the crystal orientation can be explained quantitatively using the work-hardening theory proposed by Frank and Seeger [3] [4] , where the work-hardening coefficient θ is written as

$$\theta = \frac{\delta \tau}{\delta a} = \frac{A}{L} + B \qquad\qquad (4)$$

where A is a function of the temperature and B depends on the temperature T and the strain rate $\dot{a}$, which is relatively small compared to A/L. The experimental values of θ are derived from fig.2 at a = 4 % (μ represents the shear modulus) :

$$\theta^{<100>} = 2.9 \ 10^{-3} \mu \ (\dot{a} = 2.0 \ 10^{-1} \ s^{-1});$$

$$\theta^{<110>} = 6.3 \ 10^{-3} \mu \ (\dot{a} = 3.2 \ 10^{-1} \ s^{-1}) .$$

The experimentally observed ratio $\theta^{<110>}/\theta^{<100>}$ is in good agreement with the ratio $L^{<100>}/L^{<110>}$ determined from NMR measurements (see table II). Analogously, the work-hardening theory predicts

$$\frac{\theta^{<111>}}{\theta^{<9\ 13\ 10>}} = \frac{L^{<9\ 13\ 10>}}{L^{<111>}} = 0.76 \ ,$$

whereas the NMR theory on moving dislocations predicts a value of 0.90.

        Crystals compressed along <111> do not resolve any shear stress on {110} planes (table I). They deform by a combination of {100} and {111} slip [5] [6] . The work-hardening rate (fig.3) is much higher compared to deformation along <110> which is in accordance with the observations of Franzbecker [7] . In <9 13 10> oriented crystals slip is also found in {110}<1$\bar{1}$0> systems [8] [9] , while not all {100} planes are equally stressed. This is a plausible explanation why the {110} systems cannot shorten L of <9 13 10> crystals as much as the {100} systems do for <110> crystals.

        From these analyses, the self-consistency of the work-hardening theory, NMR theory of moving dislocations and its compatibility with mechanical experiments might be concluded.

Acknowledgements   -   This work is part of the research program of the Foundation for Fundamental Research on Matter (F.O.M. - Utrecht) and has been made possible by financial support from the Netherlands Organization for the Advancement of pure Research (Z.W.O - The Hague) and the "Deutsche Forschungsgemeinschaft" (W. Germany).

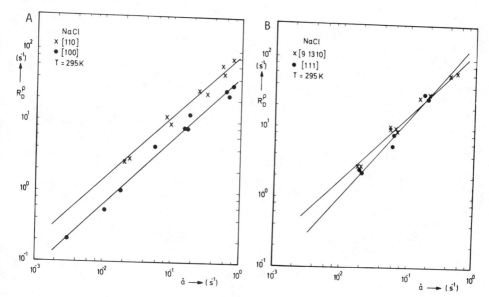

Fig.3 : Relaxation rate R_D^ρ as a function of the shear rate $\dot{a}$ resolved on $\{110\}$ (a) and on $\{100\}$ (b) planes.

REFERENCES

[1] D. Wolf, O. Kanert, Physical Review B, <u>16</u>, 4776 (1977) and references.

[2] K.H. Matucha, Physica Status Solidi, <u>26</u>, 291, (1968).

[3] W. Frank, A. Seeger, Comments on Solid State Physics, 133, (1969).

[4] W. Skrotzki, Materials Science and Engineering, <u>38</u>, 271, (1979).

[5] I.V.K. Bhagavan Raju, H. Strunk, Physica Status Solidi (a), <u>53</u>, 211, (1979).

[6] H. Strunk, private communication.

[7] W. Franzbecker, Physica Status Solidi (b), <u>57</u>, 545, (1973).

[8] O.V. Klyavin, S. G. Simashko, A.V. Stepanov, Soviet Physics - Solid State, <u>15</u>, 274, (1973).

[9] O.V. Klyavin, S.G. Simashko, Soviet Physics - Solid State, <u>16</u>, 898, (1974).

THE PHONON ASSISTED RADIATIVE TRANSITIONS IN THE

ISOELECTRONIC TRAP AgBr:I

W. Czaja

Laboratoire de Physique Appliquée
Ecole Polytechnique Fédérale

Lausanne, Switzerland

It is argued that neither the phonon assisted emission nor the
phonon assisted absorption in AgBr:I can be explained using a
configuration coordinate model. The reason is believed to be
the charge distribution of the electron which differs strongly
between ground and excited state. For the absorption process a
different model is proposed.

In a recent publication[1] it has been shown that in the
low temperature emission (luminescence) spectra of AgBr:I the
electronic part can be well represented by the model of isoelectronic traps in semiconductors. The detailed predictions of
this model are well confirmed by the experiment. However, only
about 0.1 % of the light emitted is due to the electronic
transition. The overwelming part is due to phonon assisted
transitions which, however, are not well understood.

In Fig.1a the phonon assisted emission as well as
absorption of the exciton bound to AgBr:I is shown. Fig.1b is the
result from an analysis published by Kanzaki and Sakuragi[2] based
on a configuration coordinate model of the Huang-Rhys-type[3].
Assuming lattice phonons to be responsible and in particular LO-
phonons giving rise to the fine structure one obtains an intensity
distribution marked (a) in Fig.1b. The coupling constant in this
analysis has been taken from the one-phonon part of the emission
spectrum. Neither is the experimental spectrum symmetric with
respect to the zero-phonon line nor is the maximum of the emission
at the predicted phonon number. In order to obtain better agreement
Kanzaki and Sakuragi[2] have assumed that the fine structure visible
in the absorption spectrum is again due to LO-phonons but their

121

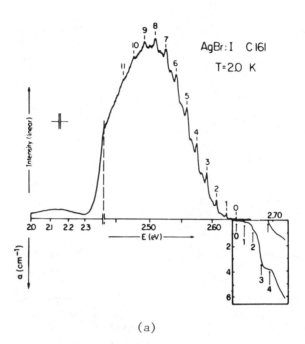

(a)

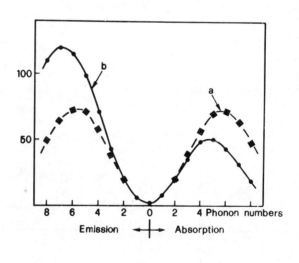

(b)

Fig.1 : (a) Emission (ref.1) – and absorption (ref.2) – spectrum
            for AgBr:I.
        (b) Relative intensities and relative absorption according
            to the C.C.-model (ref.2).

energy is measured to be about 13 meV instead of 17 meV. With
this refinement curve (b) in Fig.1b is obtained. Although the
change goes into the right direction the absorption spectrum is
still very poorly reproduced.

In ref. 1 it has been shown that the fine structure in
the emission spectrum is due to lattice LO-phonons which behave
according to the dispersion relation and the one-phonon density
of states. It is therefore very difficult to understand how a
doping of $\sim 10^{19}$ cm^{-3} can change the frequencies of lattice phonons
in the excited state only. It should be mentioned, however, that
differences in the vibrational quanta for the emission and
absorption processes might be possible if the impurity couples
mainly to impurity related vibrational modes[3].

One might argue that describing the emission spectrum
only might yield a much better agreement with the model in question.
This procedure has been tried in ref. 1. But also there strong
doubts have been raised as to the applicability of the
configuration coordinate model even in this limited case.

In trying to understand the failure of the simple
configuration coordinate model to reproduce the experimental
data one realises that a sufficiently localised impurity has
electronic states only weakly perturbed by the vibrations and
determined by approximately the same charge distribution for
the ground and the excited state. Thus the application of the
adiabatic approximation is justified and the configuration
coordinate model in the real sense of the word applies. In the
case of AgBr:I there is experimental information pointing towards
a strong violation of the requirements for the applicability of
the adiabatic approximation. Iodine being an isoelectronic impurity
has in its ground state - except for the core electrons - the same
charge distribution as bromine, i.e. with electrons localised
within one Wigner-Seitz cell. For the excited state - i.e. for
the exciton bound to iodine - recent Stark effect data published
by Marchetti and Tinti[4] yield a quadratic shift of the zero-
phonon line and therefore directly a polarisability of the bound
exciton. Neglecting polaron effects, assuming the hole to be fixed
at the impurity site and the static value for the dielectric
constant, the equivalent donor represented by the electron bound
to the "positively charged" iodine impurity - and therefore the
bound exciton - has a Bohr radius of 20 Å. This means that in the
excited state the electron is distributed over a volume containing
almost 180 cubic unit cells. Markham has already remarked[3] that
the adiabatic approximation should not be used for shallow trapped
or free electrons.

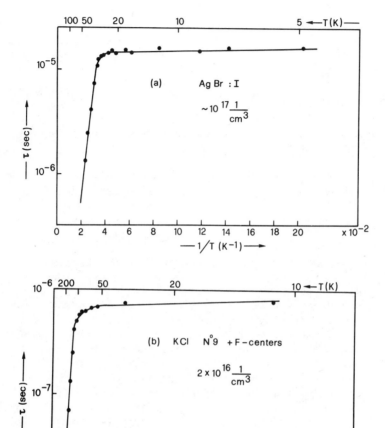

Fig.2 : Lifetimes versus temperature.
(a) AgBr:I after Moser and Lyu (ref.5).
(b) F-centers in KCl after ref.7.

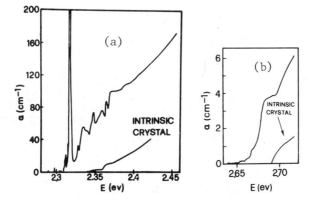

Fig.3 : Impurity related absorption.
    (a) GaP:N, 2K, 7×10^{18} cm^{-3} after ref.9
    (b) AgBr:I, 2K, $1,5 \times 10^{19}$ cm^{-3} after ref.2.

        As a consequence of the large Bohr radius the electron-
hole overlap is weak and therefore the luminescence decay time
(bound exciton lifetime) is expected to be long. This has been
found by Tsukakoshi and Kanzaki[5] and by Moser and Lyu[5], comp.
Fig.2a. It should be noted that very similar lifetimes have been
found for various isoelectronic traps, forinstance GaP:N, GaP:Bi,
ZnTe:0[6]. Furthermore the lifetimes for some F-centres (KCl, KBr,
KI and NaCl) also assume large values[7] as shown for KCl in fig.2b.
This result has been taken by Fowler as the key information which
demanded a reconsideration of the applicability of the configuration
coordinate model to certain F-centers[8].

        Finally it should be pointed out that the overall behaviour
of the iodine related absorption in AgBr[2] - not considering the fine
structure and the numerical values - is indeed similar to that
observed for nitrogen in GaP[9], see fig.3. The main effect is a
strong enhancement of the absorption in the region where the
intrinsic absorption starts. This effect has been calculated for
GaP:N by Faulkner[10]. Neglecting the electron-phonon coupling he
has found good agreement between theory and experiment. There is
hope that an application and extension of this theory to the
absorption in AgBr:I might also lead to reasonable results.

        In conclusion, it has been pointed out that the
adiabatic approximation is not applicable to transitions related
to AgBr:I and therefore configuration coordinate models fail. In
contrast a model neglecting the electron-phonon coupling might be
able to reproduce the coarse features of the absorption spectrum.

REFERENCES

1. W. Czaja and A. Baldereschi, J. Phys. C 12, 405 (1979).

2. H. Kanzaki and S. Sakuragi, J. Phys. Soc. Japan 27, 109 (1969).

3. An explication of the various configuration coordinate models has been given by J.J. Markham, Rev. Mod. Phys. 31, 956 (1959).

4. A.P. Marchetti and D.S. Tinti, Phys. Letters 69A, 353 (1979).

5. M. Tsukakoshi and H. Kanzaki, J. Phys. Soc. Japan 30, 1432 (1971).
   F. Moser and S. Lyu, J. Lumin. 3, 447 (1971).

6. J.D. Cuthbert and D.G. Thomas, Phys. Rev. 154, 763 (1967).

7. R.K. Swank and F.C. Brown, Phys. Rev. 130, 34 (1963).

8. W.B. Fowler, Phys. Rev. 135 A, 1725 (1964).

9. J.J. Hopfield, P.J. Dean and D.G. Thomas, Phys. Rev. 158, 748 (1967).

10. R. Faulkner, Phys. Rev. 175, 991 (1968).

RESIDUAL DEFECTS IN LASER ANNEALED Si

A. Goltzené, C. Schwab
Laboratoire de Spectroscopie et d'Optique du Corps
Solide, Associé au C.N.R.S. n° 232
Université Louis Pasteur
5, rue de l'Université, 67000 Strasbourg, France

J.C. Muller, P. Siffert
Centre de Recherches Nucléaires
Groupe de Physique et Applications
des Semiconducteurs (PHASE)
67037 Strasbourg Cedex, France

EPR and photocarrier cyclotron resonance investigations have been
performed on laser annealed Si wafers.
In virgin, an isotropic EPR line at g_1 = 2.0055 $\pm$ 0.0005 and a
carrier lifetime decrease, assessed by cyclotron resonance, are
correlated to the laser processing.
In p-type low-resistivity Si, the laser anneal quenches the P-
implantation induced isotropic line near $q \simeq 2$. The same treatment
is correlated to a strong hole cyclotron resonance increase.

INTRODUCTION

Recent experiments have shown that the damaged layer
resulting from an ion implantation of an amorphous layer deposited
on a crystalline substrate may be regrown under some circumstances
by using a laser beam of high power density[1]. Thus this technique
of laser annealing appears of great interest for the processing
of semiconductors in the electronic component industry.

In the case of silicon, all reports seem to indicate
that laser annealing of damaged surface layers may restore a
better crystal perfection as compared to the usual thermal
anneals.

127

Although it is known from sensitive electrical transport measurements, such as thermally stimulated currents (TSC)[2] or deep level transient spectroscopy (DLTS)[3,4] that some deep traps are still present in restored layers, only little attention has been paid to a direct determination of these remaining defects.

Electron paramagnetic resonance (EPR), which has proven a powerful method for the identification of the irradiation defects in Si, has only be used recently : in a preliminary study, EPR and cyclotron resonance (CR) results have shown that residual defects are introduced in undoped Si during the laser processing[5]. In the present paper, we extend these measurements to laser annealed phosphorus-implanted Si, using again EPR and CR parameters to assess the presence of defects.

As is well known, EPR yields information on the defect electronic structure, whereas CR is linked to the carrier properties.

If an intensity standard is measured simultaneously, EPR signals may yield the absolute spin number, i.e. the number of defects if their electronic term multiplicity is known[6]. CR may too give quantitative informations[7] : when a stationary population of carriers is excited by a constant photon flux N_{ph}, thus the CR signal is proportional to the photogenerated carriers n_c (c = e or h), which is proportional to the incoming photon flux times the lifetime τ_c of the carriers :

$$n_c \propto N_{ph}\, \tau_c$$

The linewidth may be determined by τ_c, but also by collision times[8]. In the purest silicon, the carrier-phonon scattering is predominant whereas in the doped Si impurity-carrier scattering predominates. Thus, any difference observed for different samples for similar experimental parameters (microwave power and frequency (or external magnetic field), N_{ph}, temperature) may be ascribed to the defects. Therefore, CR can be considered as a specific characterization tool, as the effects on the holes and electrons may be well separated.

EXPERIMENTAL

The experiments have been performed on high purity float zone Si and on low resistivity Si.

The former was high purity n-type of 10^4 Ωcm resistivity. Boron, the major trace impurity, present at concentration below

10^{12} cm^{-3} is compensated by phosphorus. Except oxygen and carbon, whose concentrations are unknown, all other impurities are kept at much lower concentrations. The wafers sliced and polished 1 mm thick by the manufacturer, were cut in $1 \times 3 \times 4$ mm^3 samples. Prior to any measurement, they were chemicallt etched (CP4).

Low resistivity (160 Ωcm) material was B-doped p-type Si. On one side of the [111] oriented CP4 etched wafers, doses of 5.10^{16} cm^{-2} P^{31} ions were implanted at 10 keV. The anneals were performed in air with a Q-switched ruby laser ($\lambda = 694,3$ nm) with a pulse duration of 25 ns at half power. The power density was 2.7 Jcm^{-2} for the virgin silicon and up to 10 successive shots have been employed. For the low resistivity wafers, the laser energy was set to 1.8 Jcm^{-2}, with only one shot on the implanted side.

EPR and CR measurements were performed simultaneously, at 4.2 K and 9.3 GHz. The samples dipped in liquid He, are in a finger cryostat which was introduced into a TE 102 rectangular cavity in such a way that the sample was offset by some 3 mm from the microwave electrical node, in order that both the electrical and the magnetic components of the microwave field could be applied on the sample. Care was taken to set the Si sample and the ^{61}Ni$^+$:CuGaS$_2$ EPR standard[9] in a reproducible way, to ensure reliable relative intensity measurements for CR.

RESULTS

The virgin silicon samples are EPR signal free at 4 K. After, the laser anneal, an isotropic line appears with a Landé factor $g_1 = 2.0055 \pm 0.0005$. After etching off the surface layer, or after a thermal anneal at 500 C for 30 minutes, this signal disappears again. Maximum signal intensity corresponds to some 10^{13} spins.

CR is only observed under light irradiation. Laser anneal induces only slight CR effects : the hole CR linewidth decreases, whereas the electron CR linewidth increases, but both their intensities decrease.

For the low-resistivity Si, after implantation a strong EPR line appears near g = 2. It is partially quenched, by almost 40 %, after the laser anneal.

This implantation induced line has a reproducible maximal intensity, corresponding to some 10^{14} spins, but the g_2-value varies between 1.999 and 2.004, the linewidth being about 10 Gauss.

CR intensities given in Table 2 are obtained for photo-carriers excited by photon energies $h\nu = 2.06 \pm 0.17$ eV. After implantation no strong variation is observed. After laser anneal however a strong hole CR and a definite electron CR increase are observed when the implanted side is illuminated.

Finally, a very weak CR (essentially a narrow hole CR) is observed without any photoexcitation on the laser processed wafers only.

DISCUSSION

To the defect responsible of g_1 in undoped Si, we may compare some already known signals. We add, however, the possibility of averaged g values of anisotropic centers, either by considering poor local cristallinity or motional effects. However, the latter is only effective for defects corresponding to shallow levels. Within our experimental range, we are thus left with following possibilities :

a) the anisotropic vacancy-interstitial B_1 complex $[V+O_i]^-$ which requires an averaging, with $g_{B_1} = 2.0050$ [10].

b) the unidentified defect occurring as well in crystalline or amorphous Si, with a isotropic Landé factor $2.005 < g < 2.007$ [11,12].

c) the unidentified defect observed during the baking of $SiSiO_2$ interfaces, at $g = 2.0055$ [13,14].

The second center has never been observed in purest silicon by EPR. Consequently, we are left with a) or c) hypothesis. The observation of a level at $E_c - 0.18$ eV by DLTS, ascribed to $[V+O_i]^-$ [3,4], on the same samples suggests a), however this requires a crystal disorientation to get a g averaging. Centers involving vacancy complexes trapping an oxygen seem therefore more likeable as c) shows a similar line.

CR variations show that the residual defects concentrations increase and that the mean carrier lifetime definitely decreases after laser processing.

In doped Si, the line position varies between $2.005 < g_2 < 1.999$. The latter corresponds closely to the Landé factor of the free electrons in the conduction band ($g_{CB} = 1.99875$ [15]); this could be expected if the local concentrations

Table 1 : CR line intensities I (arbitrary units) and widths ΔH (Gauss) of a stack of 3 platelets, before and after laser annealing on each side.

	Electrons		Light holes	
	I(a.u.)	ΔH(G)	I(a.u.)	ΔH(G)
Blank Stack	5.5	50	7.3	75
Laser Annealed Stack	4.2	65	4.4	70
Etched Stack	6.9	55	6.0	80

Table 2 : Light hole, heavy hole and electron CR intensities $(I_{\ell h}, I_{hh}, I_e)$ for one platelet (in a.u.).
A : light incident on implanted side.
B : light incident on opposite side.
I_e set to 1 for the blank.

	$I_{\ell h}$	I_e	I_{hh}
A Blank CP-4 Etched	0.05	1	0.07
+ Phosphorus Implant	0.01	0.86	0.11
+ Laser Anneal	0.25	1.29	1.35
B Blank CP-4 Etched	0.03	1	0.18
+ Phosphorus Implant	0.003	1.01	0.07
+ Laser Anneal	0.01	0.85	0.11

of the implanted P is high enough to allow motional narrowing effects. In implanted Si, a signal at $g \simeq 2.006$ has already been reported, whose linewidth depends on ions and implantation conditions[16]. It is, in our case, the sum of these two contributions, i.e. shallow donor and amorphization defects which yields our EPR spectrum, which is quenched after laser anneal.

CR variations are most important on the implanted side; this may be expected since, as well carrier diffusion length (~ 0.1 mm) and optical absorption length at 2.06 eV (~ 0.03 mm) are shorter than the sample thickness (1 mm).

The increase of the hole CR over electron CR intensity ratio should indicate the compensation of the p-Si by the P donors, most of them being only electricallt active after the laser anneal.

CONCLUSION

EFR and CR assessment of laser annealed virgin Si and P-implanted Si has been performed.

In undoped Si, EFR shows that residual defects appear after the laser processing, correlated to a slight but definitie decrease of the lifetimes of the carriers.

The EFR signal obtained on low resistivity p-Si after P implantation is strongly quenched by laser processing. CR shows that simultaneously the electrically active donor concentration increases.

REFERENCES

1. J.F. Gibbons, Proceedings of the Laser Effects Ion Implanted Semiconductors, Catania (1978); editor E. Rimini, p.103.

2. J.C. Muller, C. Scharager, M. Toulemonde and P. Siffert, Revue de Physique Appliquée, 14 (1979), to be published.

3. J.L. Benton, L.C. Kimerling, G.L. Miller, D.A.H. Robinson and G.K. Celler, Proceedings Catania (1978), p.543.

4. N.M. Johnson, R.B. Gold, A. Lietola and J.F. Gibbons, Proceedings of Laser Interaction and Laser Processing Symposium, A.I.P. Conf. Proc. edited by S.D. Ferris, J.H. Leamy and J.M. Poate, Boston, USA (1978), p.550.

5. A. Goltzené, J.C. Muller, C. Schwab and P. Siffert, Revue de Physique Appliquée, 15, 21 (1980).

6. C.P. Poole, Electron Spin Resonance, John Wiley, New York (1967).

7. G. Dresselhaus, A.F. Kip and C. Kittel, Physical Review, 98, 368 (1955).

8. E. Otsuka, T. Ohyama and K. Murase, Journal of Physical Society of Japan, 25, 729 (1968).

9. H.J. von Bardeleben, A. Goltzené and C. Schwab, Physics Letters, 51A, 460 (1975).

10. J.W. Corbett, Electron Radiation Damage in Semiconductors and Metals, Solid State Physics, Suppl. 7, Academic Press, N.Y. (1966), p.60.

11. D. Lepine, Physical Review, B6, 436 (1972).

12. I.Solomon, D. Biegelsen and J.C. Knights, Solid State Communications, 22, 505 (1977).

13. J. Szuber and B. Salamon, Physica Status Solidi (a), 53, 289 (1979).

14. P.J. Caplan, E.H. Poindexter, B.E. Deal and R.R. Razouk, Journal of Applied Physics, 50, 5847 (1979).

15. Y. Yafet, Solid State Physics, 4, 50 (1963).

16. G. Götz, Solid State Physics, 4, 50 (1963).
    N. Sobolev, Physica Status Solidi (a) 50, K209 (1978).

CALCULATION FOR THE ELECTRONIC STRUCTURE OF HYDROGEN

AT THE OCTAHEDRAL POSITION IN FCC METALS

M. Yussouff[*] and R. Zeller

Institut für Festkörperforschung
der Kernforschungsanlage Jülich

D-5170 Jülich, W.-Germany

KKR Green's function method developed by Beeby and others for substitutional defects is generalized to the case of interstitial defects. In this method, the ideal lattice is described by a periodic arrangement of muffin-tin potentials. The hydrogen atom is represented by an additional muffin-tin potential at the interstitial site. Charge densities and local density of states are computed self-consistently in the local density approximation to density functional theory. We present results for hydrogen at the octahedral site in Al, Ni, Cu, Pd, and Ag, and compare them with other existing calculations.

Theoretical investigations of the electronic properties of the metal-hydrogen system have become important in recent years due to their practical applications in problems related to energy storage through hydrides[1] and embrittlement[2] of metals after hydrogen is dissolved in them. One approach to the calculation of the electronic structure of hydrogen is based on the jellium model[3], where the positive charges of the host ions are smeared out uniformly to form a homogeneous background. Self consistent calculations are possible in this model[4] but the host lattice structure can appear only as a perturbation. Molecular cluster calculations[4,5] suffer from the restriction to a small number of atoms and the uncertainty about surface effects. Band structure models using the augmented plane wave (APW) method[6] are well suited for stochiometric hydrides, but not for an understanding of the properties of isolated hydrogen impurities. A Green's function method is more appropriate for such a study because it takes advantage of the periodicity of the host lattice and the short screening length of the impurity potential.

The Green's function method based on the Korringa-Kohn-Rostocker (KKR) band structure formalism was proposed by Beeby[7] and others[8]. Here the ideal lattice is described by a periodic arrangement of muffin-tin potentials and the perturbing potential due to the impurity is confined to a single muffin-tin. An important question in such calculations has been the choice of the muffin-tin potentials which are normally constructed with the Mattheiss-prescription and therefore depend on the chosen atomic configurations. A recent work by Zeller and Dederichs[9] uses the KKR Green's function approach for substitutional defects with muffin-tin potentials calculated self-consistently so that the question of atomic configurations becomes irrelevant.

Here we present a generalization of this approach for the cases of interstitial impurities and use it to calculate the electronic properties of hydrogen in the octahedral position in fcc metals. A similar, but non-selfconsistent technique has been used by Katayama et al.[10] to calculate the electronic structure of a positive muon in ferromagnetic Ni.

The one electron Green's function G of the perturbed crystal may be expanded in terms of the eigensolutions of local muffin-tin potentials[9] :

$$G(\vec{r}+\vec{R}^m, \vec{r}'+\vec{R}^n) = (E)^{1/2} \delta_{mn} \sum_L Y_L(\vec{r}) Y_L(\vec{r}') R_l^m(r_<,E) H_l^m(r_>,E)$$

$$+ \sum_{LL'} Y_L(\vec{r}) R_l^m(r,E) G_{LL'}^{mn}(E) R_l^n(r',E) Y_{L'}(\vec{r}') \qquad (1)$$

Here $\vec{r}$ and $\vec{r}'$ are restricted to the Wigner-Seitz cell, $r_<$ and $r_>$ are the smaller or larger of $|\vec{r}|$ and $|\vec{r}'|$. $\vec{R}^m$ are lattice sites except for $m = I$ when $\vec{R}^I$ denotes the octahedral site. Y_L are real spherical harmonics. The radial eigenfunctions $R_l^m(r,E)$ are regular for $r = 0$ and given by

$$\cos \delta_l^m j_l((b)^{1/2}v) - \sin \delta_l n_l((E)^{1/2}v)$$

outside the muffin-tin radius R_m in terms of spherical Bessel functions j_l and spherical Neumann functions n_l. $H_l^m = N_l^m - i R_l^m$ is the analog of the spherical Hankel function and contains the nonregular solution N_l^m which is asymptotically given by

$$\sin \delta_l^m j_l((E)^{1/2}v) + \cos \delta_l^m n_l((E)^{1/2}v)$$

The matrix elements $G_{LL'}^{mn}(E)$ are obtained from the unperturbed $G_{LL'}^{0mn}(E)$ through a Dyson equation :

$$\tilde{G}_{LL'}^{mn}(E) = \tilde{G}_{LL'}^{0mn}(E) + \sum_{L''} \tilde{G}_{LL''}^{0mn}(E)\, t_{1''}^{I}(E)\tilde{G}_{L''L'}^{In}(E) , \qquad (2)$$

with :

$$\tilde{G}_{LL'}^{mn}(E) = G_{LL'}^{mn}(E)\exp(-i\delta_1^m - i\delta_1^n).$$

It may be noted that unlike the substitutional case, here the total on shell t-matrix,

$$t_1^I = -E^{-1/2} \sin \delta_1^I \exp i\delta_1^I$$

appears in Eq. (2). Most of the impurity characteristics are contained in $\tilde{G}_{LL}$ ^{11}. There is full cubic symmetry for the interstitial at the octahedral site in an fcc lattice and $\tilde{G}_{LL'}$ ^{11} is diagonal in angular momentum for $1,1' \leqslant 2$. The Dyson equation then decomposes into four scalar equations for s, p, d-T_{2g} and d-E_g scattering. However, the interstitial impurity at the octahedral position implies a very small muffin-tin (= 0.146a, where a = lattice constant of the fcc lattice) for the impurity potential. Indeed, this is of the order of one Bohr radius in most cases and constitutes a restriction of the present calculations which we intend to remove in the future by including changes of the potentials at neighbouring host sites.

Brillouin zone integration is used to evaluate G_{LL}. By inserting the KKR Ansatz for the wave function

$$\psi_{\vec{k}\nu}(\vec{r}+\vec{R}^m) = \exp(i\vec{k}.\vec{R}^m) \sum_L i^1 \omega_L(\vec{k}\nu)Y_L(\hat{r})R_1(r,E) \qquad (3)$$

in the spectral representation of the Green's function

$$G^0(\vec{r},\vec{r}',E) = \sum_\nu \int_{BZ} d^3k \frac{\psi_{\vec{k}\nu}(\vec{r})\psi_{\vec{k}\nu}^*(\vec{r}')}{E + i0^+ - E_{\vec{k}\nu}} \qquad (4)$$

one gets

$$\mathrm{Im}\, G_{LL}^{0II}(E) = (E)^{1/2}\theta(E)$$

$$-\pi \sum_\nu \int_{BZ} d^3k\delta(E - E_{\vec{k}\nu}) \phi_L^I(\vec{k}\nu)\phi_L^I(\vec{k}\nu) \qquad (5)$$

Here $\phi_L^I(\vec{k},\nu)$ is related to the block wave components $\phi_L^0(\vec{k},\nu)$ of the ideal lattice through the equation

$$\phi_L^I(\vec{k},\nu) = \sum_{L'} G_{LL'}^I(\vec{k},E) t_1^0(E)\phi_{L'}^0(E)\alpha_{11'}(E) , \qquad (6)$$

where $\alpha_{ll'}(E) = \int_0^{Rmr} dr\, r^2 R_l(r,E) R_{l'}(r,E)$ and t_l^0 is the t-matrix

of the host muffin-tin potential. The matrix elements $G_{LL'}^I(\vec{k},E)$ are calculated as

$$G_{LL'}^I(\vec{k},E) = 4\pi \sum_{L''} D_{L''}^I C_{L'',L,L'} \tag{7}$$

where $C_{L'',L,L'}$ are the usual Gaunt numbers[11] and the interstitial structure constants[11] are

$$D_L^I = -(\frac{2}{\pi^2})(E)^{-1/2} \exp(E/\eta) \sum_n \frac{e^{i(\vec{K}_n \cdot \vec{R}^I)} |\vec{K}_n + \vec{k}|^l}{(\vec{K}_n + \vec{k})^2 - E}$$

$$\exp[\frac{-(\vec{K}_n + \vec{k})^2}{\eta}] Y_L(\vec{K}_n + \vec{k})$$

$$-\frac{2}{\sqrt{\pi}} \int_{\sqrt{\eta/2}}^{\infty} d\xi \sum_m e^{[ik\cdot(\vec{R}^m - \vec{R}^I) + \frac{E}{4\xi^2} - (\vec{R}^I - \vec{R}^m)^2 \xi^2]}$$

$$\times E^{-1/2}(2i\xi^2 |\vec{R}^I - \vec{R}^m|)^l \, Y_L(\vec{R}^I - \vec{R}^m) \tag{8}$$

where $\vec{K}_n$ denote the reciprocal lattice vectors and η is the usual Ewald parameter[11].

Finally the real parts of $G_{LL'}^{OII}(E)$ are obtained by a Kramers-Kroning integration :

$$\alpha_{ll'}(E) R_e G_{LL'}^{OII} = \pi^{-1} \int_{-\infty}^{+\infty} dE' P(E'-E)^{-1}$$

$$\alpha_{,}(E') \, ImG_{LL'}^{OII}(E') \tag{9}$$

where one has to choose a sufficiently good cut-off energy E_c for the energy integration[9]. Once $G_{LL'}^{OII}$ is known, $G_{LL'}^{II}$ can be found from Eq.(2). The electronic charge density is given by

$$\rho(\vec{r}) = -(2/\pi) \int_{-\infty}^{E_F} dE \, ImG(\vec{r},\vec{r},E) \tag{10a}$$

and the local density of states in volume V is obtained as

$$n_v(E) = -(2/\pi) \int_V d^3r \, ImG(\vec{r},\vec{r},E) \tag{10b}$$

In our calculations we have performed the Brillouin zone
integrations of Eq.(5) by the tetrahedron method[12] with 6144 tetra-
hedrons in the irreducible part of the zone. The self-consistent
potentials of Moruzzi et al.[13] are used for the host lattice. The
impurity potentials were constructed self-consistently by iteration
according to density functional theory with the local density
approximation of Hedin and Lundqvist[14]. The charge density change
is assumed to be confined to the interstitial muffin-tin during
the interactions. The final results for the impurity potential
are converged on the average to 0.01 eV.

Fig.1 shows the results of our calculations for the local
density of states for hydrogen at the octahedral site in Ni, Cu,
Pd and Ag. It shows almost no contribution from d-states, a little
contribution from p-states and predominantly s-state contributions.
In Ni and Cu, the local density of states looks similar to the s
and p local density of states for substitutional sp-impurities[9]
like Zn in Cu. There local densities of states have strong
structures in the energy range of the host d-bands due to impurity
s-electrons hybridizing with the host d-electrons. (Also the local
density of states is suppressed somewhat in this region as
compared to the energy range where the host electrons have only
s-character). This effect is quite pronounced in Pd and Ag. This
is due to the fact that in Pd and Ag the valence band has only a
narrow region ($\sim$ 1 e.v.) at the bottom where the host electrons
have s-character. So, the s-electrons of hydrogen pile up
particularly in this region.

In all cases we find a s-type bound state for the
hydrogen atom below the valence band of the host metal.

Table I : Bound state characteristic of H in the octahedral
          position. The bound state energies are given with
          respect to the Fermi energy (E_1) and with respect
          to the bottom of the host valence band (E_2). Q_b and Q_t
          are the total and bound state contributions to the
          electronic charge inside the hydrogen muffin-tin sphere.

Host	E_1 (eV)	E_2 (eV)	Q_b (e)	Q_t (e)
Al	−14.00	−2.72	0.682	1.088
Ni	−11.01	−2.12	0.785	1.089
Cu	−12.65	−3.19	0.831	1.124
Pd	−10.45	−4.30	0.817	1.101
Ag	−11.53	−4.25	0.828	1.146

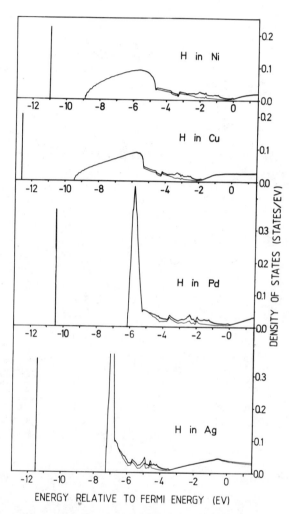

<u>Fig.1</u> : The local density of states seen by hydrogen at the octahedral interstitial site. Lower curve gives s-contribution and the upper curve gives a plus p-contribution. The vertical line gives the location of the bound state.

The bound state contribution to the charge inside the hydrogen muffin-tin sphere is of the order of 0.8 electron. Since the bound state is doubly occupied, it implies that only about 40% of the bound state charge is contained in this muffin-tin sphere. Therefore the bound state wave function extends well beyond the unit cell as was also found in cluster calculations[4]. This should be a drawback of the molecular cluster method because the cluster should be at least as large as the extension of the bound states. In our method, this is not as serious as might appear at the first sight. Although the bound state is extended, the total charge density is rather localized and only this charge density determines the range of the impurity potential. This effect, arising from cancellations of long range tails of individual states by super-position of all occupied states, was also seen in the self-consistent calculations of Baraff and Schlüter[15] and Bernholc et al.[16] for the case of a vacancy in Si.

Acknowledgement - The authors are grateful to Prof. P.H. Dederichs for suggesting this problem and for many helpful discussions. One of us (M.Y.) would like to thank Alexander von Humboldt foundation for financial support.

REFERENCES

* Alexander von Humboldt Fellow from Indian Institute of Technology, Kanpur, India.

1. J.J. Reilly, Jr in Proceedings of the International Symposium on Hydrides for Energy Storage, Geilo, Norway, 1977, edited by A.F. Anderson and A.J. Maelland (Pergamon, Oxford, 1978).

2. B.A. Kolachev, Hydrogen Embrittlement of Nonferrous Metals (Israel Program for Scientific Translation, Jerusalem, 1968).

3. W.A. Harrison, Pseudopotential in Theory of Metals (Benjamin, New York, 1966).

4. See, for example P. Jena, F.Y. Fradin and D.E. Ellis, Phys. Rev. $\underline{B20}$, 3543 (1979).

5. E.J. Baerends, D.E. Ellis and P. Ros, Chem. Phys. $\underline{2}$, 41 (1973).

6. B.M. Klein, E.N. Economou and D.A. Papaconstantopoulos, Phys., Rev. Lett. $\underline{39}$, 574 (1977).

7. J.L. Beeby, Proc. Roy. Soc. London, A $\underline{302}$, 113 (1967).

8. T.H. Dupree, Ann.Phys. (N.Y.) $\underline{15}$, 63 (1961).

9. R. Zeller and P.H. Dederichs, Phys. Rev. Lett. $\underline{42}$, 1713 (1979).

10. H. Katayama, K. Terakura and J. Kanamori, Solid State Commun. $\underline{29}$, 431 (1979).

11. F.S. Ham and B. Segall, Phys. Rev. $\underline{124}$, 1786 (1961).

12. O. Jepsen and O.K. Anderson, Solid State Commun. $\underline{9}$, 1763 (1971).

13. V.L. Moruzzi, J.F. Janak and A.R. Williams, Calculated Electronic Properties of Metals (Pergamon, New York, 1978).

14. L. Hedin and B.I. Lundqvist, J. Phys. C $\underline{4}$, 2064 (1971).

15. G.A. Baraff and M. Schlüter, Phys. Rev. Lett. $\underline{41}$, 895 (1978).

16. J. Bernholc, N.O. Lipari and S.T. Pantelides, Phys. Rev. Lett. $\underline{41}$, 895 (1978).

VACANCIES AND CARBON IMPURITIES IN IRON

A. Vehanen
Department of Physics, University of Jyväskylä
SF-40720 Jyväskylä 72, Finland

P. Hautojärvi, J. Johannson, and J. Yli-Kauppila
Department of Technical Physics
Helsinki University of Technology
SF-02150 Espoo 15, Finland

P. Moser
Section de Physique du Solide
Département de Recherche Fondamentale
Centre d'Etudes Nucléaires de Grenoble
85 X, 38041 Grenoble Cédex, France

Point defects in electron-irradiated high-purity α-iron have been studied by positron lifetime measurements. We show that the migration stage of monovacancies occurs already as low as at 220 K, which results in agglomeration of small three-dimensional vacancy clusters. Furthermore, we irradiated carbon-doped iron specimens, where formation of highly asymmetric monovacancy-carbon atom pairs was detected during the migration stage of monovacancies at 220 K.

The properties of point defects in bcc metals as well as their mutual interactions are known to a very small extent[1]. This is especially the case for iron-carbon system in spite of its substantial technical importance. We report on positron lifetime measurements on the behaviour of monovacancies as well as their interaction with interstitial carbon impurities in electron-irradiated α-iron. The characteristics of photons due to annihilations of thermalized positrons are sensitive to the presence of vacancy-type defects, whereas interstitials and their agglomerates do not affect their behaviour[2]. Positron localization at defect sites enables us to draw information on both the concentration and the internal structure of the defects. A

143

striking feature is the enormous sensitivity of the trapped
positron lifetime to vacancy condensation into three-dimensional
clusters as the cluster size ranges from one to approximately 50
vacancies[3]. Our results[4] show that the migration stage of monovacancies in iron occurs already around 220 K. Furthermore, a
strong interaction between mono-vacancies and carbon interstitials
exists leading to a highly asymmetric vacancy-carbon pair during
vacancy migration.

        The high-purity α-iron was prepared with zone-refining
methods[4]. The analyzed interstitial impurity concentration (C + N)
was less than 5 at ppm. Some samples were carbon doped resulting
in concentrations of 50 and 750 at ppm. The specimens were electron
irradiated at 20 K with 3 MeV electrons to a fluence of about
6×10^{18} e$^-$/cm^2 for pure iron and 3×10^{19} e$^-$/cm^2 for carbon doped
samples. Positron lifetime measurements[4] were carried out at 77 K
during isochronal (30 min) annealing of samples, and the
measurements were done at room temperature after 340 K annealing.
The measured spectra were fitted with two exponential lifetime
components after source-background corrections.

        The analyzed positron lifetimes τ_1 and τ_2 as well as
their relative intensities I_2 (= $1 - I_1$) are shown in Figs.1, 2
and 3 as a function of the isochronal annealing temperature. At
annealing temperatures below 200 K all spectra consist of a single
component with the lifetime τ_1 = 175 psec, much higher than that
in an annealed iron τ_{ann} = 115 psec. Thus we have 100 % positron
trapping into monovacancies produced in electron irradiation.
Annealing through 220 K (labelled as "stage III" in iron[5]) splits
the lifetime spectra in all cases into two distinct components
with the longer lifetime $\tau_2 \approx$ 300 psec and its relative intensity
I_2 varying from 10 % to 50 %. The appearance of a long lifetime
component at 220 K is a strong evidence for vacancy migration,
which results in agglomeration of small three-dimensional
microvoids[3,4,6,7]. A contradicting view has also been presented[5],
which is based on the monovacancy migration energy derived from
more indirect high-temperature data.

        The values of τ_1 in Fig.1 between 250 K and 350 K stay
at a constant level in each specimen. Moreover, the height of the
levels increases monotonically as the carbon concentration increases. This indicates that the presence of carbon impurities
enhances the survival of vacancies during their migration stage.
We ascribe this behaviour to the capture of migrating vacancies
by immobile[8] carbon atoms, which results in a bound vacancy-
carbon pair. By applying the trapping model[2] and the vacancy
concentration estimates[4] produced during irradiation, we also

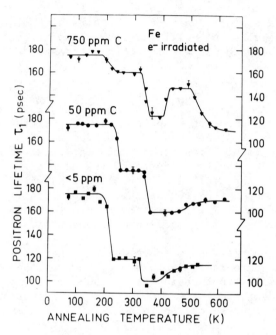

Fig.1 : The shorter positron lifetime values τ_1 as a function of isochronal annealing temperature in electron-irradiated (3 MeV, 20 K) α-iron with varying carbon concentrations. The electron dose is about 6×10^{18} e^-/cm^2 for pure iron and 3×10^{19} e^-/cm^2 for carbon-doped specimens. The error bars are statistical from the lifetime fit. Note a scale shift in each curve.

conclude that for the Fe-750 ppm C specimen the lifetime value τ_1 = 160 psec represents a saturation value; i.e. the trapped positron lifetime at the vacancy-carbon pair. Even in the undoped iron there seems to remain a residual pair concentration, since the trapping model would give τ_1 = 75 psec at 280 K assuming positron trapping only to microvoids.

The next annealing stage occurs at 350 K, where the τ_1 values decrease. At these temperatures the migration of carbon atoms is known to occur[8]. We interpret this phenomena as a further decoration of vacancy-carbon pairs with migrating carbon atoms. This results in "nullification" of positron trapping into the pairs in accord with earlier observations[9-10].

The vacancy-carbon pair formation is also reflected in the behaviour of τ_2 and I_2 in Figs. 2 and 3. The microvoid intensity I_2 is quite small in the Fe-750 C sample, due to a relatively high competing trapping rate of positrons into vacancy-carbon pairs. As this trapping decreases strongly at 350 K, the fraction of positrons annihilating at vacancy clusters increases, which results in further rise of the intensity I_2. On the other hand, the small discontinuity in I_2 values at 340 K is also due to changing the temperature of positron lifetime measurements from 77 K to room temperature : the specific trapping rate into microvoids is temperature-dependent[11].

The recovery of microvoids and other defects occurs in a broad temperature range up to 600 K. However, between 400 K and 500 K the defect structure in the Fe-750 ppm C sample rearranges as seen as a decrease in I_2 and an increase in positron trapping into vacancy-type defects. Up to 600 K all lifetime spectra resume a one-exponential form with τ_1 = 115 psec, typical for a well-annealed iron.

The most striking result of our experiment is the formation of a vacancy-carbon pair at 220 K. The migration sequence of the constituents is reversed from what has been assumed earlier. This seems to be an intrinsic feature of the lattice, and a monovacancy migration energy lower than that of interstitial carbon has been calculated by Johnson[12].

The strong positron trapping by vacancy-carbon pairs as well as the relatively long positron lifetime τ_1 = 160 psec in the pair - not far from that in pure vacancies (175 psec) - are clear indications of a high degree of asymmetry in the structure of the pair. The idea of an asymmetric pair is also consistent with early calculations of Johnson and Damask[13].

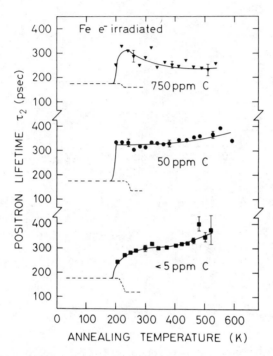

<u>Fig.2</u> : The longer positron lifetime values τ_2 as a function of
isochronal annealing temperature in electron-irradiated
iron-carbon samples. The irradiation conditions are as
in figure 1. The dashed lines denote the behaviour of τ_1
from figure 1.

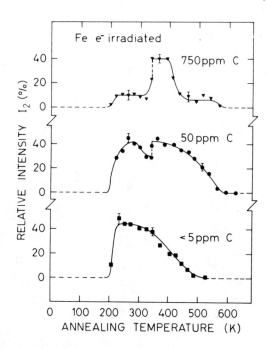

<u>Fig.3</u> : The relative intensity I_2 of the longer positron-lifetime
         component as a function of isochronal annealing temperature
         in electron-irradiated iron-carbon samples. For irradiation
         conditions see figure 1. The vertical shift is due to a
         change in the temperature of lifetime measurements.

         In conclusion our positron lifetime measurements on
electron-irradiated carbon-doped α-iron show direct evidence on
vacancy migration stage at 220 K. Furthermore, bound vacancy-
carbon atom pairs are formed as a result of migration of. vacancies.
Such pairs have a highly asymmetric structure, which is further
modified by carbon migration at 350 K. A more detailed report
will be published elsewhere.

Acknowledgement   -   We are indebted to Section des Accélérateurs
of CEN Grenoble for low temperature electron irradiations.

REFERENCES

1. Properties of Atomic Defects in Metals, Proc. Int. Conf.,
   Ed. by N.L. Peterson and R.W. Siegel, J. Nucl. Mater. 69 & 70
   (1978).

2. Positrons in Solids, Ed. by P. Hautojärvi, Topics in Current
   Physics 12, Springer (1979).

3. P. Hautojärvi, J. Heiniö, M. Manninen and R. Nieminen, Phil.
   Mag. 35, 973 (1977).

4. The data for pure iron below room temperature has been
   published by P. Hautojärvi, T. Judin, A. Vehanen, J.Yli-Kauppila,
   J. Johansson, J. Verdone and P. Moser, Solid State Commun. 29,
   855 (1979).

5. W. Decker, J. Diehl, A. Dunlop, W. Frank, H. Kronmüller,
   W. Mensch, H.E. Schäfer, B. Schwendemann, A. Seeger, H.P. Stark,
   F. Walz and M. Weller, Phys. Stat. Sol. (a) 52, 239 (1979).

6. M. Eldrup, O.E. Mogensen and J. Evans, J. Phys. F 6, 499 (1976).

7. S. Mantl and W. Triftshäuser, Phys. Rev. B 17, 1645 (1978).

8. H. Wever and W. Seith, Phys. Stat. Sol. (a) 28, 187 (1975).

9. C.L. Snead Jr., A.N. Goland, J.H. Kusmiss, H.C. Huang and
   R. Meade, Phys. Rev. B 3, 275 (1971).

10. M. Weller, J. Diehl and W. Triftshäuser, Solid State Commun.
    17, 1223 (1975).

11. R.M. Nieminen, J. Laakkonen, P. Hautojärvi and A. Vehanen,
    Phys. Rev. B 19, 1397 (1979).

12. R.A. Johnson, Phys. Rev. 134, A 1329 (1964).

13. R.A. Johnson and A.C. Damask, Acta Met. 12, 443 (1964).

ELECTRONIC IMPURITY LEVELS IN SmS COMPOUNDS

J.C. Parlebas, M.A. Khan, and C. Demangeat

Laboratoire de Magnétisme et de Structure
Electronique des Solides (LA au CNRS n° 306) U.L.P.

4, rue Blaise Pascal 67000 Strasbourg Cedex, France

We study the electronic energy levels on the nearest neighbours to a substitutional anionic atom in SmS compounds. In agreement with an experimentally observed collapse of doped crystals, the impurity is essentially supposed to introduce a drastic change in the dd hopping integrals between near neighbour conduction band orbitals of samarium. We find that a bound state extracted from the conduction band and falling below the f levels is able to drive an f quasistate above the Fermi level through fd hybridization.

Valence transitions can be chemically induced in SmS by anion substitutions and has been the subject of extensive experimental studies[1,2,3]. Recent theoretical models[4] have calculated the energy levels on the nearest samarium neighbours to a substitutional impurity. As far as the $4f^6$ and $4f^5$ levels are concerned, they are described by a non degenerate effective fermion level holding one to zero electron per cation which is the basis of the Ramirez-Falicov Model[5]. Despite the absence of direct 4f-4f overlap between nearest rare earth neighbours, it is then possible to obtain an 'f' quasiband, the band-width of which is due to a phenomenological fd hybridization. However, the previously mentioned calculations[4] have been restricted to unrealistic host band models calculations with the d band replaced by a single s band.

In this work, we present a more realistic approach by using for the spd band structure of pure SmS a generalized Slater-Koster fit to ab initio calculation[6]. Also, it can be shown easily

that the 30 nearest neighbour intersite Green functions between
f (s like) and d orbitals can be expressed in terms of two
independent functions as described in Table 1.

Table 1 : Matrix elements of the Green operator between 'f'
          (s like) and d orbitals on nearest neighbours in SmS
          compounds.

Orbital Symmetries		First nearest Sm - Sm neighbours		
		$\pm \frac{a}{2}$ ($\pm$ 1,0,1)	$\pm \frac{a}{2}$ (0,$\pm$1,1)	$\pm \frac{a}{2}$ ($\pm$1,1,0)
s 'f'	xy	0	0	$\pm$ G°(1)
	yz	0	$\pm$ G° (1)	0
	zx	$\pm$ G° (1)	0	0
	x^2-y^2	$\overline{3}$ G° (2)	$-\overline{3}$ G°(2)	0
	$3z^2-r^2$	G° (2)	G° (2)	$-$ 2G°(2)

        Before solving the impurity system, let us calculate the
electronic structure of pure semiconducting SmS compound. The
corresponding Hamiltonian·is written in the Bloch representation
from a generalization of the Anderson Hamiltonian to a lattice of
'f impurities' :

$$H° = \sum_{\alpha k\sigma} |\alpha k\sigma> \varepsilon_k^\alpha <\alpha k\sigma| + \varepsilon^f \sum_k |fk><fk|$$

$$+ 2 \left(\sum_{\alpha,k} |\alpha k> h_k^{\alpha f} <fk| + c.c. \right) \tag{1}$$

The first term represents the 'p' valence and 'sd' conduction
bands (C.B.) formed from the spd tight-binding orbitals $|\alpha\lambda\sigma>$
with spin σ and centered on samarium site λ when α = 1, xy, yz, zx,
x^2-y^2, $3z^2-r^2$ symmetries and on sulfur site λ when α = x, y, z
symmetries. The second term of eq.(1) describes a zero width
'spinless band' of energy ε^f. The last term of eq.(1) is actually
different from a one electron mixing between f and spd states
but we will consider it as a phenomenological parameter responsible
of the final non zero f band-width. In the present paper we only
retain the fd type of admixture. We remark that the admixture
between f and d states is as non negligible as the semiconducting
gap is smaller; also it is possible to show that it can be
enhanced by fd correlations[7].

Taking into account spin degeneracy for spd states, the eigen solutions $\{|nk\sigma>, E_n^\circ(k)\}$ of H° for a given k vector of the

first Brillouin zone are obtained by diagonalization of a $\{10 \times 10\}$ matrix expressed in the $\{|fk>, |\alpha k\sigma>$ set with $\alpha = $ s,p,d symmetries:

$$H^\circ |nk\sigma> = E_n^\circ(k) |nk\sigma> \qquad (2)$$

The following parameters (in two centre Slater-Koster notations) appear in eq.(2) : $ss\sigma$, $pp\sigma$, $pp\pi$, $dd\sigma$, $dd\pi$, $dd\delta$, $sp\sigma$, $sd\sigma$, $pd\sigma$, $pd\pi$, ε^s, ε^p, ε^{eg}, ε^{t2g} for α states; they are deduced from a fit to KKR bands structure of SmS[6]. For the dd hopping integrals, the following results are found : $dd\sigma = -0.9$ eV; $dd\pi = 0.4$ eV and $dd\delta = -0.08$ eV. It is important to note that part of this dd overlap parameters comes from mixing of the d orbitals with the anion outer p orbitals; moreover, part of the difference between ε^{eg} and ε^{t2g} arises from implicit renormalization through three center integrals including anion p potential. The SmS density of states (D.O.S.) displays quite a narrow p band roughly 2 eV below the bottom of a broad sd C.B. Besides, ε^f is adjusted in order to get the order of magnitude of the experimental gap between ε^f and the bottom of the C.B. Also we introduce a phenomenological parameter 'fd ' between f (s like) and d states which is consistent with a very small f bandwidth.

Now let us consider a single impurity substituted at an anion site O in SmS compound. We study the electronic impurity levels localized on the 6 nearest cation neighbouring sites of O. The impurity is essentially presumed to increase the overlap integrals $dd\sigma$, $dd\pi$ and $dd\delta$ between $|mR\sigma>$ orbitals (m = xy, yz, zx, x^2-y^2, $3z^2-r^2$) around O. Thus the perturbed Hamiltonian is given by :

$$H = H^\circ + \sum_{\substack{<R,R'> \\ m,m',\sigma}} |mR\sigma> \delta\beta_{RR'}^{mm'} <m'R'\sigma| \qquad (3)$$

where the notation $<R,R'>$ indicates that the summation of eq.(3) is only over nearest neighbouring sites among the 6 sites surrounding O; $\delta\beta_{RR'}^{mm'}$ being the dd hopping change between two such sites. For simplicity, we will only retain diagonal $\delta\beta_{RR'}^{mm}$ matrix elements which are easily expressed for the directions (0,1,1), (1,0,1) and (1,1,0) of RR' (Table 2). Also we only retain the

Table 2 : Diagonal hopping integrals between d orbitals on
nearest neighbours in SmS compounds. The following
notations are used :

$$\delta t^1 = 1/2 \ \delta(dd\pi + dd\delta); \quad \delta\tilde{t}^1 = 1/4 \ \delta(3dd\sigma + dd\delta);$$

$$\delta t^4 = 1/16 \ \delta(3dd\sigma + 4dd\pi + 9dd\delta); \quad \delta\tilde{t}^4 = \delta dd\pi;$$

$$\delta t^5 = 1/16 \ \delta(dd\sigma + 12dd\pi + 3dd\delta);$$

$$\delta\tilde{t}^5 = 1/4 \ \delta(dd\sigma + 3dd\delta).$$

Orbital Symmetries		First nearest Sm – Sm neighbours		
		$\pm \frac{a}{2}$ (±1,0,1)	$\pm \frac{a}{2}$ (0,±1,1)	$\pm \frac{a}{2}$ (±1,1,0)
xy	xy	δt^1	δt^1	$\delta\tilde{t}^1$
yz	yz	δt^1	$\delta\tilde{t}^1$	δt^1
zx	zx	$\delta\tilde{t}^1$	δt^1	δt^1
x^2-y^2	x^2-y^2	δt^4	δt^4	$\delta\tilde{t}^4$
$3z^2-r^2$	$3z^2-r^2$	δt^5	δt^5	$\delta\tilde{t}^5$

intraatomic dd host Green functions $G^{om}(E)$. Actually, in eq.(3)
we should also take into account a similar df hopping change
noted $\delta\beta^{mf}$; instead of that we only consider $\delta\beta^{mm}_{RR'}$, changes but
we consider them as being effectively enhanced by $\delta\beta^{mf}$ effects
combined with host fd mixing. The above model is then solved in
closed form after a bit of manipulations. The d and f local D.O.S.
relative to a site R nearest neighbour to 0 in the z direction is
expressed by :

$$n_R^d(E) = - \pi^{-1} \ Im \ \sum_m \ G^m(E) \qquad (4)$$

$$n_R^f(E) = - \pi^{-1} \ Im \ G^f(E) \qquad (5)$$

where

$$G^m(E) = \frac{G^{om}(E) \ \{1-2D^m(E) - 4[N^m(E)]^2\}}{1-2D^m(E) - 8[N^m(E)]^2}$$

for m = xy, x^2-y^2, $3z^2-r^2$

$$G^m(E) = \frac{G^{om}(E)\{1 - 6[N^m(E)]^2 - 2[D^m(E)]^2 - 8V^m(E)\}}{1 - 8[N^m(E)]^2 - 4[D^m(E)]^2 - 16V^m(E)}$$

for m = yz, zx

with :

$$D^m(E) = \delta \tilde{t}^m \, G^{om}(E)$$

$$N^m(E) = \delta t^m \, G^{om}(E)$$

$$V^m(E) = [N^m(E)]^2 \, D^m(E)$$

In Table 2, $\delta \tilde{t}^m$ and δt^m have been defined.

$$G^f(E) = G^{of}(E) - \frac{24 \, \delta \tilde{t}^4 [G^o(2)]^2}{1 + 2D^4(E)}$$

$$+ \frac{8[G^o(2)]^2 \{4\delta t^5 N^5(E) + \delta \tilde{t}^5\}}{1 - 2D^5(E) - 8[N^5(E)]^2}$$

We are presently engaged in a detailed numerical investigation of eqs.(4) and (5), but let us just present here preliminary (but illustrative) results, when the contributions of each of the m orbitals in eq.(5) are supposed to be equal, i.e. when we are only dealing with s type orbitals[4]. In that case, the following intriguing phenomenon may occur. A d bonding state can be extracted below C.B. and even below ε^f. As a consequence, and through the essential fd admixture, the previously considered

bonding state is able to drive a local f 'quasistate' above the Fermi level. This process causes a drastic change in the valence of the samarium ions around 0 in agreement with the experimental tendency[1,2,3]. Clearly the df mixing should play an appreciable role in all impurity problems[8] as the present.

All this calculations has found its starting point in recent experimental study of SmS compounds doped with oxygen[9]. Krill et al.[9] have advanced the idea that the replacement of S by oxygen (an isoelectronic impurity) may play an analogous role as a hydrostatic pressure. In future, we hope to extend the present model for isoelectronic substitution in SmS to the non isoelectronic impurities and include Friedel's rule.

REFERENCES

1. F. Holtzberg, O. Pena, T. Penney and R. Tournier, in
   Valence Instabilities and Related Narrow Band Phenomena,
   ed. by R.D. Parks, Plenum Press, New-York, p.507 (1977).

2. O. Pena and R. Tournier, ICM'1979, to appear in J.M.M.M.

3. D.C. Henry, K.J. Sisson, W.R. Savage, J.W. Schweitzer and
   E.D. Cater, Phys. Rev. B 20, 1985 (1979).

4. R. Camley, J.C. Parlebas, K.R. Subbaswamy and D.L. Mills,
   J. Physique C5, 372 (1978).
   J.C. Parlebas and C. Demangeat, 3èmes Journées RCP 520,
   Grenoble (1980).

5. R. Ramirez, and L.M. Falicov, Phys. Rev. B 3, 2425 (1971).

6. H.L. Davis, 9th R.E. Research Conference, ed. Field, Vol.1,
   p.3 (1971).

7. D.I. Khomskii and A.N. Kocharjan, Sol. State Commun. 18, 985
   (1976).

8. J.C. Parlebas, ICM'(1979, to appear in J.M.M.M.

9. G. Krill, M.F. Ravet, L. Abadli and J.M. Leger, 3èmes Journées
   RCP 520, Grenoble (1980).

LATTICE DYNAMICS OF SUBSTITUTIONAL ^{119m}Sn IN

COVALENT SEMICONDUCTORS

O.H. Nielsen

Institute of Physics, University of Aarhus

DK-8000 Aarhus C, Denmark

The theory of impurity vibrational amplitudes, Debye temperatures and Mössbauer Debye-Waller factors for ^{119m}Sn on substitutional positions in f.c.c., b.c.c. and zincblende lattices is described. Several models for such systems are commented on, and an extension of Mannheim's model to the zincblende structure is presented. Using Weber's adiabatic bond charge model for the host lattices, unexpectedly small impurity-host force constants relative to the host-host force constants were found for ^{119m}Sn in silicon and germanium. Possible explanations are discussed. For ^{119m}Sn in α-tin reasonable agreement between the phonon model and experiment is observed.

INTRODUCTION

The lattice vibrations of impurities were studied by, among others, Lifshitz[1], whereas the problem of impurity vibrational amplitudes was first treated by Kagan and Iosilevskii[2] within the mass defect approximation. Dawber and Elliott[3] and Elliott and Taylor[4] independently performed calculations for this case. A number of analytical models have included approximately the force constant changes around an impurity, as reviewed by Grow et al.[5] and Petersen et al.[6] The only model which presently includes force constant changes exactly is the model of Mannheim and co-workers which assumes only nearest neighbour central forces in both the pure and impure lattices. This implies that the transverse acoustical phonons have zero frequency which is a

very unphysical feature. A number of models based on numerical
Green's function calculations have also been presented[8].

We give here the essential analytical results of an
extension of Mannheim's model to the zincblende structure, thus
covering several group IV, III-V and II-VI materials within this
framework. The details are given in ref.6, and Mössbauer experiments
with which we compare our model calculations are presented in
refs. 6 and 9. We do not discuss Mössbauer results for b.c.c. and
f.c.c. metals, which were reviewed by Grow et al.[5] using
Mannheim's model for the analysis.

THEORY OF IMPURITY VIBRATIONS

The Mössbauer Debye-Waller factor, of the recoilless
fraction of γ-rays, is for cubic crystals

$$f = \exp\left(-\vec{k}_\gamma^2 \langle u'^2_x \rangle\right) \tag{1}$$

where $\vec{k}_\gamma$ is the γ-ray wavevector and u' is the Mössbauer nucleus'
displacement which depends on the temperature T. It is typically
of order 0.1 Angstrom, and is calculated by

$$\langle u'^2_x \rangle = \frac{M}{2M'} \int_0^\infty \frac{1}{\omega} g'(\omega) \coth\left(\frac{1}{2} \hbar\omega/k_B T\right) d\omega \tag{2}$$

M' is the impurity mass, and $g'(\omega)$ is the impurity "density of
states" or "dynamic response function". This is again given in
terms of a Green's function for the impure lattice

$$g'(\omega) = \frac{2M'\omega}{\pi} \lim_{\varepsilon \to 0^+} \mathrm{Im}(U_{xx}(00,00;\omega-i\varepsilon)) \tag{3}$$

The Green's functions are defined in chapter VIII of Maradudin et
al.[10], and were originally treated by Elliott and Taylor[4]. At
sufficiently high temperatures we obtain

$$\langle u'^2_x \rangle = \frac{k_B T}{M'} \int_0^\infty \omega^{-2} g'(\omega) \, d\omega \equiv \frac{k_B T}{M'} \frac{3\hbar^2}{\left(k_B \theta_D'(-2)\right)^2} \tag{4}$$

which defines the Debye temperature $\theta_D'(-2)$ corresponding to the
ω^{-2}-moment of $g'(\omega)$. It has been shown[6] that the mass defect
model yields for this quantity

$$\theta_D'(-2) = \sqrt{\frac{M}{M'}} \cdot \theta_D(-2) \tag{5}$$

where M denotes the host atom mass and $\theta_D(-2)$ is the host lattice

Debye temperature corresponding to the ω^{-2}-moment of the perfect lattice phonon density of states.

In Mannheim's model the equation for the impurity Green's function $\bar{\bar{U}}$, eq.(8.5.2) of ref.10, is solved for the f.c.c. and b.c.c. structures under the assumption of nearest neighbour central forces. We have extended[6] this model to zincblende structure, which gives a result identical to that of Mannheim. While a number of physical quantities, e.g. localized mode frequencies, can be calculated from this model, we shall here restrict ourselves to the calculation of high temperature vibrational amplitudes.

We note the general relation between vibrational Green's functions and the densities of states,

$$U_{xx}(00,00;\omega=0) = \frac{1}{M'} \int_0^\infty \frac{g'(\omega)d\omega}{-\omega^2} \qquad (6)$$

Solving the equations for $\bar{\bar{U}}$ at $\omega = 0$, eqs.(39) and (40) of ref.6, we readily arrive at the result

$$\theta_D'(-2) = \theta_D(-2) \cdot \sqrt{\frac{M}{M'}} \cdot [1 + \frac{5}{9} (\frac{\theta_D(-2)}{\theta_D(+2)})^2 (\frac{\phi_{xx}(00,00)}{\phi'_{xx}(00,00)} - 1)]^{-\frac{1}{2}}$$

$$(7)$$

Here $\theta_D(+2)$ is the Debye temperature corresponding to the ω^{+2}-moment of the phonon density of states. $\phi_{xx}(00,00)$ and $\phi'_{xx}(00,00)$ are the host-host and impurity-host force constants, respectively. This result was first derived by Grow et al.[5] by a series expansion of the Mannheim Green's function $\bar{\bar{U}}$.

The great simplicity of Mannheim's model lies in its analyticity, leading to the formula eq.(7) which can be directly compared with Mössbauer data. Apart from the masses, only the two Debye temperatures $\theta_D(\pm2)$ for the perfect lattice are needed together with one free parameter, the force constant ratio $\phi_{xx}(00,00)/\phi'_{xx}(00,00)$. In fact, the Debye temperatures can for monatomic lattices be found from an analysis of heat capacity data[11]. For many materials phonon models are also available for calculation of these quantities. $\theta_D(-2)$ may furthermore in general be found from neutron- or x-ray diffraction. It appears preferable to compare the host lattice Debye temperatures from several sources in the cases where this is possible.

COMPARISON OF EXPERIMENTAL AND THEORETICAL RESULTS

In Table 1 we show Debye temperatures calculated from Weber's adiabatic bond charge model[12,13]. The third column shows impurity Debye temperatures measured for ^{119m}Sn by Mössbauer experiments[6], and in the last column is shown the mass defect model results from eq.(5). The insufficiency of this model to

Table 1

	$\theta_D(-2)$	$\theta_D(+2)$	$\theta_D'(-2)_{EXP}$	$\theta_D'(-2)_{M.D.}$
Si	526	671	223(4)	256
Ge	297	395	191(4)	232
α-Sn	169(7)	259	161(3)	169(7)

describe the present results is clearly demonstrated since the experimental Debye temperatures fall significantly below for silicon and germanium, indicating qualitatively a decrease in impurity-host coupling strength. This decrease is quantified by means of Mannheim's model in Table 2, where the force constant ratios are calculated from Table 1 by means of eq.(7). The second column shows the Debye-Waller factor calculated from Mannheim's model via eqs.(1) and (2) and the phonon densities of states from the bond charge model. The last column gives the experimental results, where the measured α-tin $\theta_D(-2)$ was used to compute the Debye-Waller factor for this case. Silicon and germanium were measured relative to α-tin.

Table 2

	$\dfrac{\phi_{xx}(00,00)}{\phi_{xx}'(00,00)}$	f (300 K) Mannheim	Relative exp. f (300 K)
Si	1.92(15)	0.332(14)	0.34(3)
Ge	2.51(30)	0.226(20)	0.22(3)
α-Sn	=1	0.125(13)	0.13(1)

DISCUSSION

The inapplicability of the mass defect model for description of [119m]Sn vibrational amplitudes in silicon and germanium was shown. For α-tin good agreement between experiment and the bond charge model is found. The host-host model to impurity host force constant ratio is found to be 1.92(15) in silicon and 2.51(30) in germanium from Mannheim's model, indicating a drastic decrease of the impurity-host coupling. A number of possible error sources do not seen to affect this conclusion[6], which is in contrast with tight-binding calculations and experiments for the Mössbauer isomer shift[6,9]. These indicate that the host crystal has a predominant influence on the impurity electronic structure. This does not, however, a priori imply any contradiction, since vibrational properties are not directly linked to the s-electron density[6].

As mentioned above, Mannheim's nearest neighbour central force model yields zero TA-phonon frequency, and since these phonons in reality dominate the vibrational amplitudes, any conclusions drawn from Mannheim's model are invariable ambiguous. We have attempted the evaluation of a non-central force constant model, but even for this case the Green's function relations, eqs.(36)-(38) of ref.6, which make possible the analytic solution of Mannheim's model, are altered in such a way that a simple solution in terms of only the host lattice phonon density of states is hindered. This appears to leave only the possibility of doing numerical calculations for models more complex than that of Mannheim.

Even though the concluded force constant ratios may not carry the physical significances that is assigned by Mannheim's model, it is interesting to note that the ratio is larger for [119m]Sn in germanium than in silicon, but again decreases to the value 1 in α-tin, by definition. If this picture holds also for more realistic impurity models, we may point on the qualitatively very different interactions of the impurity with the host lattice TA-phonons in silicon compared to germanium[6]. For [119m]Sn in silicon a well-defined low frequency resonance mode appears, whereas in germanium somewhat more structure is found in $g'(\omega)$, especially around the host lattice TA-phonon frequencies.

CONCLUSION

Current models of vibrational amplitudes for ^{119m}Sn in covalent semiconductors were discussed. Mannheim's model encompasses all earlier analytical models for f.c.c. and b.c.c. metals, and was extended to the zincblende structure also. We have by the Mössbauer effect concluded the existence of force constant changes for ^{119m}Sn in silicon and germanium, and calculated these within the framework of Mannheim's model. We found impurity-host force constants to be weakened by about a factor of 2, less in silicon, and more in germanium. This indicates that Mössbauer Debye-Waller factors may be more sensitive to details of the impurity-host interaction than the simultaneously measured isomer shifts. Since Mannheim's model oversimplifies the physical interactions involved, a too rigid interpretation of these results is not warranted. Better, and possibly numerically based, impurity models should be developed in order to reach the same degree of sophistication as for the perfect lattice vibrations.

REFERENCES

1. I.M. Lifshitz, JETP 17, 1017, 1076 (1947); Nuovo Cimento 3 suppl. 4, 716 (1956).

2. Yu.Kagan, Ya.A.Iosilevskii, JETP 15, 182 (1962); JETP 17, 195 (1963); JETP 19, 1462 (1964).

3. P.G. Dawber, R.J. Elliott, Proc. Roy. Soc. A (GB) 273, 222 (1963); Proc. Phys. Soc. 81, 453 (1963).

4. R.J. Elliott, D.W. Taylor, Proc. Phys. Soc. 83, 189 (1964).

5. J.M. Grow, D.G. Howard, R.H. Nussbaum, M. Takeo, Phys. Rev. B 17, 15 (1978).

6. J.W. Petersen, O.H. Nielsen, G. Weyer, E. Antoncik, S. Damgaard, Phys. Rev. B, in print.

7. P.D. Mannheim, Phys. Rev. 165, 1011 (1968); P.D. Mannheim, A. Simopoulos, Phys. Rev. 165, 845 (1968); P.D. Mannheim, S.S. Cohen, Phys. Rev. B 4, 3748 (1971); P.D. Mannheim, Phys. Rev. B 5, 745 (1972).

8. E.g.: G.W. Lehmann, R.E. DeWames, Phys. Rev. 131, 1008 (1963); D. Strauch, Phys. Stat. Sol. 30, 495 (1968); M. Vandevyver, P. Plumelle, J. Phys. Chem. Sol. 38, 765 (1977).

9. G. Weyer, A. Nylandsted-Larsen, B.I. Deutch, J.U. Andersen, Hyp. Int. 1, 93 (1975).

10. A.A. Maradudin, E.W. Montroll, G.H. Weiss, I.P. Ipatova,
    Solid State Phys. suppl. 3, 3nd ed., Academic Press (1971).

11. T.H.K. Barron, W.T. Berg, J.A. Morrison, Proc. Roy. Soc.,
    London A242, 478 (1957).

12. W. Weber, Phys. Rev. B 15, 4789 (1977).

13. O.H. Nielsen, W. Weber, Comput. Phys. Comm. 18, 101 (1979).

MÖSSBAUER SPECTROSCOPY ON THE AMORPHOUS LAYER IN ION

IMPLANTED DIAMOND

M. Van Rossum, J. De Bruyn, G. Langouche[+],
M. de Potter, and R. Coussement

Instituut voor Kern- en Stralingsfysika
Leuven University

Celestijnenlaan 200 D, B-3030 Leuven, Belgium

The structure of the amorphous layer in ion implanted diamond has been investigated with the Mössbauer probes ^{133}Cs, ^{125}Te and ^{129}I. A strong resemblance is found between the nearest-neighbour coordination in amorphous diamond and ion implanted graphite.

1. INTRODUCTION

It is well known that the surface of ion implanted diamond will be amorphized when the implanted dose exceeds some critical limit. For heavy ions implanted at low energy, this transition starts somewhere between 10^{13} and 10^{14} at/cm^2 [1]. It is also widely accepted that the structure of the amorphous layer in diamond does not retain the short range order of its crystalline counterpart, as it is the case for amorphous Si and Ge. Instead, the structure of amorphized diamond exhibits many features of amorphous graphite, as has been observed by Raman spectroscopy and other techniques[2].

Recently, we have shown that the crystalline-to-amorphous transition in diamond can be followed in detail by Mössbauer Spectroscopy, using ^{133}Cs and the proble nucleus[3]. In this paper, we want to present some further evidence for the similarity between the amorphous diamond and the graphite coordination. In addition to ^{133}Cs, the Mössbauer nuclides ^{125}Te and ^{129}I (obtained from

the decay of ^{125m}Te resp. ^{129m}Te) have been used for this study.
No previous report of ^{125m}Te implanted in diamond has been published.
Implantations of ^{129m}Te in diamond have been reported by
Hafemeister and de Waard[4]. However, no special care was taken to
avoid amorphization and no comparison with graphite spectra was
made. As we shall see, this casts some doubts on the conclusions
drawn by these authors, who rely on the assumption that their
results are representative of the tetrahedral diamond lattice.

2. EXPERIMENTAL

Implantation of the parent activities ^{133}Xe, ^{125m}Te and
^{129m}Te were carried out at an accelerating voltage of 85 keV. The
implanted doses varied between 10^{14} and 10^{15} at/cm^2, thus
exceeding the amorphization limit for diamond at room temperature.
The targets consisted of natural diamond stones of type IA, and
of commercially purchased pyrolitic and polycrystalline graphite
platelets. Mössbauer absorbers were CsCl for ^{133}Cs, ZnTe for ^{125}Te
and CuI for ^{129}I. Spectra were recorded with a conventional
electromechanical transducer, both source and absorber being kept
at liquid helium temperature.

3. DISCUSSION OF THE EXPERIMENTAL RESULTS

3.1. Experiments with ^{133}Cs

The experimental results are displayed in fig. 1. The
spectrum of graphite (fig.1a) consists of a single line with
isomer shift δ = +0.03 + 0.05 mm/s and linewidth Γ = 1.0 + 0.1 mm/s.
The diamond spectrum (fig.1b) is dominated by an unsplit line
(a-line) of δ = -0.20 + 0.05 mm/s and Γ = 1.3 + 0.1 mm/s.
Arguments for the attribution of this line to Cs atoms in an
amorphous surrounding have been presented in an earlier paper[3].
The remaining components of spectrum 1b probably originate from
Cs atoms in the crystalline diamond beneath the amorphous layer.
It is noteworthy that the electronic density of Cs implanted in
pyrolytic graphite nearly coincides with the value measured by
Campbell et al.[5] for ^{133}Cs in the intercalate compound CsC$_{24}$,
where the Cs is supported to be fully ionized. In this compound
however, the Cs spectra also display a rather well resolved
quadrupole interaction, which is not found in our results.

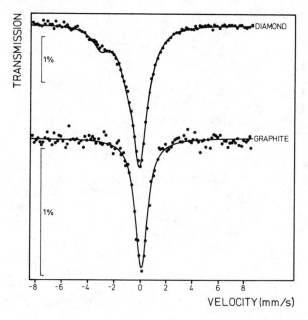

<u>Fig.1</u> : Mössbauer Spectrum of ^{133}Cs implanted in
(a) graphite   (b) diamond

<u>Table I</u> : Fitting parameters for the spectra of ^{125}Te and ^{129}I in
diamond and graphite. Isomer shift (δ), quadrupole
splitting (Δ) and linewidth (Γ) of the different
components are given.

		δ mm/s	Δ mm/s	Γ mm/s
^{125}Te	diamond	-0.32 ± 0.05	8.97 ± 0.1	5.9 ± 0.5
		$+0.04 \pm 0.4$	$-$	5.9 ± 0.5
	graphite	-0.20 ± 0.05	9.14 ± 0.1	6.8 ± 0.2
^{129}I	diamond	$-0.87 \pm 0.1^*$	$13.4 \pm 0.2^*$	1.7 ± 0.1
		$+0.74 \pm 0.2$	$-$	1.7 ± 0.1
	graphite	-0.64 ± 0.05	13.3 ± 0.2	2.0 ± 0.2

* a distribution of $\pm$ 10 % was included in the fit

Another close similarity between the Cs implanted in
pyrolytic graphite and the intercalate compounds consists of the
strong anisotropy of the Debye-Waller factor. We found the intensity
ratio of the Mössbauer resonance detected from γ-rays at 90° resp.
30° to the C planes to be about 4.4. According to Campbell, we may
use this argument to localize the implanted Cs atoms in between
the C layers.
The a-diamond line does not show this anisotropy effect, which may
follow from a random orientation of the graphitic cells in the
amorphous layer. From the total area of the a-line, it is also
clear that the Debye-Waller factor for Cs in a-diamond is much
higher than the one in graphite even at 90°; this fact has been
confirmed by subsequent f-factor measurements[6]. A more detailed
study will be necessary to understand this rather surprising
feature of the dynamical behaviour of Cs in the amorphous diamond
lattice.

3.2. Experiments with ^{125}Te and ^{129}I

The suspected similarity between the short-range order
of graphite and a-dimaond is reinforced by the results of
^{125m}Te and ^{129m}Te implantations. In graphite, the spectra show
large splittings which we interpret as arising from a strong
quadrupole interaction at the Te resp. I site. Nearly identical
splittings (both in position and magnitude) are found as the main
components of the diamond spectra (figs.2b, 3b and table I). However, the diamond spectra also contain at least one more unsplit
line. In a search for the origin of these lines, we have performed
a hot implantation of ^{129m}Te in diamond, during which the target
was kept at 600°C. EPR and Mössbauer studies have shown that under
these circumstances, the formation of the amorphous layer is
strongly inhibited, even for doses exceeding 10^{14} at/cm^2 [3,7].
Indeed, the spectrum resulting from this hot implant (fig.3c) is
very different from the previous one. It essentially displays a
strong single line with a small sidepeak, which could be interpreted either as an unsplit resonance or as part of a small
residual quadrupole interaction. The main peak should be
representative of I atoms situated in the crystalline diamond
lattice. Moreover, it turns out that the spectrum of the cold
implant (fig.3b) can be fitted quite satisfactorily by a properly
weighted combination of figs. 3a) and 3c) provided that one allows
for a small distribution of the quadrupole splitting parameter.
This quadrupole split component, which we identify with I atoms
in the a-diamond layer, accounts for about 85 % of the total
spectral intensity. Although no hot implant of ^{125m}Te in diamond
has been performed up to now, we suppose that a similar type of
analysis will also be valid for spectrum 2b).

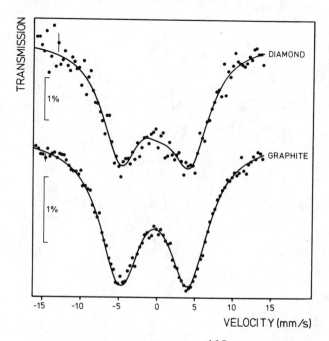

<u>Fig.2</u> : Mössbauer Spectrum of ^{125}Te implanted in
(a) graphite   (b) diamond

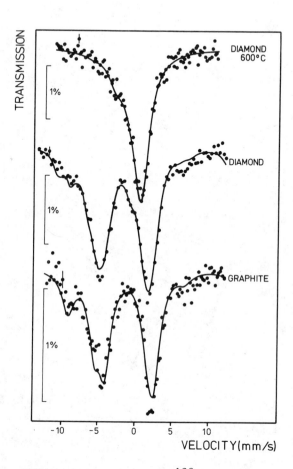

Fig.3 : Mössbauer Spectrum of ^{129}I implanted in
       (a) graphite   (b) diamond
       (c) diamond kept at 600°C during implantation

4. CONCLUSION

These experiments have brought new information for the characterization of the amorphous diamond layer. Whereas the comparison of isomer shift values for ^{133}Cs in graphite and in a-diamond only provides a rough indication for the similarity of both lattices, the identity of isomer shifts and quadrupole inter-actions observed with ^{125}Te and ^{129}I in both materials suggests a strong resemblance of their near-neighbour coordination. Moreover, this structural similarity seems now even more firmly established, as it is independent of the impurity used as a Mössbauer probe (Te, I or Cs). However, the ^{133}Cs experiments still reveal substantial differences in the vibrational properties of amorphous diamond and graphite. More experiments are in progress to clear out this situation.

Finally, we would like to point out that previous Mössbauer experiments on ^{129m}Te implanted in diamond[4] have to be reinterpreted in the light of the present results.

Acknowledgements - The authors wish to thank Dr. H. Pattyn and R. Vanautgaerden for the ion implantations. The help of E. Verbiest for the data handling is gratefully acknowledged. This work has been financially supported by the association S.C.K.-K.U.L. and by the I.I.K.W.

REFERENCES

+ Bevoegdverklaard navorsers N.F.W.O.

1) A. Talmi, E. Beserman, M. Taicher, G. Braunstein and R. Kalish, in "Defects and Radiation Effects in Semiconductors" (Ed. J.H. Albany, The Institute of Physics, London 1979), p.347.

2) M.A. Brodsky and M. Cardona, JOurnal of Non-Crystalline Solids 31, 81 (1978).

3) M. Van Rossum, J. De Bruyn, G. Langouche, M. de Potter and R. Coussement, Physics Letters 73A, 127 (1979).

4) D.W. Hafemeister and H. de Waard, Physical Review B7, 3014 (1973).

5) L.E. Campbell, G.L. Montet and G.J. Perlow, Physical Review B15, 3318 (1977).

6) M. Van Rossum, et al., to be published.

7) Y.H. Lee, P.R. Brosious and J.W. Corbett, Physica Status Solidi (a)50, 237 (1978).

THE EFFECT OF V AND Nb IMPURITIES ON THE ELECTRONIC

PROPERTIES OF TiSe$_2$

F. Lévy and H.P. Vaterlaus

Laboratoire de Physique Appliquée, EPEL

Lausanne, Switzerland

TiSe$_2$ is a semimetallic compound undergoing a second order structural phase transition at about 200 K. At room temperature, the electrical conductivity $\sigma = 10^3 \ ^{-1}\text{cm}^{-1}$ is dominated by positive charge carriers. In relation with the phase transition a change of sign of the Hall coefficient has been measured and the crystal becomes n type at low temperature with an absolute value of the Hall constant depending on the stoichiometry. This paper presents the results of the investigation of the electrical properties of TiSe$_2$ doped with V and with Nb. A strong increase of the electrical resistivity is observed at 5 K for 1 at %, respectively. In the V doped samples, a tendency of remaining p-type at lower temperature with respect to TiSe$_2$ emphasizes the relative importance of the electron and hole densities in these semimetallic compounds.

Titanium selenide belongs to the class of layered materials crystallizing in the CdI$_2$ type of structure. The semimetallic character of this compound is suggested by the small value of its resistivity at room temperature $\rho(295) \cong 10^{-3} \ \Omega\text{cm}$ [1] and is emphasized by the result of recent band structure calculations[2]. At low temperature a distorted phase is stable, characterized by a doubling of the lattice parameters on the three crystallographic axes. The origin of the phase transition has been interpreted in several ways[3] and appears to be related with the quasi soft-mode behaviour of certain vibrational branches.[4]

The role of impurities in $TiSe_2$ has been mainly investigated with respect to their influence onto the phase transition. As in similar systems, the principal effect of the mixing of cations (Ta) or anions (S) is to shift the phase transition towards lower temperature[1,5]. This progressive suppression of the phase transition is linked with the increasing disorder due to the foreighn ions. In $TiSe_2$, however, the effect of non-stoichiometry is even more important to wipe out the phase transition. Ti in excess is intercalated between the slabs and donates electrons to the crystal. The resulting variation of the electron density at the Fermi surface appears to stabilize the undistorted phase without strong variation of the resistivity at room temperature.

This paper reports on the investigation of the influence of incorporated impurities onto the electrical properties of $TiSe_2$. The most drastic effects have been observed in crystals doped with transition metal ions and especially with V and with Nb[6,7]. At low temperature, a strong increase of resistivity is observed. At room temperature, however, the resistivity does not change much even if the crystal becomes n type in comparison to pure stoichiometric $TiSe_2$ which is p type. Moreover, magnetic moments have been measured at low temperature as well as a large negative magneto-resistance[8].

Single crystals of $Ti_{1-x}V_xSe_2$ and $Ti_{1-x}Nb_xSe_2$ have been grown by chemical transport reaction from the elements mixed with selenium in excess. In order to improve the stoichiometry the crystal growth was done at the lowest possible temperature (T_g = 650°C). However, for concentrations $x_{nom} > 0.5$, higher growth temperatures had to be used. In the system $Ti_{1-x}Nb_xSe_2$ for example, single phase crystals for $0.3 < x_{nom} < 0.95$ had to be grown at $T_g > 800°C$. For $x > 0.8$, crystals grown at about 1000°C were cooled rapidly to the room temperature. Such crystals grown at high temperature evidence a decrease of the room temperature resistivity which can be due to a deficiency of stoichiometry.

In the system $V_xTi_{1-x}Se_2$ it appears to be difficult to grow well formed crystals for $0.5 < x < 0.95$ even at high temperature. This effect could be expected from the large difference between the ionic radii : V^{5+} is smaller than Ti^{4+}.

How much of the added impurity is actually incorporated in the $TiSe_2$ matrix? This crucial question can be partially answered with the help of electron microprobe analysis; a 20 keV electron beam was used and X-ray emissions were detected. The results of these investigation are represented in Fig.1 where

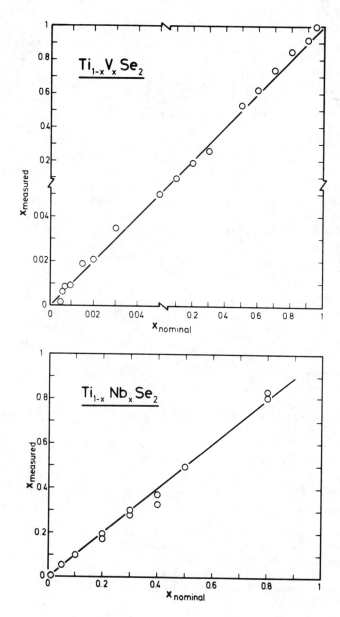

<u>Fig.1</u> : Actual concentration x in single crystals of the systems
Ti$_{1-x}$V$_x$Se$_2$ and Ti$_{1-x}$Nb$_x$Se$_2$ measured by electron microprobe
analysis and plotted as a function of the nominal
concentration.

the actual concentration is represented as a function of the
nominal concentration for a large number of crystals. Except at
very low concentration, the nominal and actual concentrations
agree within about 10 %. A strong difference appears between the
two systems. The homogeneity of the samples containing V is more
reliable than in the Ti/Nb system. It can be noted that the
existence of Nb in an octahedral configuration is not evident
since the only niobium selenide phase with Nb in octahedral
configuration is the $4H_d-NbSe_2$ polytype stable above 980°C [9].

      The second crucial question is to discern the structural
position of the impurity ions. X-ray and electron Laue diagrams as
well as powder diagrams have given no evidence of interstitial
ions. This assertion is based on the interpretation of the diffuse
streaking observed in the electron diffraction patterns as due to
the short range order prepearing the superlattice formation[4]. However, the presence of intercalated cations and in particular of
intercalated Ti cannot be excluded, especially not in crystals
grown at high temperature, as indicated by several checks of the
stoichiometry.

      The electrical resistivity of typical crystals is
represented as a function of the temperature in Fig.2. The strong
increase of resistivity occuring with decreasing temperature is
reminiscent of $1T - TaS_2$ [10]. This increase is interpreted in
terms of an Anderson localization, as in the case of $TaSe_2$ doped
with Fe. However, the importance of the electron correlations has
been emphasized for these kind of transitions[11].

      Even if no saturation of the resistivity has been
measured with decreasing temperature, its value at liquid Helium
temperature represented as a function of the alloy concentration
gives a good notion of the behaviour of these systems (Fig.3).
The maximum resistivity at low temperature occurs for $x \cong 0.01$ in
$Ti_{1-x}V_xSe_2$ and for $x \cong 0.2$ in $Ti_{1-x}Nb_xSe_2$. It can be added that
no such increase of resistivity has been observed in $Ti_{1-x}Ta_xSe_2$ [6].
The difference between these three alloys indicates that the
tendency of building extended states increases from V to Ta and
is connected with the relative spread of d electrons. This is
consistent with the maximum resistivity in $Ti_{1-x}Nb_xSe_2$ which is
notably lower than in $Ti_{1-x}V_xSe_2$.

      The Hall coefficients of several $Ti_{1-x}Nb_xSe_2$ samples are
represented as a function of the alloy concentration x in Fig.4.
The value of R_H at room temperature decreases with increasing Nb
concentration x. In a single band model this would correspond
to an increase of the hole density from about 2×10^{20} for $x = 0.01$

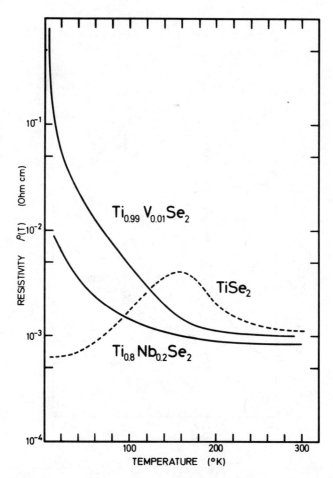

Fig.2 : Electrical resistivity as a function of the temperature
for TiSe$_2$ containing V or Nb impurities in the critical
concentration range and compared with stoichiometric TiSe$_2$
(broken line).

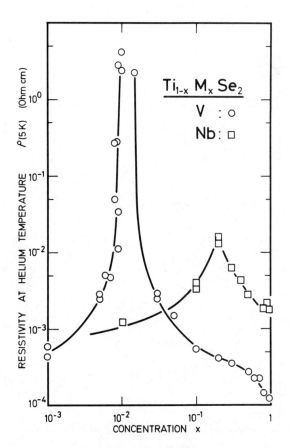

Fig.3 : Electrical resistivity measured at liquid helium
temperature versus concentration in the systems
$Ti_{1-x}V_xSe_2$ and $Ti_{1-x}Nb_xSe_2$.

to 6×10^{21} cm^{-3} for x = 0.95. However, the Hall coefficient R_H = $r(p - b^2n)/e(p + bn)^2$ is more likely related with the increase of the electron density n. More over, the mobility ratio b = μ_n/μ_p [13] cannot be neglected. It appears from the bandstructure of TiSe$_2$ that a shift of the Fermi level towards higher energy would be accompagnied by a decrease of hole effective mass and by an increase of the electron mass, that is by a sensitive decrease of b. For R_H this decrease can compensate the variations of the carrier densities so that for small x R_H remains nearly constant especially in Ti$_{1-x}$V$_x$Se$_2$. The large value of $R_H \cong 10^{-3}$ cm^{3}A^{-1}s^{-1} for x = 0.95 contrasts with value R_H = 4×10^{-4} cm^{3}A^{-1}s^{-1} for NbSe$_2$ [12] and is characteristic of the 1T form in Ti$_{1-x}$Nb$_x$Se$_2$. The Hall coefficient at room temperature remains positive in Ti$_{1-x}$Nb$_x$Se$_2$. However, in Ti$_{1-x}$V$_x$Se$_2$ it becomes smaller for x $\overset{\sim}{>}$ 0.1 and then negative for higher values of x. This effect reminds the influence of non stoichiometry in TiSe$_2$ where excess Ti renders the crystal n-type.

The Hall coefficient at liquid Helium temperature is negative for small values of x as in the TiSe$_2$ crystals. But in the concentration range corresponding with the high value of the resistivity, a positive R_H (5 K) is measured. This behaviour is probably related with the existence of magnetic moments associated with the localization of electrons. At low temperature the distorted phase is stable. For example, the structural phase transition typical of TiSe$_2$ has been observed by electron diffraction and in IR reflectivity spectra in Ti$_{0.97}$V$_{0.03}$Se$_2$ [14]. Assuming that portions of the Fermi surface corresponding to holes are lost at the phase transition, an additional contribution of the hole to the typical value R_H (5 K) = 2×10^{-2} cm^{3}A^{-1}s^{-1} for x = 0.03 is hardly to be expected.

In summary this paper reports on the synthesis of single crystals and on the electrical properties of Ti$_{1-x}$V$_x$Se$_2$ and Ti$_{1-x}$Nb$_x$Se$_2$. The resistivity anomaly of TiSe$_2$ connected with the structural phase transition is progressively shifted towards lower temperature by the incorporated impurity ions.

A strong increase of the low temperature resistivity is observed in both systems and can be related with a disorder-induced localization.

The complex behaviour of the galvanomagnetic effects in the critical concentration range is related with the existence of

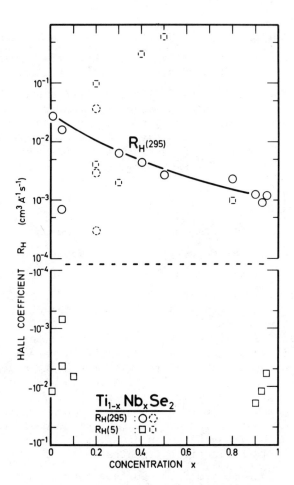

<u>Fig.4</u> : Hall coefficient versus concentration in the system
$Ti_{1-x}Nb_xSe_2$; R_H (295 K) : o, R_H (5 K) : □ . The dotted
marks correspond to critical measurements.

of magnetic moments associated with the substituted ions.

        The sensitive variations in the electronic properties
compared with the relative stability of the low temperature
distorted phase support the interpretation of the phase transition
driven by lattice dynamical effects.

Acknowledgements - The technical assistence of H. Berger and of
G. Burri is greately acknowledged. We should like to thank
Professor E. Mooser for his constant interest and the Swiss
National Fundation for Scientific Research for financial support.

REFERENCES

1. Di Salvo F.J., Moncton D.E. and Waszcak J.V., Phys. Rev: B14, 4321 (1976).

2. Zunger A. and Freeman A.J., Phys. Rev. B17, 1839 (1978).

3. Liang W.Y., Lukcovsky G., Mikkelsen J.C. and Friend R.H., Phil. Mag. B39, 133 (1979).

4. Brown F.C., Physica 99B, 264 (1980).

5. Lévy F. and Froidevaux Y., J. Phys. C : Sol. State Phys. 12, 473 (1979).

6. Di Salvo F.J. and Waszcak J.V., Phys. Rev. B17, 3801 (1978).

7. Lévy F., J. Phys. C 12, 3725 (1979).

8. Uchida S., Tanabe K., Okajima K. and Tanaka S., Solid State Commun. 31, 517 (1979).

9. Kadjik F. and Jellinek J., Less Common Metals 23, 437 (1971).

10. Di Salvo F.J. and Graebner J.E., Solid State Commun. 23, 825 (1977).

11. Fazekas P. and Tosatti E., Physica 99B, 183 (1980).

12. Huntley D.J. and Frindt R.F., Physics and Chemistry of Materials with Layered Structures, Lee. P.A. Ed. (Reidel, Dordrecht 1976), pp.385.

13. Zunger A. and Freeman A.J., Phys. Rev. B17, 1839 (1978).

14. Vaterlaus H.P., Ansermet S., Py M. and Lévy F., to be published.

ON THE BAND GAP NARROWING IN IMPURE SILICON:

EFFECTS OF IMPURITY SCATTERING

B.E. Sernelius and K.F. Berggren

Dept of Physics and Measurement Technology
Linköping University

S-581 83 Linköping, Sweden

The presence of impurities in heavily doped semiconductors changes
positions and shapes of the conduction and valence bands of the
host. From experiments it is known that the value of the band gap
is reduced as compared to that of the pure material. We present a
calculation of this reduction in n-type Si. In the calculations we
take into account the effects of electron-electron and electron-
donor ion interaction on the energies of the valence and conduction
band states within a dynamical RPA treatment.

1. INTRODUCTION

The value of the energy gap, E_g, of heavily doped semiconductors is an important parameter in transitor design. It is
experimentally known that E_g changes noticably at such high doping
levels that are currently being considered in connection with e.g.
the miniaturization of electronic components[1,2]. In addition the
problem of band gap narrowing is of theoretical interest.

When a semiconductor is doped with donors (acceptors),
electrons (holes) and positive (negative) ions are inserted into
the system. At high doping level, i.e. above the Mott critical
density, the added electrons (holes) will in a first approximation
occupy states at the bottom of the conduction band (top of the
valence band). A homogeneous gas of interacting particles is formed.
The high polarizability of this gas will change the initial
conduction and valence bands. In addition these bands will be
perturbed by the presence of the impurity ions. The change

of the band structure of a semiconductor with doping has attracted
considerable interest. Experiments and theory have recently been
reviewed by Keyes[1] and Abram, Rees and Wilson[2] to whom we refer
for details and more extensive lists of references to previous
work.

In the present work we will investigate the case of n-
type silicon. The reason for this is that detailed calculations
for this material are lacking. We will in this paper, because of
short-age of space, only briefly describe our calculations and
confine ourselves to the change in band gap and optical threshold
with doping.

2. THEORETICAL CALCULATIONS FOR n-SILICON

We restrict ourselves to such high doping level that
the system is metallic which for the specific example of Si:P
means a donor density above about $3.2 \ 10^{18}$ donors per cm^3. We
assume that the electrons occupy the states at the bottom of
the host conduction band, i.e. they occupy the states at the
bottom of the six equivalent, anisotropic valleys of the silicon
conduction band. Intervalley processes are important at very high
donor concentrations but can be neglected for doping concentrations
considered here, i.e. for concentrations less than the solubility
limit which is somewhat above 10^{20} donors per cm^3. Thus we have a
situation with six separate Fermi systems. It can be shown that
the anisotropy is less important; thus we adopt an isotropic
approximation in which the electrons occupy six spheres in
reciprocal space and the energy dispersion for the electrons is
spherically symmetric and characterized by the density of states
effective mass $m_d = (m_t^2 m_l)^{1/3}$. Hence the polarizability from the
conduction band electrons is the sum of the independent
contributions from six isotropic Fermi spheres.

The top of the valence band consists of two double
degenerate hole bands plus one band below, split off by spin
orbit interactions. This band is sufficiently below in energy
to be ignored. The two hole bands have parabolic energy
dispersions in all directions but have a complicated cubic warping.
We follow Combescot and Nozieres[3] and Kohn and Luttinger[4] and use
a spherical approximation in which the energy dispersions of these
bands are spherically symmetric. The matrix elements are, however,
angle dependent and different for processes between like and unlike
bands. The energy dispersions for the bands are characterized by
the heavy and light hole masses, m_{hh} and m_{lh}.

Furthermore we assume that the host contributes with a back ground screening constant, κ, independent of the doping level $(e^2 \to e^2/\kappa)$.

The donor ions are assumed to be randomly distributed and we treat the interaction between the electrons and ions in perturbation theory to second order in the Coulombic ion potential. These are the basic assumptions and approximations we are relying on and the calculation is performed within the fully dynamic Random Phase Approximation.

The model describes the electrons in the conduction band minima as a collection of independent quasi-particles. The quasi-particle dispersion, defining the perturbed conduction band, is

$$E_c(\vec{k},\omega) = \varepsilon_c^o(\vec{k}) + \hbar\Sigma_c(\vec{k},\omega) \tag{1}$$

where $\varepsilon_c^o(\vec{k})$ is the unperturbed conduction band energy and $\hbar\Sigma_c(\vec{k},\omega)$ the self-energy associated with electron-electron and electron-impurity scattering. Thus, if $G_c^o(\vec{k},\omega)$ is the unperturbed Green's function for the conduction band and $\varepsilon(q,\omega)$ the Lindhard dielectric screening function modified to the present case of six valleys, the electron-electron interactions give rise to the conduction band shift

$$\hbar\,\Sigma_c^{ee}(\vec{k},\omega) = \frac{i}{(2\pi)^4} \int d\vec{q} \int d\nu \, \frac{4\pi e^2}{q^2\, \varepsilon(q,\nu)}$$

$$G_c^o(\vec{k}+\vec{q},\omega+\nu) \tag{2}$$

From second order perturbation theory in the screened impurity potential one gets the shift associated with electron impurity scattering

$$\hbar\,\Sigma_c^{ei}(\vec{k}) = n \int \frac{d\vec{q}}{(2\pi)^3} \left| \frac{4\pi e^2}{q^2\, \varepsilon(q,0)} \right|^2 / (\varepsilon_c^o(\vec{k}) - \varepsilon_c^o(\vec{k}+\vec{q})) \tag{3}$$

where $\underline{n}$ is the density of donors. For a valence band with index one obtains the somewhat more complicated expression

$$\hbar\,\Sigma_{v,\alpha}^{ee}(\vec{k},\omega) = \frac{i}{(2\pi)^4} \sum_{\alpha'} \int d\vec{q} \int d\nu \, \frac{4\pi e^2}{\kappa q^2} \{\frac{1}{\varepsilon(q,\nu)} - 1\}$$

$$\Lambda_{\alpha\alpha'}(\vec{k},\vec{k}+\vec{q}) \times G_{v,\alpha'}^o(\vec{k}+\vec{q},\omega+\nu) \tag{4}$$

where the 'vertices' $\Lambda_{\alpha\alpha'}$ are the same as defined in ref.3. Eq.(4) takes into account that the unperturbed energy $\varepsilon^o_{v,\alpha}(\vec{k})$ of the valence band already contains a Hartree-Fock exchange contribution[5]. This contribution is therefore subtracted in Eq.(4). For the valence band the shift due to impurities is

$$
\hbar\Sigma^{ei}_{v,\alpha}(\vec{k}) = n\sum_{\alpha'} \int \frac{d\vec{q}}{(2\pi)^3} \left| \frac{4\pi e^2}{q^2\,\varepsilon(q,0)} \right|^2 \Lambda_{\alpha\alpha'}(\vec{k},\vec{k+q}) /
$$

$$
/(\varepsilon^o_{v,\alpha}(\vec{k}) - \varepsilon^o_{v,\alpha'}(\vec{k+q})) \tag{5}
$$

The band gap of the perturbed system is defined as the difference in energies for quasi-particles at the bottom of the conduction band and the top of the valence band, i.e. the shift is

$$
\Delta E_g = \hbar\Sigma_e(k=0) - \hbar\Sigma_v(k=0) \tag{6}
$$

The change in the interband optical threshold, $\Delta E'_g$ is obtained as

$$
\Delta E'_g = \varepsilon^o_c(k_F) - \varepsilon^o_c(0) + \hbar\Sigma_c(k=k_F) - \hbar\Sigma_v(k=0) \tag{7}
$$

Numerical results are shown in fig.1 and are based on the input parameters $\kappa = 11.4$ $m_l = .9163$, $m_t = .1905$, $m_{hh} = .523$ and $m_{lh} = .154$.

3. CONCLUDING REMARKS

The band gap shrinkage in n-Si has previously been treated by, among others, Inkson[5] and Abram et al.[2] In his calculation Inkson neglected the ion contribution and used a jellium model. He also used a static Thomas Fermi approximation which grossly overestimates the shift $|\Delta E_g|$. Abram et al. made the improvement to use the Lindhard Approximation but ignored the valence band structure as well as impurity scattering. The shift is then reduced. In the present work we have tried to explore a simple model as far as possible, i.e. dynamic screening effects, a valence band structure, and impurity scattering are explicitly considered. One then finds that the valence band structure further reduces the shift $|\Delta E_g|$. On the other hand this reduction is roughly compensated by the increase due to impurity scattering. One notes that the contribution to the band gap narrowing from the shifts of the conduction and valence bands are of the same order of magnitude.

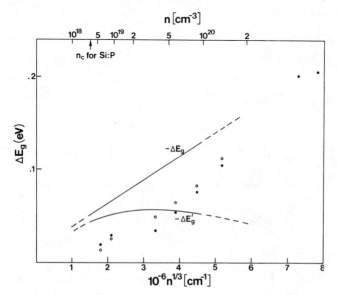

<u>Fig.1</u> : The change in band gap, ΔE_g, and in the interband optical
threshold, $\Delta E_g'$, for n-type Si as a function of donor
concentration, n. The circles show the results for the
band gap shrinkage obtained by Balkanski et al. (ref.6)
(open circles refer to 35°K and filled to 300°K). The
critical Mott density is denoted n_c.

As extensively reviewed by Abram et al.[2] there is considerable scatter in experimental data in ΔE_g. When comparing with experiments we have chosen the optical data of Balkanski et al.[6] because experiments with devices seem to be strongly affected by e.g. doping profiles. At any rate our comparison with experiments in fig.1 should be regarded with caution; the deduction of the experimental ΔE_g involve several drastic assumptions - so does theory. With these precautions we note that there is order of magnitude agreement between theory and experiments. In the concentration regime considered both theory and experiments follow approximately a $n^{1/3}$-law. Fig.1 seems to indicate that the agreement improves with increasing concentration. In view of the perturbation method used here this seems reasonable; our method is basically a high density expansion.

REFERENCES

1. R.W. Keyes, Comments in Solid State Physics 7, 149 (1977).

2. R.A. Abram, G.J. Rees, B.L.H. Wilson, Advances in Physics 27, 799 (1978).

3. M. Combescot, P. Nozieres, Solid State Communications 10, 301 (1972).

4. J.M. Luttinger, W. Kohn, Physical Review 97, 869 (1955).

5. J.C. Inkson, Journal of Physics C : Solid State Physics 9, 1177 (1976).

6. M. Balkanski, A. Aziza, E. Amzallag, Physica Status Solidi 31, 323 (1969).

MEASUREMENT OF THE QUADRUPOLE INTERACTION FOR ^{111}Cd

IN As, Sb AND Te

H. Barfuß, G. Böhnlein, P. Freunek, R. Hofmann,
H. Hohenstein, W. Kreische, H. Niedrig, and A. Reimer

Physikalisches Institut
Universität Erlangen-Nürnberg

Erwin-Rommel-Straße 1, 8520 Erlangen

The quadrupole interaction for ^{111}Cd in As, the system $Sb_{1-x}In_x$ and Te has been investigated. For the system $Sb_{1-x}In_x$ unexpected dependences on temperature as well as on concentration have been observed. The quadrupole frequency of the semiconductor Te increases with increasing temperature. This behaviour of the coupling constants is probably due to changes in the charge carrier densities. For Arsenic a $T^{3/2}$-temperature dependence was obtained.

INTRODUCTION

The electric field gradient (efg) in metals was
extensively studied by perturbed angular correlation methods.
It could be shown, that for metals not only the lattice parameters but also the conduction electrons are essential for the
efg. Special attention was directed to the temperature
dependence of the quadrupole coupling constant. In some cases
the pressure dependence has been studied as well. The intention
of such investigations was to distinguish between contributions
of the positive ion cores and the non uniform charge density of
conduction electrons. A review is given in ref. 1).

The characteristic feature of a semimetal is that the
conduction band edge is slightly lower than that of the valence
band. In a semiconductor there is an energy gap between the
valence and the conduction band. In table 1 the values for As,

189

Table 1

	Energygap/eV	Ref.
As	− 0.5	2)
Sb	− 0.16	2)
Te	+ 0.33	3)

Sb and Te are given at low temperatures. The negative quantities mean a band overlap. In solid solutions of such materials the energy gap depends on the composition. A suitable choice of the latter allows preparation of materials with a small band overlap or small gap.

Measuring the efg in semimetals or semiconductors offers the possibility to change the electron density either by adding impurities or by changing the temperature.

METHOD

The applied method was the time differential perturbed angular correlation method (TDPAC). The experimental arrangement was a conventional four detector set-up. Details are described elsewhere[4]. For getting the quadrupole frequency, defined by (for I = 5/2)

$$\omega_o = 6\omega_Q = \frac{3\pi}{10} \cdot \nu_Q \; ; \qquad \nu_Q = \frac{e^2 qQ}{h} \; ,$$

the following formulas were used, which are valid for the angular correlation of ^{111}Cd displayed by a powder source :

$$W(\theta,t) = 1 + A_{22}^{eff}(t) \, P_2 \, (\cos \theta) \; .$$

A_{22}^{eff} is the effective angular correlation coefficient and $G_{22}(t)$ is the time dependent perturbation factor, which is given for an axially symmetric, randomly oriented, static electric field gradient by

$$G_{22}(t) = \sum_n \exp\left[-\frac{1}{2}(n\delta t)^2\right] s_{2n} \, \cos(n\omega_o t) \; .$$

The parameter δ describes a possible Gaussian distribution of the values of the efg. For the fit procedures always the ratio of the coincidence counting rates $N(180°,t)$ and $N(90°,t)$ is used :

$$R(t) = 2 \cdot \frac{N(180°,t) - N(90°,t)}{N(180°,t) + 2N(90°,t)} = A_{22}G_{22}(t) \; .$$

The measurements were performed with the 173 keV - 247 keV γ-γ-cascade including the $5/2^+$ state of ^{111}Cd at 247 keV with a lifetime τ = 121 nsec. The upper $7/2^+$ level was populated by EC from ^{111}In. A typical time-spectrum is shown in fig.1.

^{111}Cd IN ARSENIC

The ^{111}In has been electroplated on an Arsenic-tube and afterwards the As has been sublimated together with the ^{111}In into a glass tube at about 770 K condensation temperature. The quadrupole frequency ν_Q for ^{111}CdAs has been investigated in the temperature range 77 K $\leqslant$ T $\leqslant$ 763 K. The measured values for ν_Q are plotted versus the temperature T in a $T^{3/2}$ scale in fig.2. The drawn line is a fit to the data according to the relation :

$$\frac{eq(T)}{eq(0)} = 1 - BT^{3/2}$$

Such a behaviour is mainly due to the influence of lattice vibrations on the efg. It was first discovered as an experimental rule[5], latter on there were theoretical explanations [6,7,8]. Since As has a relatively large band overlap, this metallic like behaviour is not unexpected.

^{111}Cd IN THE SYSTEM $Sb_{1-x}In_x$

For the first measurement we used ion-implantation to get the radioactive ^{111}In into the anitmony matrix. The quadrupole frequency ν_Q is plotted versus the temperature in a $T^{3/2}$ scale in fig.3 (filled circles). Up to 400 K the frequency increases with temperature. Since the deviation from a $T^{3/2}$ behaviour is very obvious, we added different amounts of In impurities to change the density of the conduction electrons.

For all following measurements we used samples prepared by electroplating the radioactive ^{111}In on antimony or indium carrier foils. Subsequently the samples were melted and quenched to room temperature to produce a metastable solution of In in Sb. To get considerable amplitudes the sources were tempered between 600 K and 700 K for some hours.

For the Indium concentration x = 0.005 and x = 0.002 we got a $T^{3/2}$ relation, but at lower concentrations we got deviations. For x = 0 there are only small differences between the measurements done after ion-implantation and those after electroplating. These differences may be due to unknown impurities in both cases. The

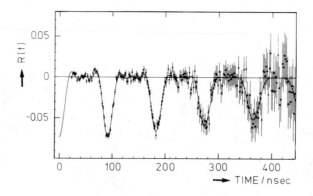

Fig.1 : Typical time spectrum of ^{111}CdSb.

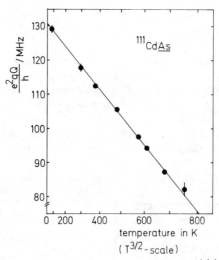

Fig.2 : Temperature dependence of ^{111}CdAs.

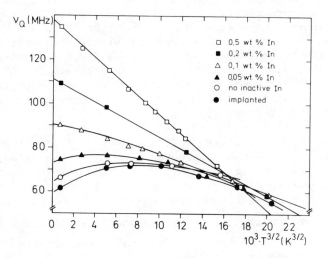

Fig.3 : Temperature dependence of ^{111}Cd in $Sb_{1-x}In_w$.

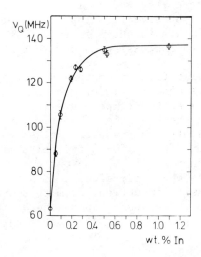

Fig.4 : Concentration dependence of ν_Q at T = 77 K.

results for the quadrupole coupling constant ν_Q at 77 K for
different Indium concentrations are shown in fig.4. The quadrupole
frequency ν_Q increases by about a factor of two for x = 0.005.
After that the frequency remains constant within the experimental
errors up to the highest measured value of 5 wt. % Indium, which
is not shown in fig.4. X-ray diffraction experiments indicate no
change in the c/a ratio within experimental errors. Hence the
drastic increase of the quadrupole frequency cannot be explained
by the change of lattice parameters.

From the phase-diagram of In and Sb [9] there is no
evidence of different phases, moreover from fig.4 the steady
increase of the quadrupole frequency ν_Q excludes different phases.
The segregation of In-clusters would cause the quadrupole frequency
ν_Q = 17.5 MH$_z$ (T = 293 K) of CdIn which was not observed. Since
the effective amplitudes A_{22}^{eff} remain constant over the whole
range of concentration an influence of grain boundaries on the
coupling constant can be excluded.

Preliminary experiments for the system ^{111}Cd in $Sb_{1-x}Cd_x$
show similar results.

^{111}Cd IN TELLURIUM

The samples are prepared by ion implantation of ^{111}In
into the Te-matrix. The observed quadrupole frequency ν_Q as a
function of the temperature is shown in fig.5. The quadrupole
frequency increases with increasing temperature.

To get a qualitative understanding, it is useful to plot
$\hbar\nu_Q$ versus 1/T. This is done in fig.6. One observes an increase of
ν_Q at about 350 K. This is the temperature region where intrinsic
conductivity dominates the electric properties of the semiconductor
Te. Comparing the data with the temperature dependence of the
electric charge carrier density for a typical semi-conductor (given
in the right upper corner of fig.6) one can see, that the efg
increases in the region where the electric conductivity is
governed by the intrinsic conduction. From this comparison one
can conclude, that the temperature dependence of the efg for ^{111}Cd
in Te is mainly given by changes of the charge carrier density.

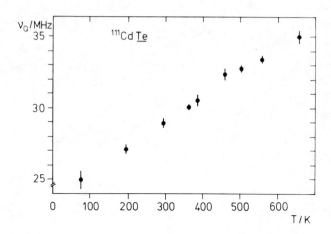

Fig.5 : Temperature dependence of [111]CdTe.

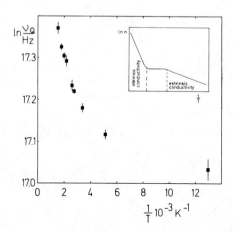

Fig.6 : Quadrupole frequency ν_Q versus 1/T in a semilogarithmic
scale.

SUMMARY

The measurements of the quadrupole frequency on semi-metals and semiconductors show new and unexpected results. A strong concentration dependence for the system $Sb_{1-x}In_x$ has been observed. For $^{111}CdTe$ the efg increases with increasing temperature. This behaviour is probably due to changes in the electron density.

Acknowledgement - We are very indebted to Dr. G. Schatz and his co-workers for doing the ion-implantation and to Dr. W. Irler for the X-ray analysis. This work was supported by the Bundesministerium für Forschung und Technologie.

REFERENCES

1) E.N. Kaufmann and R.J. Vianden, Review of Modern Physics 51, 161 (1979).

2) G.A. Saunders, Contemporary Physics 14, 149 (1973).

3) P. Grosse, Die Festkörpereigenschaften von Tellur, Springer Tracts in Modern Physics 48, 147 (1969).

4) U. Frey, H. Hohenstein, W. Kreische, M. Meinhold, H. Niedrig, U. Pechtl and K. Reuter, Zeitschrift für Physik B33, 141 (1979).

5) P. Heubes, H. Hempel, H. Ingwersen, R. Keitel, W. Klinger, W. Loeffler and W. Witthuhn, Contributed Papers, International Conference on Hyperfine Interactions studied in Nuclear Reactions and Decay, Uppsala, Sweden, June 10-14, 1974, ed. by E. Karlsson and R. Wäppling, published by Upplands Grafiska AB, Uppsala 1974, p.208.

6) D. Quitmann, K. Nishiyama and D. Riegel, Proceedings of the 18th Ampere Congress, Nottingham, 9-14 September 1974, in : Magnetic Resonance and Related Phenomena, Vol.2, ed. by : Allen P.S., Andrew E.R., Bates C.A., Amsterdam, North Holland 1975, p.349.

7) K. Nishiyama, F. Dimmling, Th. Kornrumpf and D. Riegel, Physical Review Letters 37, 357 (1976).

8) P. Jena, Physical Review Letters 36, 418 (1976).

9) M. Hansen and K. Anderko, Constitution of Binary Alloys, ed. by Mc.Graw.Hill Book Company, New York 1958, p.859.

ENERGY LEVELS OF ELECTRONS BOUND TO SHALLOW DONORS IN GaP

B. Pödör

Research Laboratory for Inorganic Chemistry
of the Hungarian Academy of Sciences

H-1502 Budapest, P.O. Box 132, Hungary

Energy levels of electrons bound to S, Te and Si donors in GaP
were determined from Hall effect and infrared absorption
measurements. From the Hall data ground state ionization
energies of S, Te and Si donors, and valley-orbit splitting
energies of the 1s states of S and Te donors were deduced.
Infrared absorption lines due to transitions from the 1s state
of the donors to p-like excited levels were observed. Several
excited states were assigned on the basis of the hydrogenic
effective mass theory. Thermal and optical ionization energies
of S, Te and Si donors were found.

Most of the information about impurity levels in semiconductors found in the literature refers to the elementary
semiconductors Ge and Si. Much less is known about impurity
energy levels in the compound semiconductors of III-V type.
The case of substitutional donors in GaAs seems to be the simplest,
hydrogenic effective mass theory /HEMT/[1] giving a good description
of the energy level structure. The case of GaP is more difficult,
both experimentally and theoretically. Experimental ionization
energies of the donors, as well as the positions of the excited
levels show a wide scatter[2,3,4]. The difficulties in the interpretation of ionization energies derived from electrical
measurements come perhaps from using data obtained on ill-
characterized or noncontrollably doped samples. The theoretical
interpretation of the observed energy levels is marred by the
strong anisotropy and non-parabolicity of the conduction band
minima of GaP, leading to the assertion of only limited
applicability of HEMT to shallow donors in GaP[2].

Below we summarize the results of a study of energy
levels and ionization energies of S, Te and Si donors in GaP, using
Hall effect data and infrared absorption spectra obtained
simultaneously on the same series of well-characterized GaP single
crystals and epitaxial layers. These results were obtained on the
following groups of samples : i./ GaP liquid phase epitaxial /LPE/
layers containing either S or Si as a dominant donor, ii./ Te
doped bulk GaP single crystals, iii./ Si doped solution grown GaP
single crystals. Using electrical methods and low temperature
photoluminescence spectra measurements it was ascertained that
in each group the electrical properties were determined only by
one donor species within experimental error. Neither S nor Si
donors were found in GaP:Te crystals moreover Si contamination
did not influence appreciably the electrical properties of GaP:S
LPE layers, the same applies to the possible S contamination of
GaP:Si layers. In the purest samples the sum of donor and acceptor
concentration was as low as 3×10^{16} cm^{-3}, electron mobility at
77 K exceeded 1900 cm^2/Vs [5].

Hall effect, conductivity and mobility were measured
by the Van der Pauw method from 77 to 400 K. The estimated total
error of the measured parameters was less than 8 to 10 per cent.
The results presented here are based on data taken on more than
30 samples.

Infrared absorption spectra were obtained using infrared
transmission spectra recorded on a Perkin-Elmer 225 grating
spectrophotometer from 1000 to 400 cm^{-1} with a resolution of about
1 cm^{-1}.

In the infrared spectra besides strong two-phonon
absorption bands some weaker lines attributed to transitions
of electrons from the donor ground state to the upper lying p-
like excited states were observed[6], see Table 1.

The assignments, using HEMT[2,6,7] are based on the
assumption that the strongest lines correspond to the 1s-2p/$\pm$/
transition. In the spectra of Si donors, besides lines due to
transitions originating from the 1s/Γ_8/ ground state, two lines
/Nos 2' and 4'/ are thought to originate from transitions from
the split off 1s/Γ_7/ state. The reported value of this splitting[2]
is about 4.0 cm^{-1}. Lines Nos 5, 6 and 7 of Si obtained here differ
from those reported by Carter et al.[2].

The analysis of the Hall data was based on a six-valley
ellipsoidal model of the conduction band minima, incorporating
the effect due to valley-orbit splitting of the ground state of
P-site donors /i.e. S and Te/, in an analogous way as was done
by Long and Myers in the case of P doped Si[8]. The ground state

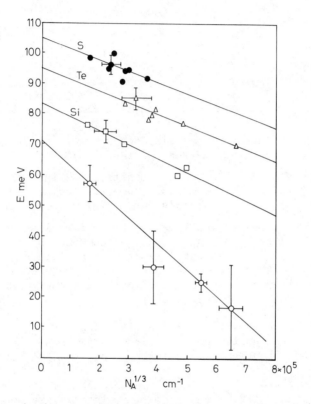

Fig.1 : Donor ionization energies of S, Te and Si, and the energy
level of the presumed quinte state versus the cubic root
of acceptor concentration.

Table 1 : Observed absorption lines for S, Te and Si donors in
GaP. Initial state is 1s/A_1/ for S and Te, 1s/Γ_8/ for
Si except lines Nos 2' and 4' where it is 1s/Γ_7/.

Line	Assignment /Final state/	Excitation energy /cm^{-1}/ S	Te	Si
7	5p/o/	–	–	653
6	3p/+/	826	–	647
5	4f/$\overline{o}$/	808	–	630
4	2p/+/	781	666	603
4'	2p/$\overline{+}$/	–	–	599
3	4p/$\overline{o}$/	774	–	–
2	3p/o/	–	–	539
2'	3p/o/	–	–	534
1	2p/o/	571	454	–

of the six-fold degenerated P-site donors /not including the
degeneracy due to spins/ is the split off 1s/A_1/ state with a
degeneracy of 1 [7,9]. This state exhibits a substantial "chemical
shift" relative to the level position predicted by HEMT[1,7,9]. The
remaining triplet 1s/T_2/ and doublet, 1s/E/ levels, being close
to each other and to the HEMT value[7,9], were considered to
coalesce into a quintet level[8], designated here as 1s/E and T_2/.
For Ga-site donors /Si in our case/ the ground state splits into
two states with degeneracies of 3 each, but the energy separation
between them is much less than the ionization energy itself, so
it was neglected in the analysis.

Fitting the charge neutrality equation to the measured
concentration versus temperature curves using a nonlinear-least-
squares method, four parameters, i.e. ionization energy, E_D, donor
and acceptor concentration, N_D and N_A, and valley-orbit splitting
energy, Δ, defined here as the difference between the ground
state energy and that of the presumed quintet state were determined.
The sensitivity of the fitting procedure for these parameters
decreased approximately in the above order, resulting in typical
standard errors amounting to 2-3 meV, 10-20 per cent, 20-50 per
cent and 10-50 meV respectively.

Plots of the donor ionization energies, E_D versus the
cubic root of N_A are presented in Fig.1 [5,10]. The ground state
ionization energies deduced from electron concentration versus
temperature curves should follow a linear dependence on the cubic
root of the ionized donor concentration /i.e. the concentration
of compensating acceptors/ due to the usual concentration broadening.

From the data presented in Fig. 1 we have[10]

$$E_D/S/ = /105.0 \pm 1.8/ - /3.78 \pm 0.66/ \times 10^{-5} \; N_A^{1/3}$$

$$E_D/Te/ = /94.7 \pm 1.4/ - /3.82 \pm 0.33/ \times 10^{-5} \; N_A^{1/3}$$

$$E_D/Si/ = /83.4 \pm 1.2/ - /4.56 \pm 0.15/ \times 10^{-5} \; N_A^{1/3}$$

Similar relationships are not found in the literature except the one by Vink et al.[11] for Si donors, which is in a good agreement with our result. Data for GaP:Te reported by Toyama et al.[12] obtained by a similar analysis are also close to our ones.

Table 2 summarizes the donor ground state ionization energies extrapolated to infinite dilution obtained above and compares them with the relevant data from the literature, as well with the optical ionization energies deduced for the same set of samples from the infrared absorption spectra to be discussed below.

Table 2 : Ground state ionization energies of donors in GaP.
          T = thermal, O = optical.

S meV	Te meV	Si meV	Method	Reference
105.0 + 1.8	94.7 + 1.4	83.4 + 1.2	T	this work
104*	92 + 2**	83 + 2***	T	
110.5 + 0.7	96.2 + 0.7	88.4 + 0.7	O	this work
108		86.7	O	2
107 + 1	92.6 + 1	85 + 1****	O	4
107.1	92.8	85.1	O	13
104.2	89.8	82.1	O	14

*   deduced by Vink et al.[11]
**   deduced by us from the data of [12]
***   deduced by us from the data of [11]
****   corrected upward by 3 meV from the data of Vink et al.[14] as suggested in [13]

From the valley-orbit splitting energies, Δ, obtained
for S and Te donors, the position of the presumed quintet level
was deduced. This resulted in a level position which is independent
of the donor species within experimental errors. The data /see
Fig.1/ could also be described with a linear relationship

$$E_{1s}/E \text{ and } T_2/ = (71 \pm 8/ - /8.5 \pm 2/ \times 10^{-5} N_A^{1/3}$$

The such greater slope is not surprising because one can
presume that for levels lying nearer to the conduction band edge
the concentration broadening should be stronger. This results in
a valley-orbit splitting energy at infinite dilution of 34 ± 8 meV
for S and 24 ± 8 meV for Te donors.

The position of the quintet level agrees reasonably with
the binding energy of the 1s/T_2/ state of about 70 meV for S, Se
and Te donors deduced from the data of Carter et al.[2].

The positions of the p-like excited levels relative to
the position of the 2p/+/ lines do not depend on the donor species
within experimental errors. The fit of the HEMT[7] using the multi-
ellipsoidal model of the conduction band minima resulted in the
theoretical line separations given in Table 3 together with the
experimental data. The fit was obtained with the following parameter values : effective mass anisotropy factor $\delta^{-1} = m_l/m_t = 22 \pm 2$,
effective Rydberg constant $R_t^x = 37.3 \pm 1.3$ meV /i.e. 301 ± 10 cm^{-1}/.
These values differ substantially from that ones deduced by Carter
et al.[2] from the spectra of Si donors. As seen from the data of
Table 3 all of the identified experimental line separations agree
with the theoretical predictions.

Table 3 : Separation between the excited states of donors

Assignment	HEMT cm^{-1}	Experiment /cm^{-1}/ S	Te	Si
4p/+/ – 2p/+/	63 ± 3			
5f/o/ – 2p/+/	60 ± 3			
5p/o/ – 2p/+/	54 ± 3			50
3p/+/ – 2p/+/	44 ± 2	45		44
4f/o/ – 2p/+/	31 ± 3	27		34
2p/+/ – 4p/o/	4 ± 3	7		
2p/+/ – 3p/o/	67 ± 6			65
2p/+/ – 2p/o/	211 ± 5	210	212	

The theoretical fit to the HEMT resulted in a binding energy of the 2p/+/ state a value of 13.6 $\pm$ 0.5 meV /i.e. 110 $\pm$ 4 cm^{-1}/. This adds up with the observed 1s - 2p/+/ transition energies of the different donors to result the following values of the optical ionization energies of S, Te and Si donors, respectively: 110.5 $\pm$ 0.7 meV /891 $\pm$ 5 cm^{-1}/, 96.2 $\pm$ 0.7 meV L776 $\pm$ 5 cm^{-1}/, and 88.4 $\pm$ 0.7 meV /713 $\pm$ 5 cm^{-1}/. These values are about 2 meV higher than the currently suggested optical ionization energies[2,4] /c.f. Table 2/, and are 2 to 5 meV higher than the thermal ionization energies obtained in this work. This latter difference could be interpreted as resulting from a Franck-Condon shift[1].

Acknowledgement   -   Part of the work reported here /electrical measurements and Hall data analysis/ was performed in the Research Institute for Technical Physics of the Hungarian Academy of Sciences, Budapest, where the author was affiliated to. The GaP epitaxial layers were grown by Dr. J. Pfeifer at the same institute. Her cooperation in this work as well the help of Mrs. N. Nádor, Mrs. Z. Püspöki and Dr. Z. Laczkó is gratefully acknowledged.

REFERENCES

1. S.T. Pantelides, Rev. of Mod. Phys. 50, 797 /1978/.

2. A.C. Carter, P.J. Dean, M.S. Skolnick and R.A. Stradling, J. Phys. C : Solid State Physics 10, 5111 /1977/.

3. W. Scott, J. Appl. Phys. 48, 3173 /1977/.

4. A.A. Kopylov and A.N. Pikhtin, Phys. and Tech. of Semi-conductors 11, 867 /1977/.

5. J. Pfeifer, B. Pódör, L. Csontos and N. Nádor, Revue de Physique Appliquée 13, 741 /1978/.

6. B. Pódör and Z. Laczkó, Acta Physica Hungarica, in press.

7. R.A. Faulkner, Phys. Rev. 184, 713 /1969/.

8. D. Long and J. Myers, Phys. Rev. 115, 1119 /1959/.

9. W. Kohn, in Solid State Physics, ed. F. Seitz and D. Turnbull, Academic Press, New York, 1957, Vol.5, p.257.

10. B. Pódör, J. Pfeifer, L. Csontos, N. Nádor and F. Deák, to be published.

11. A.T. Vink, A.J. Bosman, J.A.W. Van der Does de Bye and R.C. Peters, J. of Luminescence 5, 57 /1972/.

12. M. Toyama, K. Unno and A. Kasami, Jap. J. Appl. Phys. 7, 1418 /1968/.

13. P.J. Dean, D. Bimberg and E. Mansfield, Phys. Rev. B15, 3906 /1977/.

14. A.T. Vink, R.L.A. Van der Heyden and J.A.W. Van der Does de Bye, J. of Luminescence 8, 105 /1973/.

ANOMALOUS RELATIVE ENHANCEMENT OF THE INTENSITY OF PHONON

SIDEBANDS IN GaP:N

H. Chang*, C. Hirlimann, M. Kanehisa, and M. Balkanski

Laboratoire de Physique des Solides, associé au C.N.R.S.
Université Pierre et Marie Curie

4, Place Jussieu, 75230 Paris-Cedex 05, France

The temperature variation of the integrated intensity of
luminescence bands due to radiative recombination of excitons
trapped at different nitrogen pair sites in GaP shows a maximum.
This maximum shifts toward higher temperatures for decreasing N-N
distances. The activation energy for dissociation of the $(NN_i-$
exciton) complex is very close to the binding energy for
each different NN_i configuration. The LO-side bands corresponding
to NN_i configuration show a quite different temperature dependence.
The maximum of the integrated intensity for all the LO-side bands
corresponding to different NN_i configurations occurs at the same
temperature and the activation energy for the dissociation of $((NN_i-$
exciton)-LO)complex is the same for all different NN_i configurations.

INTRODUCTION

The radiative recombination of exciton bound to nitrogen
in GaP has been intensively studied since the pionnering work of
D.G. Thomas and J.J. Hopfield[1,2]. These authors pointed out that
nitrogen substitutes phosphorus and has a greater electronegativity
so that it can trap an electron which binds a hole through
coulombian interaction. The net result of these combined interactions is a nitrogen trapped exciton whose radiative recombination
gives the A-B lines in the luminescence spectra. When the nitrogen
concentration in the crystal is increased (typically up to 10^{19}
cm^{-3}) one finds that pairs of nitrogens act as traps for the
excitons. The binding energy of the trapped exciton decreases
when the distance between two nitrogen atoms increases. It is
maximum and of the order of 143 meV for two atoms in nearest

205

Table (1)

Band	E_B (meV)	E (meV)	A	B ($\times 10^2$)
NN_1	141.7 ± 2	138 ± 15	$(8.68 \pm 1.36) \times 10^6$	2428
NN_3	63.1 ± 1	59 ± 3	$(6.77 \pm 0.34) \times 10^5$	10560
NN_4	38.2 ± 1	29 ± 2	$(7.76 \pm 2.79) \times 10^4$	7902
NN_5	30.5 ± 1	22 ± 5	$(2.54 \pm 0.16) \times 10^4$	4793
NN_6	24.6 ± 1	15 ± 6	$148 \pm 44 \times$	78
$NN_1^\star$		64 ± 13	$(8.6 \pm 3.4) \times 10^3$	1059
$NN_3^\star$		60 ± 3	$(8.88 \pm 1.08) \times 10^5$	3277
$NN_4^\star$		60 ± 18	$(1.36 \pm 0.15) \times 10^6$	2765
$NN_5^\star$		59 ± 2	$(1.01 \pm 0.02) \times 10^6$	1319
$NN_6^\star$		58 ± 5	$(3.5 \pm 2.7) \times 10^5$	227

neighbour position. The luminescence lines resulting from the radiative recombination of nitrogen pairs trapped excitons are refered to as the NN_i lines (i indicates the distance between the two nitrogen atoms). In this work we have investigated the temperature variation of the intensity of the $LO(\Gamma)$ phonon side bands associated to the nitrogen pair lines when the temperature of the sample is varied between 5 and 200 K. These phonon side bands will be labeled NN_i^* hereafter.

EXPERIMENTAL RESULTS

We have performed a systematic experimental study of the intensity variation of the luminescence due to the recombination of exciton bound to various nitrogen pairs in GaP:N ($n_N = 3 \times 10^{19}$ cm^{-3}) in the temperature range from 5 K to 200 K. We used as an exciting source the 458 nm line, of an Ar^+ laser, with a power of 130 mw for all measurements. The emitted light was analysed through a Coderg PHo double-monochromator (800 cm focal length) and detected by a FW/30 I.T.T. photomultiplier (S-20 cathode). A photon counting system was used for the analysis of the signal. The samples were cooled by one helium vapor flow in a cryostat. The temperature was measured using a platinum resistor and was regulated with an accuracy of one K (at 77 K).

In figure 1 is shown a series of luminescence spectra at different temperatures. Two series of lines are observed in each spectrum, they are labeled NN_i and NN_i^* (NN_1, NN_2 and NN_1^*, NN_2^* are not included in the figure.

Each NN_i^* band is shifted from the corresponding NN_i band towards lower energy side by an energy which is about 4 cm^{-1} less than the $LO(\Gamma)$ phonon energy 404 cm^{-1}. At low temperature A-B lines are seen. They are 11 meV below the intrinsic exciton in GaP:N[2]. The energy difference between intrinsic exciton and each NN_i band gives the trapped exciton binding energy (E_B) for each site i. The values of E_B are given in table (1). It should be noticed that each NN_i band has a complex fine structure, at low temperature, with two main peaks separated by an energy of 7 cm^{-1}. The NN_i^* bands show always only two peaks with a splitting of about 14 cm^{-1}.

The temperature variation of the integrated intensity (I) of each NN_i and NN_i^* band is studied. When it is necessary the

PHOTON ENERGY (ev)

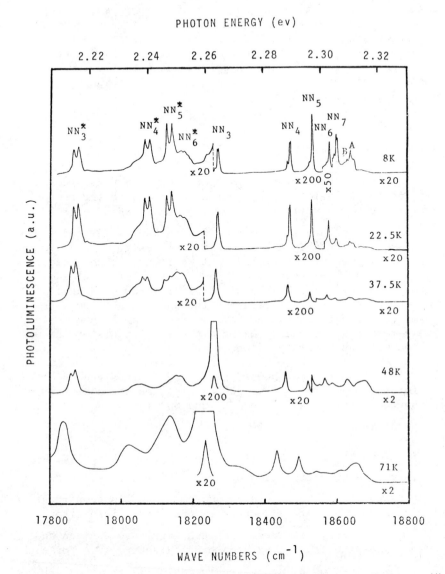

Fig.1 : Luminescence spectra of nitrogen doped GaP ($n_N = 3 \times 10^{19}$ cm^{-3}) for various sample temperatures. The exciting wave length is 458 nm with a power of 130 mW. A-B lines are the isolated nitrogen trapped exciton recombination lines. NN_i labels the nitrogen pairs trapped excitons recombination bands, while NN_i^* shows their phonon sidebands. NN_1 and NN_2 are not included.

spectra are decomposed into single bands. Figure 2 shows the temperature dependence of the integrated intensity of NN_i and NN_i^* bands. In the low temperature range, the integrated intensity has been carefully studied. It is found that the luminescence increases up to a maximum, with increasing temperature, before it is drastically quenched. NN_i bands, for i = 3, 4 and 5 have their maximum intensity at about 35 K, 21 K and 14 K respectively. For NN_6 we did not observe any maximum of the intensity down to a temperature of 5 K. All NN_i^* bands for i = 3, 4, 5 and 6 show a maximum of intensity at almost the same temperature of about 25 K.

The log I versus inverse temperature curves for NN_i and NN_i^* bands in the temperature range from 20 to 120 K are shown in figure 3. It can be seen from figure 3 that log I versus 1/T curves are different for different NN_i configurations, while they remain almost the same for different NN_i^*. The quenching process of NN_i bands depends on their configuration (i), while the quenching process of NN_i^* bands is independent of the configuration (i).

The ratio of integrated intensities of NN_i to NN_i^*, $R(NN_i^*/NN_i)$ is shown in figure 4. For i = 4, 5 and 6 $R(NN_i^*/NN_i)$ has a maximum value, of the order of 20, at about 50 K and for i = 1 and 3 $R(NN_i^*/NN_i)$ remains less than one and almost unchanged with temperature.

DISCUSSION

The fact that the energy difference between NN_i and NN_i^* almost equals LO phonon energy implies that one LO phonon is involved in the recombination process which occures at NN_i site and gives the NN_i^* line. We suggest that two types of complexes are present on the nitrogen pair sites. The first one is the usual exciton bound to the pair (NN_i-exciton), the other one is the trapped exciton coupled to one LO phonon : the ((NN_i-exciton) – LO) complex. Each type of complex gives respectively the NN_i and NN_i^* band in the luminescence spectra.

The experimental results mentioned above indicate that NN_i bands have quite different characters from NN_i^* bands in several respects, such as : fine structure, low temperature behaviour and high temperature process. NN_i bands have a behaviour dependent on

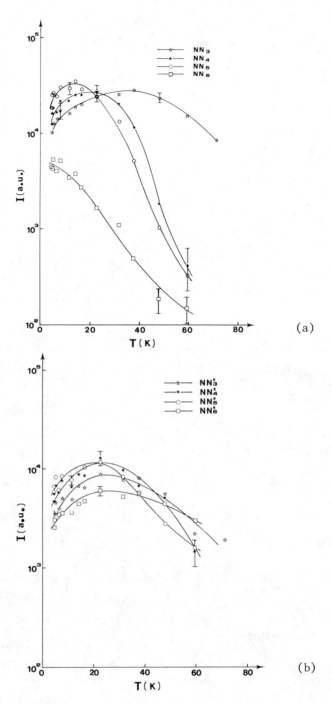

(a)

(b)

Fig.2 : (a) Low temperature variation of the integrated intensity
of various NN_i bands

(b) Low temperature variation of the integrated intensity
of the corresponding phonon sidebands NN_i^*.

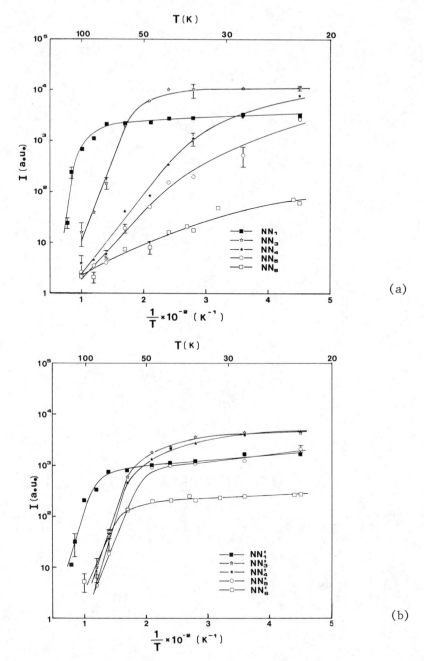

Fig.3 : Logarithm of integrated intensity versus invers temperature:
(a) for nitrogen pairs trapped exciton bands NN_i

(b) for their phonon sidebands NN_i^*.

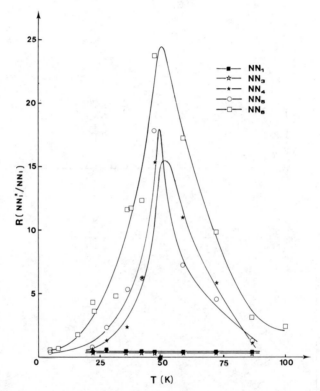

<u>Fig.4</u> : Temperature dependence of the ratio (R) of the integrated
intensity of the phonon sidebands NN_i^* to that of the zero
phonon bands NN_i.

their configuration (i). On the contrary, all the NN_i^* bands have the same behaviour. The integrated intensity of luminescence is given by[3]

$$I \propto \frac{R}{R + N} = \frac{1}{1 + \dfrac{N}{R}} \tag{1}$$

where R is the probability of radiative transitions, N is the probability of nonradiative transitions. When the temperature is high enough, the thermal quenching will be predominant, thus,

$$\frac{N}{R} = A \; e^{-E/kT} \tag{2}$$

where, E is the activation energy of the thermal quenching process, A is a constant which is equal to the ratio of N/R in the limit of very high temperature, k is the Boltzmann constant. From (1) and (2), we obtain

$$I = \frac{B}{1 + A \; e^{-E/kT}} \tag{3}$$

where, B is the saturation of luminescence when temperature tends to 0 K.

We use the least square method to fit the experimental results shown in figure 3 with formula (3). Parameters E (i.e. slopes of $\log I = f(T^{-1})$ plot at high temperatures) A and B are obtained. Table (1) gives the values of these parameters.

E as a function of the binding energy E_B is shown in figure 5. The differences between NN_i bands and NN_i^* bands is again clearly seen. The activation energies of NN_i bands are approximately equal to the binding energy of the corresponding configuration (i). The values of E are a little less than the corresponding binding energy E_B (figure 5). This is because the temperature is not high enough and the quenching is not complete. While the activation energy of NN_i^* bands is a constant. Therefore we consider that the thermal quenching of NN_i bands is due to the dissociation of the complex (NN_i-exciton). The activation energy is equal to the binding energy at different NN_i sites. The constant value E of NN_i^* gives, in fact, the thermal activation energy of the dissociation of the ((NN_i-exciton) – LO) complex. The quenching of NN_i^* is due to the dissociation of this complex.

It is also reasonable to think that the coupling of bound exciton and LO lattice phonon is so strong that all the NN_i^* bands

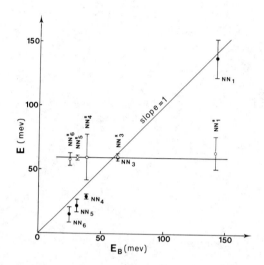

<u>Fig.5</u> : Activation energy versus the binding energy for the NN_i^* and NN_i bands. The binding energies are measured on the emission spectrum at 5 K. The activation energy is obtained by fitting the experimental results shown in Fig.4 with formula (3).

will not be dependent on the configuration (i) and they are "controlled" by the lattice vibration, therefore show the same character.

CONCLUSION

In the luminescence spectra of GaP:N, two series of bands NN_i and NN_i^* are found. They have different quenching processes. NN_i bands result from the recombination of exciton bound to different NN_i sites which have different binding energies. The temperature quenching of NN_i bands depends strongly on their binding energies. While NN_i^* bands have a clearly identical character. Their thermal quenching process has the same activation energy. How the $((NN_i\text{-exciton}) - LO)$ complex decays thermally is yet to be investigated.

REFERENCES

(1) D.G. Thomas, J.J. Hopfield and C.J. Frosch, Phys. Rev. Lett. 15 (22) 857 (1965).

(2) D.G. Thomas, J.J. Hopfield, Phys. Rev. 150 (2) 680 (1966).

(3) For example, P.J. Dean, Phys. Rev. 157 (3) 655 (1967).

* Permanent address : Changchun Institute of Physics, CAS Changchun, China.

INFLUENCE OF THE LOCAL FIELD EFFECTS ON THE DETERMINATION

OF THE EXCHANGE ENERGY OF EXCITONS

R. Bonneville
Centre National d'Etudes des Télécommunications
196 rue de Paris, 92220 Bagneux (France)

G. Fishman
Groupe de Physique des Solides de l'Ecole Normale
Supérieure, 4 Place Jussieu, 75005 Paris (France)

One investigates the influence of the local field effects on the interpretation of the excitonic polariton curves in cubic semiconductors. It is shown that in real compounds where the background dielectric constant is greater than one, and where the electron and the hole are weakly bound, the usual description of the polariton effect can be a great deal altered. A careful examination of the effective field acting on the exciton is needed to deduce the exchange energy Δ from such an experiment as resonant Brillouin scattering on excitonic polaritons; taking the local field effects into account can modify the value of Δ up to an order of magnitude.

The knowledge of the transverse energy E_T and of the longitudinal transverse splitting (LTS) $E_{LT} = E_L - E_T$ is sufficient to characterize the polariton dispersion curve in the case of phonon–polaritons[1]. In the case of excitonic polaritons[2,3], a third parameter is necessary, namely the translational mass M of the exciton; besides in some cases more information is needed : when the linear $\vec{k}$ terms (CuBr)[4] or the existence of both heavy and light excitons (ZnSe)[5] play a part, the exchange energy Δ appears to be an important parameter. In a more general way, although the formal definition of the exchange energy is well known, its connection to the measurable parameters of the polariton effect is not a trivial problem. For instance in zinc-blende compounds, the short range part of the exchange interaction splits the $\Gamma_6 \times \Gamma_8$ exciton[4] into a triplet level Γ_5 (J = 1) and a quintuplet level ($\Gamma_3 + \Gamma_4$) (J = 2); the splitting between the J = 1 and the

J = 2 levels is precisely the so-called exchange energy Δ; the
long range part of the interaction splits the J = 1 level into
a longitudinal non degenerate level E_L (M = 0) and a two-fold

degenerate level E_T (M = $\pm$ 1), whose splitting is the LTS.

It is worth noting here that the two measurable para-
meters are E_{LT} and the splitting δ between E_T and the quintuplet

level. Up to now, the exchange energy was assumed to be either
equal to the splitting between the center of gravity of the triplet
level and the quintuplet level,[4] i.e.

$$\Delta = \delta + \frac{E_{LT}}{3}$$ (1)

or merely equal to the splitting between the transverse energy
and the quintuplet level[6], i.e.

$$\Delta = \delta$$ (2)

Although the reason is not quite clear, it appears that δ is often
much weaker than E_{LT}[4,5]; in this case, it is obvious that the

magnitude of Δ depends crucially upon the relations between Δ, δ,
and E_{LT}. It is precisely the goal of this paper to study the kind

of relations which exist in real compounds and more precisely to
show how the expression of the effective electric field undergone
by an exciton influences the value of Δ as it is deduced from the
experiments.

In the latter case (Eq.2), the basic equation of the
polariton effect can be written as

$$\varepsilon = 1 + \frac{4 \pi n_o \alpha_o^{(o)}}{1 - \omega^2/\omega_o^2}$$ (3)

where $\varepsilon = k^2 c^2/\omega^2$ is the dielectric constant, $\alpha^{(o)}$ the
polarizability of the exciton, n_o the density of cells; $\hbar\omega_o$ is the
energy of the triplet (J = 1) level when the long range interaction
is not taken into account. One easily deduces

$$\omega_L^2 = \omega_o^2 (1 + 4\pi n_o \alpha_o^{(o)})$$

(4)

$$\omega_T = \omega_o$$

The assumption leading to Eqs.(2), (3), (4), is that the exciton
undergoes the macroscopic field in the medium. On the contrary,
the former case (Eq.(1)) results from the exciton undergoing a
Lorentz local field, which yields

$$\varepsilon = 1 + \cfrac{\cfrac{4\pi n_o \alpha_o^{(o)}}{1 - \cfrac{4\pi}{3} n_o \alpha_o^{(o)}}}{1 - \cfrac{\omega^2}{\omega_o^2 (1 - \frac{4\pi}{3} n_o \alpha_o^{(o)})}} \qquad (5)$$

It is easy to show that Eq. (5) is equivalent to Eq. (29) of Ref. (3), and that it involves

$$\left.\begin{array}{l} \omega_T^2 = \omega_o^2 \left(1 - \dfrac{4\pi}{3} n_o \alpha_o^{(o)}\right) \\[2ex] \omega_L^2 = \omega_o^2 \left(1 + \dfrac{8\pi}{3} n_o \alpha_o^{(o)}\right) \end{array}\right\} \qquad (6)$$

Those equations can be summarized by

$$2 \omega_T^2 + \omega_L^2 = 3 \omega_o^2 \qquad (7)$$

Since the LTS is much weaker than $\hbar\omega_o$, Eq. (7) becomes

$$2 \omega_T + \omega_L \simeq 3 \omega_o \qquad (8)$$

which is the result of the perturbation calculation worked out by Knox[7].

If one introduces a background dielectric constant ε_∞, the following equation is currently used :

$$\varepsilon = \varepsilon_\infty + \cfrac{4\pi\beta}{1 - \cfrac{\omega^2}{\omega_T^2}} \qquad (9)$$

together with Eq. (1) as in Ref. (4) or with Eq. (2) as in Ref. (6).
We are going to show that
i) Eq. (1) is never verified in any compound where the background dielectric constant is larger than one : if one deals with Frenkel excitons, ε_∞ is explicitely needed to know the relation between Δ, δ, and E_{LT}.
ii) Eq. (2) originates from the great average distance (compared to the dimensions of the unit cell) of the electron and the hole forming a Wannier exciton; in this last case, the background dielectric constant does not play any part.
iii) In an intermediate case, it is possible to give the relation between Δ, δ, E_{LT} if one knows the delocalization factor γ which
expresses the delocalization of the electronic charges in the crystal[8] ($\gamma = 1$ in the Frenkel case, and $\gamma = 0$ in the Wannier case).

We now start dealing with the problem of Frenkel
excitons. Hopfield's calculation[3] taking dipole-dipole interaction
of the excitons is performed with a background dielectric constant
equal to one and results in a formulation (Eq.(5) of this paper)
which is identical to a classical calculation, the only quantum
mechanical feature being that the polarizability $\alpha_o^{(o)}$ depends on
a dipole matrix element. Thus we shall henceforth consider that
an exciton has the same properties as a classical oscillator
immersed in the background dielectric constant ε_∞ of the solid.
The total dielectric polarization P originates from a set of
polarizable oscillators, each of them being characterized by its
eigenfrequency $\hbar\omega_i(\vec{k})$; $\hbar\omega_i$ is equal to $\hbar\omega_i(\vec{0}) + \hbar^2 k^2/2M$ in the
effective mass approximation[2,3]. The polarizability of the ith
oscillator is given by

$$\alpha_i(\vec{k},\omega) = \frac{\alpha_i^{(o)}}{1 - \dfrac{\omega^2}{\omega^2(\vec{k})}} \tag{10}$$

For the sake of simplicity, we assume that the resonance
frequency of the ith oscillator is far enough from the other
ones so that $\alpha_i^{(o)}$ does not depend upon $\vec{k}$. The problem of two
coupled oscillators will be reported elsewhere. If the field
undergone by the ith oscillator is the Lorentz local field

$$F_i = E + \frac{4\pi}{3} P \tag{11}$$

then the dielectric function is given by

$$\varepsilon(\vec{k},\omega) = 1 + \frac{4\pi n_i \alpha_i(\vec{k},\omega)}{1 - \dfrac{4\pi}{3} n_i \alpha_i(\vec{k},\omega)} \tag{12}$$

Since we are interested by the oscillator of frequency ω_o, then
in the vicinity of ω_o, one has

$$\varepsilon_\infty \simeq 1 + \frac{4\pi \sum\limits_{\omega_i > \omega_o} n_i \alpha_i^{(o)}}{1 - \dfrac{4\pi}{3} \sum\limits_{\omega_i > \omega_o} n_i \alpha_i^{(o)}} \tag{13}$$

and a tedious but straightforward calculation shows that

$$\varepsilon = \varepsilon_\infty + \cfrac{\cfrac{4\pi n_o \alpha_o^{(o)} (\frac{\varepsilon_\infty + 2}{3})^2}{1 - \frac{4\pi}{3} n_o \alpha_o^{(o)} (\frac{\varepsilon_\infty + 2}{3})}}{1 - \cfrac{\omega^2}{\omega_o^2(\vec{k})[1 - \frac{4\pi}{3} n_o \alpha_o^{(o)} (\frac{\varepsilon_\infty + 2}{3})]}} \qquad (14)$$

Only for $\varepsilon_\infty = 1$, Eq.(14) is identical to Eq.(5). From Eq.(14), it is easy to see how strongly the background dielectric constant influences the oscillator strength

$$4\pi\beta = \cfrac{4\pi n_o \alpha_o^{(o)} (\frac{\varepsilon_\infty + 2}{3})^2}{1 - \frac{4\pi}{3} n_o \alpha_o^{(o)} (\frac{\varepsilon_\infty + 2}{3})} \qquad (15)$$

If $4\pi n_o \alpha_o \ll 1$ (typically $10^{-2} - 10^{-3}$ in semiconductors), and if $\varepsilon_\infty \gg 1$ ($\varepsilon_\infty \sim 10$ in usual compounds), one has $4\pi\beta/4\pi n_o \alpha_o^{(o)} = [(\varepsilon_\infty + 2)/3]^2$. Let us now come back to the determination of ω_T and ω_L from Eq.(14); using $\varepsilon(\omega_L) = 0$ and $\varepsilon^{-1}(\omega_T) = 0$, we find

$$\omega_T^2 = \omega_o^2 [1 - \frac{4\pi}{3} (\frac{\varepsilon_\infty + 2}{3}) n_o \alpha_o^{(o)}]$$

$$\omega_L^2 = \omega_o^2 [1 + \frac{8\pi}{3\varepsilon_\infty} (\frac{\varepsilon_\infty + 2}{3}) n_o \alpha_o^{(o)}] \qquad (16)$$

These two equations, which must be compared with the usual Eq.(4) of this paper and to Eq.(3-19) and (3-20) of ref.7, lead to the solution of our problem :

$$\varepsilon_\infty \omega_L^2 + 2 \omega_T^2 = (\varepsilon_\infty + 2) \omega_o^2 \qquad (17)$$

or, since $E_{LT} \ll \hbar\omega_o$

$$\varepsilon_\infty \omega_L + 2 \omega_T = (\varepsilon_\infty + 2) \omega_o \qquad (18)$$

which demonstrates that the center of gravity of the triplet Γ_5 level is not conserved when the long range part of the exchange interaction is taken into account in the case of a Frenkel exciton. From Eq.(18) and the definition of Δ, one finds

$$\Delta = \delta + \frac{\varepsilon_\infty}{\varepsilon_\infty + 2} E_{LT} \qquad (19)$$

This shows that the "well-known" Eq.(1) used up to now when the background dielectric constant is larger than one is not valid : if ε_∞ is much larger than one, then one obtains $\Delta = \delta + E_{LT}$. We shall see later on the influence of Eq.(19) upon the order of magnitude of the exchange energy.

When the electrons are very strongly delocalized the Sellmeier expressions of the dielectric constant

$$\varepsilon(\vec{k},\omega) = 1 + 4\pi \sum_{i} n_i \alpha_i (\vec{k},\omega) \qquad (20)$$

is more convenient[9,10] than the Lorentz formula Eq.(12). Most generally, the local field F_i is

$$F_i = E + \frac{4\pi}{3} \gamma P \qquad (21)$$

Table I : Influence of the effective field expression on the calculated exchange energy. (The experimental data for CuBr and ZnSe are extracted from ref.4 and 5).

Energies (meV)		CuBr $(\varepsilon_\infty = 5,4)$	ZnSe $(\varepsilon_\infty = 9,0)$
Experimental data	E_{LT}	12	1.4
	δ	1,5	$\lvert\delta\rvert < 0.2$
Δ	$\Delta = \delta + E_{LT}/3$ $(\varepsilon_\infty = 1, \gamma = 1)$	5,5	0.5
	$\Delta = \delta + E_{LT}(\varepsilon_\infty/\varepsilon_\infty + 2)$ $(\varepsilon_\infty > 1, \gamma = 1)$	10.3	1.2
	$\Delta = \delta$ $(\gamma = 0)$	1.5	$\lvert\Delta\rvert < 0.2$

where $\gamma = 1$ for a Frenkel exciton, and (although there is no quantum-mechanical description) very likely $\gamma = 0$ for a Wannier exciton. From this starting point, if we put in the latter case

$$\varepsilon_\infty \simeq 1 + 4\pi \sum_{\omega_i > \omega_o} n_i \alpha_i^{(o)} \qquad (22)$$

we find

$$\varepsilon = \varepsilon_\infty + \frac{4\pi\, n_o \alpha_o^{(o)}}{1 - \dfrac{\omega^2}{\omega_o^2 (\vec{k})}} \qquad (23)$$

so that $\omega_T = \omega_o$, and therefore we obtain Eq.(2), i.e. $\Delta = \delta$.

Table I shows the influence of using Eq.(1), (2), or (19) on the calculation of the exchange energy Δ from the attainable parameters δ and E_{LT}. The value of Δ can be increased by an order of magnitude!

In some cases, it is known from the Bohr radius of the exciton that one deals with an intermediate case between a Frenkel exciton and a Wannier exciton[7]. Then, one must use Eq.(21) with $0 < \gamma < 1$. It is a relatively straightforward matter to find

$$\varepsilon = \varepsilon_\infty + \frac{\dfrac{4\pi\, n_o \alpha_o^{(o)} \left[\dfrac{\gamma(\varepsilon_\infty - 1) + 3}{3}\right]^2}{1 - \dfrac{4\pi}{3} n_o \alpha_o^{(o)} \gamma \left[\dfrac{\gamma(\varepsilon_\infty - 1) + 3}{3}\right]}}{1 - \dfrac{\omega^2}{\omega_o^2(\vec{k})\left\{1 - \dfrac{4\pi}{3} n_o \alpha_o^{(o)} \gamma \left[\dfrac{\gamma(\varepsilon_\infty - 1) + 3}{3}\right]\right\}}} \qquad (24)$$

and then, since $E_{LT} \ll \hbar\omega_o$,

$$\gamma \varepsilon_\infty \omega_L + (3 - \gamma)\omega_T = [\gamma(\varepsilon_\infty - 1) + 3]\omega_o \qquad (25)$$

so that

$$\Delta = \delta + \frac{\gamma\, \varepsilon_\infty}{\gamma(\varepsilon_\infty - 1) + 3} E_{LT} \qquad (26)$$

Of course, $\gamma = 1$ and $\gamma = 0$ give back the previously investigated cases.

Up to now, we have shown how it is possible to deduce Δ if δ, E_{LT}, ε_∞ and γ are known. However, from an experimental point of view, γ is not directly measurable, and indeed it is more reasonable to try to obtain an independent measurement of Δ^{11}. This experiment is quite difficult to perform with accuracy, except may be in CuBr where the order of magnitude of the exchange interaction is relatively large.

REFERENCES

1. Born M. and Huang K., Dynamical Theory of Crystal Lattice (Oxford University Press 1954).

2. Pekar S.I., Soviet Physics, JETP 6, 785 (1958).

3. Hopfield J.J., Physical Review, 112, 1555 (1958).

4. Suga S., Cho K., and Bettini M., Physical Review, B13, 943 (1976).

5. Sermage B., and Fishman G., this conference and references therein.

6. Ekart W., Lösch K., and Bimberg D., Physical Review, B20, 3303 (1979).

7. Knox R.S., Theory of Excitons, Solid State Physics, Supplement 5, Academic Press, New-York (1963).

8. Bonneville R., Physical Review, B21, 368 (1980).

9. Darwin C.G., Proceedings of the Royal Society, A146, 13 (1934).

10. Nozières P. and Pines D., Physical Review, 109, 762 (1958).

11. Fishman G., and Lampel G., Physical Review, B16, 820 (1977).

INVESTIGATIONS ON THE BIEXCITONS BY OPTICAL RESONANT

DEGENERATE MIXING

A. Maruani, D.S. Chemla, and E. Batifol

Centre National d'Etudes des Télécommunications

92220 Bagneux (France)

A previously described method of active nonlinear spectroscopy, applied to CuCl, has very clearly displayed a quite general peculiarity of the biexciton states in semiconductors : their autoionizing character. This method has been used to investigate the Γ_1 biexciton in CdS, where the theoretical and experimental conditions are by far more complicate. In that uniaxial compound it is possible, through a proper choice of the polarization of the impinging beams to discriminate among the various intermediate levels involved in that third order process. Then, the leading phenomenon which gives rise to the observed signal can be selected to be 3 or 2 times resonant, with a correspondingly "large" or "low" absorption coefficient. A theoretical analysis is presented, which accounts for the competition between on the one hand the non-linear gain and optical Kerr effect and on the other hand the one and two-photon absorptions.
A comparison between our analytical formulae and the whole set of our data enables us to extract, amidst other physical quantities, the "Fano's q parameter" which drives the decay of the biexciton into-mostly-two polaritons. We give a comparison between this phenomenological parameter and theoretical estimates. Finally, we present preliminary data on the successive higher order mixings and briefly discuss the difference between direct and cascade optical nonlinear processes.

We have used the method described in (1) to investigate the Γ_1 exciton in CdS at low temperatures. The principle of that method is that two beams (pump and test) with same energy $\hbar\omega$, and wave vectors k_p and k_t, recombining in a thin slab of mate rial give rise, through successive coherent scatterings, to signal at energy $\hbar\omega$ and wave vectors $k(n,n') = nk_p - n'k_t$ with $|n-n'| = 1$. In that communication, we shall focus on the first order ($n = 2$, $n' = 1$) scattering in CdS. The observed signal results from a forward four wave degenerate frequency mixing, in which a nonlinear polarization $P(\omega, k_s)$ is created through the third order process :

$$P(\omega, k_s) = \varepsilon_o \chi^{(3)} (-\omega; \omega, \omega, -\omega) |E(\omega, k_p) E(\omega, k_p) E^*(\omega, k_t)$$

Since the process is coherent no change of populations of the crystal excited states occurs, all the transitions are virtual, the initial and the final state are the ground state. Nevertheless, the third order susceptibility can exhibit resonant behaviour on the intermediate states, either one photon or two photon transitions. The mixing is expected to be two photon resonant if $2\hbar\omega$ is around W_{x2}, the formation energy of the Γ_1 biexciton in CdS, the most efficient one photon transition being quasi resonances through the A_1 exciton for a polarization perpendicular to the crystal optic axis $\vec{c}$ and through the B_1 exciton for a polarization parallel to it.

We have explored two configurations : one with all the beams polarized perpendicular to $\vec{c}$ and another one with the pump beam polarized parallel to $\vec{c}$, the test and the signal beams beeing perpendicular to it. In the former case the dominant term in $\chi^{(3)}$ has a triple resonant character with one two-photon resonant denominator on the biexciton and two one photon quasi-resonant denominators on the A_1 exciton :

$$\chi^{(3)} \propto [(W_{A1} - \hbar\omega - \Gamma_{A1}/2)^2 (W_{x2} - 2\hbar\omega - \Gamma_{x2}/2)]^{-1}$$

In the latter case one of these denominators involves a B_1 exciton and is less contributing because of the higher energy of B_1.

$$\chi^{(3)} \propto [(W_{A1} - \hbar\omega - \Gamma_{A1}/2)(W_{x2} - 2\hbar\omega - \Gamma_{x2}/2)(W_{B1} - \hbar\omega - \Gamma_{B1}/2)]^{-1}$$

Then the biexciton can be considered as virtually formed by two virtual B excitons. Typical line shapes are shown in fig.(1) for a 25 μm thick sample. But for a $\lambda/2$ plate for ensuring the proper polarization of the pump beam, the experimental conditions were strictly identical for both of those spectra. Their salient features include :

a) efficiency of the process : note the very low values of the intensity of the exciting beams

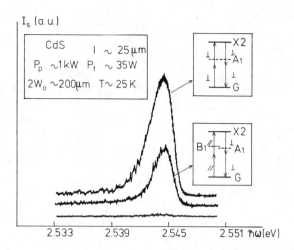

Fig.1 : Upper spectrum : First scattering spectrum in the all
        perpendicular configuration.
        Lower spectrum : First scattering spectrum when the pump
        polarization is parallel to the c axis.
        Those spectra have been vertically shifted for clarity.
        The lowest, nearly straightline is the signal detected
        after a small disalignment of the exciting beams, it
        represents the value of all residual scatterings.

b) the two spectra have about the same maximum intensity
c) the energy for that maximum is more than 3 meV red shifted
with respect to the expected two photon resonance on the
biexciton, indicated with an arrow on fig.(1)
d) the lower spectrum is slightly blue shifted with respect to
the upper one
e) the spectra manifest the autoionizing character resulting from
the coupling between the biexciton and a continuum, which is most
likely the pair of free polaritons. While that effect reveals in
a dissymmetric line shape for thinner samples, at lower intensities,
it appears in a more evident way, in thicker samples and at higher
intensities where the spectrum presents two maxima with very
different values (see fig.(1)) of ref.(3)).

        The first point has enabled us to work in the condition
where a very simple analytical solution of the propagation
equations system is operational with a large accuracy, i.e. :
$I_t(o)/I_p(o) = 1/3$ (notations of ref.(1) will be used throughout).

The second point shows there is some kind of compensation between the two following configurations :

a) $I_p /\!/ I_t \perp c$: $\chi_{\perp\perp}^{(3)}$ is triply resonant and α "high"

b) $I_p /\!/ c$, $I_t \perp c$: $\chi^{(3)} /\!/ \perp$ is doubly resonant and $\alpha_{/\!/}$ low.

The third and fourth point show that propagation effects due to the dispersion of n and α around excitonic resonances are most important for the line shape and position of the signal : by putting $W_{x2} \simeq$ Twice the energy of the maximum of the signal, one would derive $W_{x2} \simeq 5.088$ eV, in strong disagreement with all other published data[4]. By putting, following the method of ref.(2) $W_{x2} \simeq 5.096$ eV as deduced from the spectrum at a thicker sample excited by higher intensities (see ref.(3)) one would deduce a more satisfactory result.

One is thus led to carefully process the data. We have accounted for all sample thickness and all incident intensities by introducing in our analytical expressions, as fixed data, our measured absorption coefficients, the refractive index as given by Mahr (5), the energies and oscillator strengths of excitons and, as fitting parameters the oscillator strengths for the exciton-biexciton transitions, the Fano parameter, q, which is the signature of the autoionizing character of biexcitons, and the line width of the exciton and biexciton states. The extreme non linearity of the signal implies that the solution for the best fit is indeed unique, we checked this unicity by putting, as said above Γ_{x2} as a fitting parameter : that values could have been deduced actually by a simple inspection of the spectra : analytical solutions show that Γ_{x2} is the linewidth of the first order scattering spectrum. It was found that our fitted Γ_{x2} agreed with its expected value within 0.2 meV. For what concerns the nearly equal signal intensities in the two configurations, we have investigated, we deduce from our model that the ratio of the matrix elements $|\langle x^2|Z|B1\rangle|^2/|\langle x2|X|A1\rangle|^2$ is about 4, where the direction of c axis is Z. This has to be connected with the fact that the oscillator strengths from the ground state to A and B excitons and for polarizations respectively perpendicular and parallel to the c axis are equal to within some per cent.

A typical fit is shown in fig.(2). The formation energy of the biexciton was indeed found to be 5.096 eV. The fit is very acceptable for the low energy part of the spectrum, it is less so for the high energy one. The reasons for that include :
i) we found a rather low value for q (q = 6) showing that the coupling of the Γ_1 biexciton with - most probably - the two polariton pairs is strong, then, the assumption that q is constant is a rather crude one.
ii) we have neglected possible contributions from bound exciton

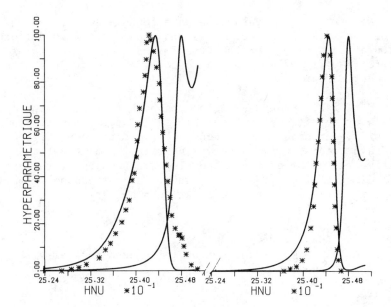

Fig.2 : Comparison of the theoretical results (full line) with
the experimental data (stars). The left part corresponds
to a sample 90 µm thick and the left one to a sample 25 µm
thick. The following values have been used :
W_{x2} = 5.096 eV, Γ_{x2} = 1.8×10^{-3} eV, Γ_{A1} = 2×10^{-4} eV
q = 6.

The curves accompanying those spectra are respectively
$|\chi^{(3)}|$ and $Im(\chi^{(3)})$. Note the shift between the position
of the two photon resonance and the maxima of the signals.

complexes, whose oscillators strengths are known to be giant.
iii) we have neglected intensity dependent dampings, and also
the polariton nature of the intermediate virtual states.
iv) we have not accounted for the B contribution in the all
perpendicular configuration, although it is doubtless it
manifests on the high energy side of the spectrum. Nor have we
included single or non resonant terms (background) in our model
$\chi^{(3)}$.

        Reasons ii) and iii) are for the sake of computational
facility; reason iv) would have added additional fitting parameters, thus leadings to some loss in the physics.

Nevertheless, the fact we have found a same reasonable quality of the fits for all crystals and experimental conditions, in particular accounting for the broadenings and the shifts of the spectra (see fig.(2)) leads us to consider we have not missed any important effect.

Work is in progress concerning higher order scatterings. As preliminary results we can indicate that, on a computational point of view, our experimental conditions (relatively thick samples) favor cascade processes versus direct mixing paths. On the experimental point of view, the slight narrowing of the spectra indicates also the same tendancy. The corresponding results will be the subject of a separate publication.

Acknowledgements - Mrs. Bonnouvrier is thanked for her very helpful programming work.

REFERENCES

(1) A. Maruani, J.L. Oudar, E. Batifol and D.S. Chemla, Physical Review Letters, 41, 1372 (1978).

(2) D.S. Chemla, A. Maruani and E. Batifol, Physical Review Letters, Vol.42, 16, 1075 (1979).

(3) A. Maruani, D.S. Chemla and E. Batifol, Solid State Communications, Vol.33, p.805 (1980).

(4) R. Planel and C. Benoît à la Guillaume, Physical Review B15, 1192 (1977).

(5) J. Jackel and H. Mahr, Physical Review 17, 8, 3387 (1978).

COULOMB GREEN'S FUNCTION FOR HYDROGENIC EXCITON

M. Suffczynski and A. Lusakowski

Institute of Physics, Polish Academy of Sciences

Lotnikow 32, Warsaw 02-668 Poland

The excited state reduced Coulomb Green's function in coordinate space is calculated and expressed in closed form. The reduced Green's function enables a calculation, in the first order of the perturbation theory, of the correction to the bound state wavefunction and, in the second order, correction to the energy. Since the Green's function is independent of the particular perturbation problem it can be calculated once and for all and then used in computation of various physical effects. Matrix elements for perturbation hamiltonian terms, transforming according to definite angular momentum representations, especially those practically important in cubic semiconductors with degenerate bands are computed analytically and presented in a succinct form.

The reduced Green's function is the coordinate space representation of the sum over the intermediate states encountered in the bound state perturbation theory. The reduced Green's function enters in the calculation of the first order corrections to the bound state wave functions belonging to the discrete energy eigenvalue, and also in the second order corrections to the energy of the bound state. The reduced Green's function can be calculated once and for all and then used for computation of particular perturbation problems [1,2].

The reduced Coulomb Green's function for the n-th hydrogenic bound state can be expressed [3] :

$$G^n(\underline{r}_2, \underline{r}_1) =$$

$$= \frac{e^{-(u-v)/2}}{4\pi |\underline{r}_2 - \underline{r}_1|} \{ L_n(u)Q_n(v) - \frac{u}{n} L_{n-1}^1(u) \; [\dot{Q}_n(v) + L_{n-1}^1(v)] \}$$

$$+ \frac{e^{-(u+v)/2}}{4\pi |\underline{r}_2 - \underline{r}_1|} \{ [uL_{n-1}^1(u)L_{n-1}(v) - vL_{n-1}(u)L_{n-1}^1(v)] \times$$

$$\times [\ln u - g(v) + \gamma - 2 \sum_{k=1}^{n} \frac{1}{k} - \frac{1}{2n}] + \frac{u-v}{n} L_{n-1}^1(u)L_{n-1}^1(v)$$

$$- \frac{u-v}{2} [L_n(u)L_{n-1}(v) + L_{n-1}(u)L_n(v)]$$

$$+ K_n(u)L_{n-1}(v) - L_{n-1}(u)K_n(v) + \frac{u}{n} L_{n-1}^1(u)\dot{K}_n(v) - \frac{v}{n} \dot{K}_n(u)L_{n-1}^1(v) \}$$

Here :

$$u = (r_2 + r_1 + |\underline{r}_2 - \underline{r}_1|)/na_1 \; ,$$

$$v = (r_2 + r_1 - |\underline{r}_2 - \underline{r}_1|)/na_1 \; ,$$

with the ground state Bohr radius a_1 ,

$$g(v) = \int_0^v \frac{e^t - 1}{t} \, dt \; ,$$

$\gamma = 0.5772...$ is the Euler-Mascheroni constant.

$$Q_n(v) = -1 - v \sum_{k=1}^{n-1} \frac{(-1)^k}{k} L_{n-k-1}^{k+1}(v) \sum_{p=0}^{k-1} \frac{(-v)^p}{p!} \; , \quad n = 2,3,...$$

$$Q_1(v) = -1 \; ,$$

$$K_n(z) = 2z \sum_{k=1}^{n-1} \frac{1}{k} L_{n-k-1}^1(z) - L_{n-1}(z) \; , \quad n = 2,3,...$$

$$K_1(z) = -1.$$

For $n = 0,1,2,3,...$ and for any m

$$L_n^m(z) = z^{-m} \frac{e^z}{n!} \frac{d^n}{dz^n} (z^{m+n} e^{-z}) , \qquad L_n(z) = L_n^0(z) .$$

The dot denotes the derivative with respect to the argument of the function.

When the perturbation hamiltonian can be decomposed into terms of definite angular momentum about an axis it is useful to have the partial wave expansion of the Green's function [4,5]

$$G^n(\underline{r}_2,\underline{r}_1) = \frac{1}{4\pi} \sum_{l=0}^{\infty} (2l+1) P_l(\cos \theta_{21}) \, G_l^n(r_2,r_1)$$

where P_l are Legendre polynomials of $\cos \theta_{21} = \underline{r}_2 \cdot \underline{r}_1 / r_2 r_1$.

The contributions in the second-order perturbation theory of Baldereschi and Lipari [6] from the p and d type anisotropy of the degenerate valence band to the energy levels of the ns hydrogenic exciton states can be evaluated by integrating the reduced Coulomb Green's function partial wave component with the normalized hydrogenic radial wave functions $\varphi_{ns}(r)$:

$(H_p G^n H_p)$:

$$-\pi \int_0^\infty dr_2 \int_0^\infty dr_1 \, r_2^2 [\frac{d}{dr_2} \varphi_{ns}(r_2)] G_{l=1}^n(r_2,r_1) r_1^2 \frac{d}{dr_1} \varphi_{ns}(r_1)$$

$$= 3/16 \quad \text{for} \quad n = 1,2,3;$$

$(H_d G^n H_d)$:

$$-\pi \int_0^\infty dr_2 \int_0^\infty dr_1 \, r_2^2 [(\frac{d^2}{dr_2^2} - \frac{1}{r_2} \frac{d}{dr_2}) \varphi_{ns}(r_2)] G_{l=2}^n(r_2,r_1)$$

$$\cdot r_1^2 (\frac{d^2}{dr_1^2} - \frac{1}{r_1} \frac{d}{dr_1}) \varphi_{ns}(r_1) = -\pi^2/8 + 35/24 \quad \text{for} \quad n = 1 ,$$

$$= \pi^2/8 - 55/48 \quad \text{for} \quad n = 2 ,$$

$$= 577/14400 \qquad \text{for} \quad n = 3 .$$

This completes a purely analytical evaluation of the perturbation corrections to the lowest exciton energy levels in cubic semiconductor [7,8] .

REFERENCES

1. Hostler, L., Phys. Rev. 178, 126 (1969).

2. Hostler, L., J. Math. Phys. 16, 1585 (1975).

3. Swierkowski, L. and Suffczynski, M., Bull. Acad. Polon. Sci.,
   Ser. Sci. Math. Astronom. Phys. 21, 285 (1973).

4. Swierkowski, L., Bull. Acad. Polon. Sci., Ser. Sci. Math.
   Astronom. Phys. 22, 1083 (1974).

5. Suffczynski, M. and Swierkowski, L., Bull. Acad. Polon. Sci.,
   Ser. Sci. Math. Astronom. Phys. 22, 1273 (1974).

6. Baldereschi, A. and Lipari, N.O., Phys. Rev. B 3, 439 (1971).

7. Suffczynski, M., Phys. Stat. Sol. (b) 52, K51 (1972).

8. Suffczynski, M. and Swierkowski, L., Bull. Acad. Polon. Sci.,
   Ser. Sci. Math. Astronom. Phys. 23, 807 (1975).

EXCITONS IN MOTT INSULATORS

M. Héritier and P. Lederer

Université de Paris-Sud, Laboratoire de Physique des

Solides, 91405 Orsay, France

In a Mott insulator, excess charge carriers in narrow half-filled degenerate bands form spin polarons. An electron-hole pair forms an exciton embedded in a ferromagnetic volume, the radius of which is large compared to the exciton Bohr radius. If a wide conduction band lies in the Hubbard gap, the exciton radius can be large compared to the hole-polaron radius. Then, depending on the exciton concentration, a gas of excitonic molecules of an electron-hole plasma can be formed.

1. MAGNETIC POLARONS IN NARROW BANDS

Correlation effects on the electronic properties of narrow band Mott insulators can be investigated, in a first approximation, in a simple Hubbard model. The properties of excess charge carriers have been studied by various authors. The simplest case of one hole in a single narrow half-filled s band has been treated in details in the limit $U/T \to \infty$, where U is the intraatomic Coulomb interaction and t is the nearest neighbour transfer integral[1,2,3,4]. In this limit, an excess in an alternate lattice induces a fully ferromagnetic ground state[1] : any departure from the complete alignment of spins reduces the number of paths for the hole propagation and thus increases the kinetic energy of the hole. For large but finite values of U/t, the Anderson superexchange couples nearest neighbour spins with a coupling constant $J = 2t^2/U$. In the absence of the excess hole (exactly one electron per site) the ground state is antiferro-magnetic. In the presence of the excess hole, there is a competition between the superexchange coupling and the Nagaoka's coupling . The hole forms a spin polaron[2,5,6] : to lower its kinetic energy, it induces a ferromagnetic polarization over a volume limited by exchange (and entropy at finite temperature), in which it becomes self-trapped. Recently, we have studied in details the conditions of existence and the properties of these ferromagnetic spin polarons[7,8]. The first condition is that the lattice is alternate. In non alternate lattices, it may happen that the lowest kinetic energy of the hole is not obtained in the ferromagnetic configur- ation[1,9] (as in f.c.c., h.c.p. or triangular lattices). The second condition is a favourable balance between, on one hand, the hole kinetic energy lowering in a ferromagnetic medium and, on the other hand, the hole localization energy in the ferro- magnetic volume and the exchange energy cost of the polarization. This condition is fullfilled if the ratio U/t exceeds a minimum value $(U/t)_m$. In the simplest approximation, we assume a perfect localization of the hole in the ferromagnetic sphere. Then, the estimation of $(U/t)_m$ is about 460 for a b.c.c. lattice. If we take into account the finite depth of the effective potential well binding the hole to the ferromagnetic volume (i.e. the kinetic energy lowering), the variational hole wave function may have exponential tails in the antiferromagnetic region. This effect decreases the localization energy within the ferromagnetic region in an important way, and lowers the estimation of $(U/t)_m$ to about 170. Taking into account exchange interactions at the polaron surface decreases the exchange cost of the polarization and still lowers the estimation of $(U/t)_m$ to about 60. This value is large. We conclude that polaron formation in single s band is possible but requires quite narrow bands.

However, spin polaron formation is easier in narrow degenerate half-filled bands, a case of practical interest in d bands with partially lifted orbital degeneracy. In that case, one should take into account intraatomic exchange. We consider now a simple Hamiltonian incorporating this effect

$$H = \sum_{im\sigma} t_m c^+_{im\sigma} c_{jm\sigma}$$

$$+ U \sum_{\substack{i \\ mm'\sigma\sigma'}} c^+_{im\sigma} c^+_{im'\sigma'} c_{im\sigma} c_{im'\sigma}$$

$$+ K \sum_{\substack{i \\ mm'\sigma\sigma'}} c^+_{im\sigma} c^+_{im'\sigma'} c_{im\sigma} c_{im'\sigma} \qquad (1)$$

K is the intraatomic exchange integral and m labels the various degenerate states of the atoms. We consider here the doubly degenerate case, with two electrons per atoms, in the limit $U \gg t$ and $K \gg t$. We still assume that the lattice is alternate. The transfer term introduce an antiferromagnetic superexchange

coupling between nearest neighbour spins $J = 2t^2/(U+K)$, and therefore, the ground state for exactly halffilled bands is antiferromagnetic. In the presence of an excess hole, this antiferromagnetic
coupling has to compete with two effects favouring ferromagnetism.
The first is a sort of generalized double exchange in the
degenerate bands. On each site, the two electrons are coupled
according to Hund's rule into triplet state. Consider an atomic
site occupied by one hole and one electron with, say, spin up.
The hole can hop to a neighbouring site with a matrix element t
if the neighbouring spin is in the state $S_z = +1$, but with reduced
matrix element if it is in the state $S_z = 0$ or $S_z = -1$. This effect
decreases the bandwidth in non ferromagnetic configurations. Moreover, the reduction of paths described by Nagaoka in the non
degenerate case increases in the same way the hole kinetic energy
in the degenerate bands. The overall effect on the antiferromagnetic
bandwith is a narrowing[2] by a factor of 2. Therefore, a hole tends
to induce a ferromagnetic polarization to decrease its kinetic
energy, as in the non degenerate case, but the condition on t/J =
$(U+K)/t$ for the existence of this spin polaron is less severe. In
the approximation of complete hole localization t/J must be larger
than 200, a figure lowered to 80 in the finite potential well model.
When surface interactions are taken into account, we may expect
spin polarons to exist for $U/2t \sim 1$, at least in a b.c.c. lattice.
While both effects discussed above favour ferromagnetism in
alternate lattices, they may have to compete in non alternate
lattices (as in the triangular lattice). Spin polarons with more
complex magnetic structure may exist. In what follows, we shall
consider only alternate lattices.

        Although spin polarons may form not far from the Mott
transition in the degenerate case, we restrict ourselves to the
case of very narrow bands. In this limit $(t/(U+K) \gg 1)$ the
polaron radius at temperature T much higher than J/k_B can be
written in the paramagnetic phase

$$R = a(\pi t/2k_B T Ln2)^{1/5} \qquad\qquad (2)$$

a is the lattice spacing. Expression (2) holds for $J \ll k_B t \ll t$.
At very low temperature, $k_B t \ll J$, in the antiferromagnetic phase

$$R = a(\pi(U+K)/2zt)^{1/5} \qquad\qquad (3)$$

For $k_B T \sim t$, the spin polaron can diffuse by propagation of spin
waves at its surface[10]. However, for $k_B T \ll J$, these diffusive
processes are exponentially damped. Then, at low enough T, the
spin polaron can be considered as a heavy particle propagating
coherently, with a high effective mass $m^* = m \exp - 0.2 N_s$, where
m is the ferromagnetic effective mass and N_s the number
of spins at the surface.

2. EXCITONS IN NARROW BANDS

The Hubbard model describes properly the formation of spin polarons in narrow bands. However, when one deals with exciton states, one must include explicitly a 1/r interaction between carriers. In this section, we consider states with one excess hole and one excess electron, and we add to Hamiltonian (1) a Coulomb term

$$H_C = - \sum_{\substack{i \neq j \\ mm'}} (e^2/\kappa r_{ij}) n_{im\uparrow} n_{im\downarrow} (1-n_{jm'\uparrow}) \cdot (1-n_{jm'\downarrow})$$

κ is the static dielectric constant, r_{ij} is the distance between site i and site j. In a ferromagnetic configuration, an exciton is formed with Bohr radius

$$a_o = 2at/(e^2/\kappa a) \tag{4}$$

Except for very large values of κ, a case which we do not consider here, $a_o \ll R(T)$. In fact, at the low temperature considered here ($k_B T \ll t$), each carrier, when isolated, forms a spin polaron with radius $R(T)$. They can lower their energy by sharing the same ferromagnetic volume of radius $R' = 2^{1/3} R(T)$. In this volume, they form an exciton, with radius $a_o \ll R'$. This exciton, which becomes self-trapped in a large polaron, has a binding free energy :

$$F_B = (2^{4/5}-1) \frac{10\pi^2}{3} t \ (a/R)^2 + \frac{1}{4} (e^2/\kappa a)^2/t$$

The self-trapping condition, even in the approximation of perfect localization is that $(U+K)/t > 20$ in the doubly degenerate bands, and >40, in the non degenerate band.

Now consider a dilute gas of these excitonic polarons. At fixed number of excitons, a lower free energy is obtained in a two phase system : an antiferromagnetic phase without exciton and a ferromagnetic phase in which free excitons are condensed at a concentration[11]

$$x_G(T) = [\frac{10}{8\pi^2} \frac{(4\pi)^{2/5}}{3} \frac{kTLn2}{t}]^{3/5} \tag{5}$$

for $J \ll k_B T \ll t$. In this ferromagnetic drop the excitons are still identical with the free carrier ones.

In most cases of Mott insulators, only one kind of carriers, for example the holes, propagate in a narrow band,

while the conduction band is large. Therefore we consider now the
case of one hole in a half-filled Hubbard band and one electron
in a broad s band. The hole forms a spin polaron, which for
$k_B T \ll J$ can be considered as a heavy particle propagating

coherently. We consider only small s-d exchange coupling so that
the light electron does not form a spin polaron. An exciton is
formed with radius :

$$a_1 = a \; \frac{\hbar^2/2m_e a^2}{e^2/\kappa a} \qquad (6)$$

m_e is the conduction electron effective mass. This radius may
be much larger than R(T) for light enough electron and large
enough dielectric constant. In fact our treatment is only valid
in this limit and we shall restrict ourselves to such a situation.
Then the exciton is quite analogous to a Hydrogen atom, with a
heavy positive "nucleus" and a light spin 1/2 electron. However
the "nuclear" magnetic moment is quite large. For example for
$U/t = K/t = 100$, $m^* \sim 10^3 m$, and the nuclear magnetic moment $\sim 10^5$
proton Bohr magnetons.

Our treatment assumes a coherent propagation of the
polaron. In fact, it is justified by an adiabatic approximation
when the polaron diffusion coefficient is small enough, i.e. for
$k_B T \ll t$.

3. EXCITONIC MOLECULES AND ELECTRON-HOLE PLASMA

The excitons studied in section 2 may form various
phases which we discuss in this section. First consider the case
where both carriers propagate in narrow Hubbard sub-bands. The
exciton gas in its ferromagnetic drop behaves like the free
carrier one. Therefore two free excitons can be found to form an
excitonic molecule with a binding energy[11] of 0.02-0.03 $e^2/2\kappa a_0$.
At fixed number of molecules, there is a segregation of an antiferromagnetic phase without molecule and a ferromagnetic drop
with a molecule concentration

$$x_M(T) = x_G(T) \cdot 2^{3/5}$$

As argued by Binkman and Rice[14], the intermolecular Van der Waals
attraction is probably unable to induce a gas-liquid transition
because of the large zero-point motion of these light molecules.
In practice, the exciton concentrations which can be obtained
are small in units of a_o^{-3}. Therefore, a condensation in electron-

hole plasma droplets[13,14] is not possible.

When excitons are formed from heavy hole polarons and
light electrons, we can make use of the analogy with Hydrogen
atoms to discuss the various possible phases. Their stability
depends on the charge carrier mass ratio $\alpha = m_h/m_e$, which can be
varied by applying a pressure or a magnetic field. It is convenient
to express the energies in exciton Rydberg units and the lengths
in exciton Bohr radius units. The binding energy E_B of the
excitonic molecule is much larger than in the previous case.
For large α, Wehner[12] found : $E_B = 0.35-0.76\alpha^{-12}$. The intermolecular potential is similar to that of Hydrogen. Then, following
de Boer's analysis[15], we expect a gas-liquid transition whenever
the molecular mass exceeds about 90 m_o, (m_o is the free electron
mass). In the high density limit, the stability of an electron-
hole plasma has been compared to that of the exciton gas in broad
band semiconductors[13,14]. The result depends on the details of the
band structure. For isotropic and non degenerate bands, the
plasma is never stable[13]. However, it is favoured by band
degeneracy and anisotropy. In Ge or Si, a detailed study taking
into account a realistic band structure shows that the plasma is
stable, as observed experimentally[13,14]. Such a study is out of
the scope of this work. We want simply to point out that
condensation in electron-hole droplets is easier in a Mott
insulator when α is large and may occur even in the less favourable
band model, i.e. isotropic and non degenerate bands. The energy of
the plasma per electron-hole pair can be found from the work of
Wigner and Huntington on metallic Hydrogen[13,16] :
$E_o = -1.05 + 0.8\alpha^{-1/2}$. It is lower than that of the exciton gas
when $\alpha > 260$, but larger than that of the molecular gas for any α.
Therefore, we expect, at low T, the following sequences of phases
when the exciton density increases   : at low density, a molecular
gas, then if the molecular mass >80 m_o, a mixed phase of molecular
gas and molecular liquid then a molecular liquid, then if $\alpha > 260$
a mixed phase of molecular fluid and electron-hole plasma, then
an electron-hole plasma alone.

This work is not an extensive study of the exciton
phase diagram. It is only a frame indicating the main lines for
further theoretical investigations. A more detailed work will
appear elsewhere.

REFERENCES

1. Y. Nagaoka, Physical Review 147, 392 (1966).

2. W.F. Brinkman and T.M. Rice, Physical Review B2, 1324 (1970).

3. N. Ohata and R. Kubo, Journal of the Physical Society of Japan, 28, 1402 (1970).

4. M. Héritier, Thèse d'Etat, (Orsay 1975).

5. N.F. Mott, Metal-Insulator Transitions (Taylor and Francis, 1974).

6. C. Herring (unpublished).

7. M. Héritier and P. Lederer, Journal de Physique Lettres 38, L209 (1977).

8. G. Montambaux, P. Lederer and M. Héritier, Journal de Physique Lettres 40, 499 (1979).

9. M. Héritier and P. Lederer, Physical Review Letters 42, 1068 (1979).

10. S.V. Iordanskii, Journal of Experimental and Theoretical Physics Letters, 26, 171 (1977).

11. M.V. Feigelman, Journal of Experimental and Theoretical Physics Letters, 27, 491 (1978).

12. R.K. Wenner, Solid State Communications, 7, 457 (1969).

13. W.F. Brinkman and T.M. Rice, Physical Review B7, 1508 (1973).

14. M. Combescot and P. Nozières, Journal of Physics C5, 2369 (1972).

15. J. de Boer, Progress in Low Temperature Physics (C.J. Gorter Editor, North-Holland), 2, 1 (1957).

16. E. Wigner and H.B. Huntington, Journal of Chemical Physics, 3, 764 (1935).

LUMINESCENCE DECAY IN SILICON AT LOW TEMPERATURES

P. Voisin, C. Benoît à la Guillaume, and M. Voos

Groupe de Physique des Solides* de l'ENS
Université Paris VII, Tour 23

2 Place Jussieu, 75005 Paris, France

We present a study of the luminescence decay of free excitons and electron-hole drops performed in pure Si in a rather broad temperature range. We get the exciton and drop lifetimes but, to interpret our results, we have to consider that the exciton-drop system is not spatially homogeneous. We propose also to take into account the exciton recombination at the sample surface to explain the observed non exponential exciton decay.

It is now well known that the equilibrium between electron-hole drops (EHD)[1] and the free exciton (FE) gas in Ge or Si is satisfactorily described in terms of condensation and thermoionic-like evaporation of FE associated with the intrinsic recombination of the two species. It follows that the lifetimes of EHD (τ_g) and FE (τ_{ex}) are two important parameters in any nucleation study[2], and must be carefully studied. As far as we know, only few explicit and detailed investigations[3,4] of the decay of the FE-EHD system following a pulsed excitation at liquid helium temperatures were reported in Si. In particular, using short exciting pulses (18 ns) and a rather high excitation peak power ($\sim$10 W) provided by a cavity-dumped Ar^+ laser, Schmid[4] concluded from such a study carried out at 4.2 K in 4000$\Omega \times$ cm p-type Si that EHD were formed directly from a dense plasma rather than from a FEgas, and that the gaseous and liquid phases were spatially separated.

We wish to report here a systematic investigation of the FE and EHD luminescence decay performed in very pure Si in a wide temperature range. Furthermore, we used a lower excitation

peak power and longer exciting pulses to allow FE and EHD to decay
from a steady-state regime. The analysis of our data indicates that
the spatial inhomogeneities of the FE-EHD system should certainly
be taken into account. However, in our experimental conditions,
we have no evidence that EHD are formed from a dense plasma rather
than from a FEgas. We get also the FE and EHD lifetimes, and to
analyse the observed non-exponential FE decay we propose to take
into account the recombination of FE at the sample surface.

In these experiments we used a $(2 \times 5 \times 10) mm^3$ pure Si
sample $(1.2 \times 10^5 \Omega \times cm$, p-type), and the experimental set-up was
the same as in Ref.5. We wish to emphasize that the excitation
source, which was a cavity-dumped Ar^+ laser (5145 Å), provided
long pulses (40 μs) whose frequency repetition and maximum peak
power were respectively 10^4 Hz and 1 W. Finally, the response time
of our experimental system was 15 ns.

Fig.1 shows the experimental FE and EHD decays at T =
9.6 K. Identical curves were obtained for EHD at any temperature
between 2 and 12.5 K, exhibiting an exponential decay over at
least 1.5 order of magnitude with a time constant $\tau = (170 \pm 10)$ ns.
Fig.2 presents the observed FE decays at different temperatures.
If we assume that the system is spatially homogeneous and that
the number of EHD remains constant, the equations governing the
FE and EHD decays are[1] :

$$\frac{dn_{ex}}{dt} = - \frac{n_{ex}}{\tau_{ex}} - (n_{ex} - n_{ex}^T \exp(\frac{2\sigma}{Rn_o kT})) v_{ex} n_g S \P R^2$$

$$\frac{d}{dt} (\frac{4}{3} \P n_o R^3) = - \frac{4}{3} \P n_o \frac{R^3}{\tau_g} + \qquad (1)$$

$$(n_{ex} - n_{ex}^T \exp(\frac{2\sigma}{Rn_o kT})) v_{ex} S \P R^2$$

where[6] n_{ex} and v_{ex} are the FE density and mean thermal velocity,
S the FE sticking coefficient[5-7] at the EHD surface, n_g the EHD
density, R their radius, σ their surface energy, and n_o the
electron-hole pair density in EHD. Besides,
$n_{ex}^T = g(h^{-2} 2\P mT)^{3/2} \exp(-\phi/kT)$ where g is the FE ground state
degeneracy, m their translation mass, and ϕ the EHD binding energy.
Using the initial equilibrium condition $n_{ex}(o) = 4n_o R(o)/3v_{ex} ST_g +$
$n_{ex}^T \exp (2\sigma/R(o)n_o kT)$, we get easily :

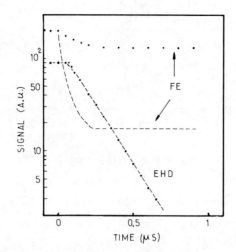

Fig.1 : Experimental FE and EHD luminescence decays at 9.6 K (full dots). Theoretical fits (dashed lines) using equations (2) with C = 4 and $n_{ex}(o) = 1.8 \times 10^{15}$ cm^{-3}. The other parameters involved in these calculations are given in Refs. 5 and 6.

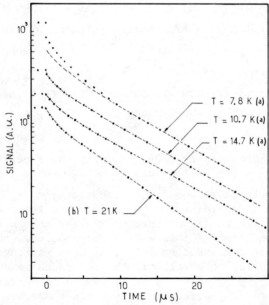

Fig.2 : Experimental FE luminescence decays at different temperatures (full dots). Theoretical fits (dashed lines) with the values of τ_{ex} s/L given in the text.

$$\frac{dx}{dt} = -\frac{x}{\tau_{ex}} - C\frac{y^2}{\tau_g}\left(\frac{x - n_{ex}^{-1}(o)n_{ex}^{T}\exp(2\sigma/yR(o)n_{o}kT)}{1 - n_{ex}^{-1}(o)n_{ex}^{T}\exp(2\sigma/yR(o)n_{o}kT)}\right)$$

(2)

$$\frac{dy}{dt} = -\frac{y}{3\tau_g} + \frac{1}{3\tau_g}\left(\frac{x - n_{ex}^{-1}(o)n_{ex}^{T}\exp(2\sigma/yR(o)n_{o}kT)}{1 - n_{ex}^{-1}(o)n_{ex}^{T}\exp(2\sigma/yR(o)n_{o}kT}\right)$$

with $x = n_{ex}(t)/n_{ex}(o)$, $y = R(t)/R(o)$ and $C = 4\pi n_g n_o R(o)^3/3n_{ex}(o)$.
At low temperatures ($\lesssim 7$ K), n_{ex}^{T} is quite negligible[5] ($2.7 \times 10^7 cm^{-3}$
at 4 K, while the threshold for EHD formation is $n_{ex}^{th} = 1.5 \times 10^{15} cm^{-3}$).
Then, for a given set of τ_{ex} and τ_g, the solution of equations (2)
depends only on C, i.e. on the initial ratio of the liquid to the
gaseous phase. Under rather high excitation conditions, which are
usual in time resolved experiments, C is currently much larger
than 1, and calculations[6] from equations (2) show that the FE
signal should relax faster than the EHD one, in agreement with
Ref.4. They show also that the larger C, the faster the relaxation,
and that the EHD signal tends rapidly towards an exponential decay
with a time constant $\tau = \tau_g = (170 \pm 10)$ ns. At higher temperatures
(>7 K), n_{ex}^{T} is not negligible and the solution of equations (2)
depend now on two parameters (C and $n_{ex}(o)$). However, for large
variations of these parameters, as long as $n_{ex}(o) \gg n_{ex}^{T}$ and $C \gtrsim 2$,
similar theoretical results are obtained : a fast initial drop in
the FE signal which relaxes towards $\sim n_{ex}^{T}$ and an exponential decay
for EHD with $\tau = \tau_g = (170 \pm 10)$ ns, as shown in Fig.1 for T = 9.6 K.
Therefore, this model is in agreement with the experimental
results for EHD, and we get a value of τ_g which is consistent
with previous determinations[1,4,5,7]. However, it leads to a
large discrepancy regarding the FE decay since the observed
initial drop in the FE signal is much smaller than what is
expected from equations (2) (see Fit.1). We believe that this
indicates that in our experimental conditions (strongly
inhomogeneous surface excitation), the EHD cloud is smaller than
the volume occupied by the FE gas, so that a large part of the
initial FE signal arises from regions where there are no EHD.
We wish to point out that this effect depends on n_{ex}^{th} and is
thereby temperature dependent, because the FE density outside
the EHD cloud is necessarily smaller than n_{ex}^{th}. Note however that
in this temperature range n_{ex}^{th} is fairly high[5]. Though our

experimental conditions are very different, this is consistent
with the results of Ref.4, but we do not have to postulate that
EHD are formed from a dense plasma. We think that this can also
explain the FE decay observed by Collet et al[3] and casts some
doubt on the value of τ_g (105 ns) reported by these authors.

Indeed, to analyse the EHD signal they used the observed FE
decay, which in our model is mainly due to FE outside the EHD
cloud. In fact, from their data we would rather find $\tau_g \sim$ 180 ns.

We can now study the FE decay for t $\gtrsim$ 1 μs, i.e. in
such conditions that EHD have mostly disappeared. Taking into
account FE diffusion and recombination at the sample surface,
on gets :

$$\frac{d\, n_{ex}(x,t)}{dt} = - \frac{n_{ex}(x,t)}{\tau_{ex}} + D \frac{\partial^2 n_{ex}(x,t)}{\partial x^2} - s n_{ex}(o,t)\delta(x)$$

$$(3)$$

where D is the FE diffusion constant, and s a coefficient
describing the surface recombination. Equation (3) can be solved
numerically, the solution depending only on τ_{ex} s/L with
$L = (D\tau_{ex})^{1/2}$. In figure 2 are shown the calculated FE signal
vs. t which is proportional to the number of FE, i.e. to
$\int_x n(x,t)\, dx$. One can see that the agreement between experiment
and theory is satisfying up to 15 K if we take τ_{ex} s/L : 3. For
this value, the time constant of the FE decay is $\tau = 0.83\ \tau_{ex}$,
yielding $\tau_{ex} = (11.5 \pm 1.5)$ μs between 7 and 15 K. At higher
temperatures (21 K) we find τ_{ex} s/L = 2.5 and a somewhat smaller
value for τ_{ex} (9.2 μs). For T < 7 K, our signal-to-noise ratio
is not good enough to allow us to study reliably τ_{ex}, but we can
nevertheless conclude that τ_{ex} does not depend strongly on T.
Considering the data reported [1,3,7] previously on τ_{ex}, it seems
that this parameter is sample dependent.

To summarize, we have presented a detailed study of
the FE-EHD system in pure Si in a rather wide range of temperature.
We think that these results give support to the FE-EHD equilibrium
model, provided we consider that the system may be spatially inhomogeneous. Furthermore the FE luminescence decay has been
satisfactorily interpreted by taking into account the
recombination of FE at the sample surface, an effect which is in
fact quite likely.

REFERENCES

* Laboratoire associé au CNRS.

1. For a review, see J.C. Hensel, T.G. Phillips and G.A. Thomas
   in Solid State Physics (Edited by F. Seitz, D. Turnbull and
   H. Ehrenreich), Vol.32, p.88. Academic Press, New York (1978);
   M. Voos and C. Benoit à la Guillaume, Optical Properties of
   Solids, New Developments (Edited by B.O. Seraphin), p.143.
   North-Holland, Amsterdam (1976); Ya. Pokrovskii, Phys. Status
   Solidi (a) 11, 385 (1972); and references therein.

2. For a recent review on drop nucleation, see B. Etienne,
   J. Luminescence 18/19, 525 (1979).

3. J. Collet, J. Barrau, M. Brousseau and H. Maaref, Solid State
   Commun. 19, 1141 (1976).

4. W. Schmid, Solid State Commun. 19, 347 (1976); Proc. 13th Int.
   Conf. Phys. Semicond. Roma, 1976 (Edited by F.G. Fumi), p.898
   Tipografia Marves, Rome (1976).

5. P. Voisin, B. Etienne and M. Voos, Phys. Rev. Lett. 42, 526
   (1979).

6. The values of the parameters involved in the calculations done
   in this study are given in Ref.5 and in : P. Voisin, B. Etienne
   and M. Voos, Solid State Commun. 33, 541 (1980).

7. R.B. Hammond and R.N. Silver, Phys. Rev. Lett. 42, 523 (1979);
   Appl. Phys. Lett. 36, 68 (1980), and references therein.

ON THE HYDRODYNAMICS OF THE ELECTRON-HOLE PLASMA PHASE

TRANSITION IN HIGHLY EXCITED SEMICONDUCTORS*

S.W. Koch

Institut für Theoretische Physik der
Universität Frankfurt
Robert-Mayer-Str. 8
D-6000 Frankfurt-Main, Fed. Rep. Germany

The dynamics of the phase separation in the system of electronic
excitations in highly excited semiconductors is treated within
a hydrodynamical model, taking into account diffusion of the
electron-hole pairs due to the local chemical potential and
generation and recombination of electron-hole pairs. The equation
of motion for the two point correlation function of the local
density of electronic excitations is established and the Fourier
transformed correlation function, i.e. the structure factor, is
calculated.

The dynamics of the phase separation in highly excited
semiconductors, especially the build up of electron-hole droplets
(ehd), has been studied extensively within the framework of
nucleation theory[1-5]. This theory is very succesful in describing
the formation and decay of ehd in indirect gap semiconductors.
For direct gap semiconductors it was shown very recently[6], that
nucleation theory predicts the build up of e-h clusters to be a
nonequilibrium phase transition of second order, due to the very
short life-time of the electronic excitations.

Another approach to the e-h plasma phase transition is
the hydrodynamical approach[7-9]. Within the hydrodynamical
description one treats the system of electronic excitations in
highly excited semiconductors as a spatially inhomogeneous system,
characterized e.g. by a local density field $n(\vec{r},t)$ and a local
velocity field $\vec{v}(\vec{r},t)$. The equations of motion for these variables
are the continuity equation and the Euler equation, respectively.
Due to the finite life-time of the electronic excitations in the

system, the generation of e-h pairs by external laser excitation and
the scattering of e-h pairs with phonons, one has to include in the
equations of motion dissipative terms[7-9] and the corresponding
fluctuations. This constitutes the basic set of Langevin equations
which describes the system of electronic excitations in highly
excited semiconductors.

However, it has been shown[10], that one can derive the
equations of the nucleation theory from the hydrodynamical
equations by integrating the generalized continuity equation. The
equation of motion of the velocity field seems to be less important.
Therefore, to simplify the present treatment an adiabatic
elimination procedure is used for the velocity field. This yields
a generalized diffusion equation for the density of electronic
excitations

$$\frac{\partial n(\vec{r},t)}{\partial t} = M \vec{\nabla}^2 \{-K \vec{\nabla}^2 n + \mu(n)\} + \frac{\bar{n} - n}{\tau} + F_1 + F_2 \qquad (1)$$

Here, the term $-K \vec{\nabla}^2 n$ stems from the surface energy, $\mu(n)$ is the
chemical potential of the e-h pairs and M is proportional to the
mobility. The term $\{\bar{n}(t) - n(\vec{r},t)\}/\tau$ describes pumping and
recombination and F_1 and F_2 are fluctuation terms due to thermal
fluctuations and shot noise, respectively. The second moments
are given by

$$<F_1(\vec{r},t)F_1(\vec{r}',t')> = -2 k_B T M \vec{\nabla}^2 \times \delta(\vec{r}-\vec{r}')\delta(t-t') \qquad (2a)$$

and

$$<F_2(\vec{r},t)F_2(\vec{r}',t')> = \frac{\bar{n} + n}{\tau} \times \delta(\vec{r}-\vec{r}')\delta(t-t') \qquad (2b)$$

Fourier transformation of the linearized mean eq.(1) yields

$$\frac{\partial u(k,t)}{\partial t} = -\omega(k) u(k,t) , \qquad (3)$$

where $u = n - \bar{n}$

and $\omega(k) = M k^2 (K k^2 + \frac{\partial \mu}{\partial n}\Big|_{n=\bar{n}}) + \frac{1}{\tau}$ \qquad (4)

A comparison of the eigenfrequency $\omega(k)$ with the eigenfrequencies
of the full hydrodynamical equations[9] (i.e. generalized continuity
equation and generalized Euler equation) shows, that the previously
mentioned adiabatic elimination of the velocity field is equivalent
to an elimination of that branch of the eigenfrequency spectrum,
which is always positive, i.e. to the elimination of the damped
mode.

The aim of the following calculations is to establish
an equation for the structure factor, which is the Fourier
transformed two points density correlation function. The details
of the derivation will be omitted here. Mainly, the mathematical
treatment follows a theory given by Langer[11], who treated similar
equations for a system with infinite lifetime of the constituent
particles. Therefore, only the main steps of the calculation are
given here. Firstly, one establishes a functional Fokker–Planck
equation, which is statistically equivalent to the Langevin
equation (1).

Then one multiplies this Fokker–Planck equation with
$u(r)u(r')$ and integrates over all configurations u. This procedure
yields an equation of motion for the two point correlation
function

$$S(|\vec{r}-\vec{r'}|,t) = \int Du\ u(\vec{r})u(\vec{r'})\rho\{u\} = <u(\vec{r})u(\vec{r'})> , \qquad (5)$$

where $\rho\{u\}$ is the distribution functional. The Fourier-transform
of $S(\vec{r},t)$ is $S(k,t)$, the structure factor. Its equation of motion
contains also higher order correlation functions. As a first
approximation one may neglect all this higher order correlations and
one ends up with a linear equation for the structure factor

$$\frac{\partial}{\partial t}\ S(k,t) = -\ 2\omega(k)S(k,t) + 2M(k_B T)k^2 + \frac{2\bar{n}}{\tau} . \qquad (6)$$

This equation can be solved easily for various mean densities $\bar{n}$,
i.e. for various generation rates $G(t)$ of the exciting laser. The
generation rate $G(t)$ and $\bar{n}$ are linked by

$$\frac{\partial}{\partial t}\ \bar{n} = -\ \frac{\bar{n}}{\tau} + G(t) \qquad (7)$$

To get preliminary results, eqs.6 and 7 are solved for the
example of CdS at a temperature T = 30 K. The chemical potential
is taken from a microscopic calculation of Rösler and Zimmerman[12]
and shows a Van-der-Waals like instability region. The
generation rate is choosen to stimulate pulsed excitation, with
a small pulse width in order not to allow more than the beginning
of a phase separation in the excited crystal. It is known from
Langers[11] calculations on the spinodal decomposition in metallic
systems, that the linear analysis only yields a qualitative correct
description of the beginning of the phase separation. The results
of the present calculations are summarized in figs. 1 and 2.
Fig.1 shows the generation rate G and the corresponding mean
density (eq.7) as a function of time. In fig.2 the resulting
structure factor is plotted as a function of k for various times.
As long as the mean density $\bar{n}$ is below the critical value $\bar{n}_c$
belonging to the instability point of the chemical potential
$(\partial\mu/\partial n_c = 0)$ the structure factor is very small. For $\bar{n} > \bar{n}_c$, the

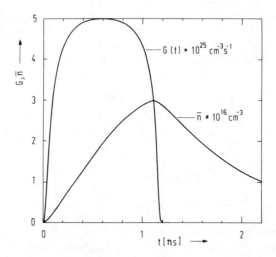

<u>Fig.1</u> : Laser generation rate G(t) of e-h pairs and the
         corresponding mean e-h pair density n̄(t) as functions
         of time for CdS. (τ = 1 ns).

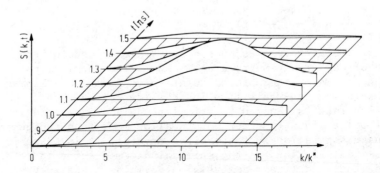

<u>Fig.2</u> : Structure factor S(k,t) for CdS as function of k/k^* at
         vareous times ($k^* = 8.10^4$ cm^{-1}) for the generation rate
         of fig.1.

derivative $\partial u/\partial \bar{n}$ becomes negative and $\omega(k)$ (eq.4) becomes positive
for some k values. Therefore, for this k values $S(k,t)$ grows.
However, the linear analysis yields an unrestricted growth of S
with increasing time. It can easily be shown[11], that this artifact
is prevented by the inclusion of higher order correlations. In
order to stay in the regime where the linear analysis is appropriate,
in the present calculations $G(\underline{t})$ was chosen so that $\bar{n}$ exceeds $\bar{n}_c$
only for a short time. Later $\bar{n}$ decreases again causing a decrease
of $S(k,t)$. The obtained results yield a reasonable description of
the expected behaviour of the structure factor. However, to improve
the calculations it will be necessary to include the higher order
correlations. This will be done in a future work.

In principle, the structure factor could be measured
in highly excited direct and indirect gap semiconductors e.g. by
x-ray scattering experiments. If these experiments are done with
and without simultaneous high excitation of the respective sample
one could get the structure factor of the system of electronic
excitations as the difference spectrum.

In indirect gap semiconductors these experiments could
be done for rather high excitation intensities, i.e. for densities
of the electronic system which belong to the instability region
of the chemical potential ($\partial \mu/\partial \bar{n} < 0$). In this regime a phase
separation in the system of electronic excitations via spinodal
decomposition should take place because the plasma is spontaneously
unstable.

On the other hand, as mentioned in the beginning of this
paper, in direct gap semiconductors the formation of e-h clusters
is expected to be a nonequilibrium phase transition of second
order. Therefore, no potential barrier to nucleation exists (i.e.
no critical droplet size exists) and the phase separation in this
system is assumed to occur for densities inside the instability
regime of the chemical potential. Therefore, the mechanism of
droplet formation in highly excited direct gap semiconductors
should correctly be described by the theory outlined in the
present paper. For the example of GaAs and CdS as typical direct
gap semiconductors, nucleation theory predicts the existence of
very small e-h clusters with a size of about 100 Å[13]. Up to now
these predictions are not experimentally verified, perhaps
because of the extremly small expected cluster size. Measurements
of the structure factor, however, in comparison with a more
elaborated version of the present theory could eventually help
to clarify this situation.

REFERENCES

* Project of the Sonderforschungsbereich Frankfurt/Darmstadt,
  financed by special funds of the Deutsche Forschungsgemeinschaft.

1. For a recent review see : T.M. Rice, J.C. Hensel, T.G. Phillips
   and G.A. Thomas, Solid State Physics $\underline{32}$, 1 (1977).

2. V.S. Bagaev, N.V. Zamkovets, L.V. Keldysh, Soviet Physics-J.
   exper. Theor. Phys. $\underline{43}$, 783 (1976).

3. J.L. Staehli, Physica Status Solidi (b) $\underline{75}$, 451 (1976).

4. M. Combescot, Time Evolution of the Electron-Hole Plasma
   Nucleation : An Analytic Approach, Preprint, Ecole Normale
   Supérieure, Paris (1979).

5. S.W. Koch and H. Haug, Physica Status Solidi (b) $\underline{95}$, 155 (1979).

6. S.W. Koch and H. Haug, Physics Letters $\underline{74A}$, 250 (1979).

7. R.N. Silver, Physical Review $\underline{B17}$, 3955 (1978).

8. M. Combescot, Hydrodynamics of an Electron-Hole Plasma,
   Created by a Pulse, Preprint, Ecole Normale Supérieure,
   Paris (1979).

9. H. Haug and S.W. Koch, Physics Letters $\underline{69A}$, 445 (1979).

10. H. Haug and S.W. Koch in : Dynamics of Synergetics (Ed.
    H. Haken, Springer Verlag, Berlin) p.57 (1980).

11. J.S. Langer, Annals of Physics (New York) $\underline{65}$, 53 (1971).

12. M. Rösler and R. Zimmermann, Physica Status Solidi (b) $\underline{83}$,
    85 (1977).

13. S.W. Koch, to be published.

ELECTRON DENSITY DISTRIBUTIONS IN Ti_2O_3 AND V_2O_3

EXPERIMENT AND THEORY

J. Ashkenazi
Université de Genève, Section de Physique, DPMC
24, Quai Ernest-Ansermet
1211 Geneva, Switzerland

M.G. Vincent[*] and K. Yvon
Université de Genève
Laboratoire de Cristallographie aux Rayons X
24, Quai Ernest-Ansermet, 1211 Genèva, Switzerland

Electron density deformation maps of Ti_2O_3 and V_2O_3 at room temperature have been obtained by Fourier analysis of carefully measured X-ray diffraction intensities from spherical single crystals. They have been compared with theoretical maps based on a band calculation, including self-consistently, the effect of intra-atomic electron-electron interactions. The results agree concerning the different natures of the occupied 3d sub-bands, and the corresponding metal-metal bonds in these substances.

The sesquioxides Ti_2O_3 and V_2O_3 are similar, having both the corundum structure at room temperature (see Fig.1), and undergoing a metal-insulator transition (MIT). However, they possess distinctly different structural and MIT features : The hexagonal c/a ratio at room temperature is anomalously low for Ti_2O_3, and anomalously high for V_2O_3. Ti_2O_3 has a gradual transition, while V_2O_3 has two transitions, a first-order one at 170 K followed by lattice distortion and antiferromagnetism, and a gradual high temperature one. Interesting modifications of these transitions have been found by doping V_2O_3 with Cr or Ti, and Ti_2O_3 with V [1, 2 , 3] .

The theoretical works on these compounds are mostl, divided between two schools. The school of Mott [4] , which attributes the MIT to electron-electron interactions, and the school of Goodenough [5] , which attributes the transitions to

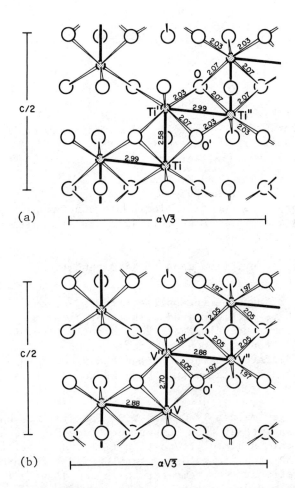

(a)

(b)

Fig.1 : The structure of (a) Ti$_2$O$_3$ and (b) V$_2$O$_3$, viewed
perpendicular to the (11$\bar{2}$0) plane. Inter-atomic distances
are given in Å.

the structural anomalies, with emphasis on directional metal-metal bonds. A band structure calculation of Ti_2O_3 and V_2O_3 [6], followed by a self-consistent calculation of the effect of intra-atomic electron-electron interaction parameters [7], has synthesized ideas from both these schools. The band structures obtained are consistent with Goodenough's model; however, neither the 3d subband arrangement, nor the MIT's could be explained without the effect of electron-electron interactions. This band and bond picture is tested here directly by comparing experimental and theoretical electron density deformation maps (i.e. maps showing the difference between the density distribution in the crystal, and that calculated by a superposition of spherical neutral atoms) of these substances.

The details of the experimental and theoretical methods are published elsewhere [8]. We present here only the essential points.

The measurements were done on spherical crystals of V_2O_3 and Ti_2O_3 of mean radius 79 μm, ground from fragments of single crystals supplied by J.M. Honig. X-ray intensity data were collected with graphite monochromated $AgK\alpha$ radiation ($\lambda = 0.5608$ Å) in two quadrants of reciprocal space on a 4-circle diffractometer out to a limit of 1.36 in $\sin \theta/\lambda$. The data were corrected for absorption and anisotropic extinction. The internal consistency factor between symmetry related reflection was 1.1 %. A least-squares refinement of the positional and thermal parameters and the scale factor using high angle data ($\sin \theta/\lambda > 0.65$ Å^{-1}) converged at R(F) = 1.2 %. The parameters obtained were used to calculate deformation maps based on 75 low angle reflexions in the region $0.0 < \sin \theta/\lambda < 0.65$ Å^{-1}. Errors in the deformation maps were computed as described elsewhere [8].

The theoretical maps were calculated using a cluster of 23 metal atoms and 27 oxygen atoms. Deformation densities due to the 3d band were calculated taking bonding and anti-bonding combinations of directed 3d atomic orbitals, using the results of ref.7. Deformation densities due to the metal oxygen bond were approximated taking spherical atomic oxygen and metal orbitals, with occupations deduced from ref. 6. It is noted that the band structure [6,7] is based on the tight-binding method, with emphasis on the 3d band.

The experimental density deformation maps in two sections are represented in Figs. 2, 3. The first section corresponds to the (11$\overline{2}$0) plane, parallel to the c-axis, and containing the metal atom sites M, M', M" (M = Ti, V) defined in Fig.1. The second section is the "basal plane" perpendicular to the c-axis, and bisecting the M' - M" line. Experimental

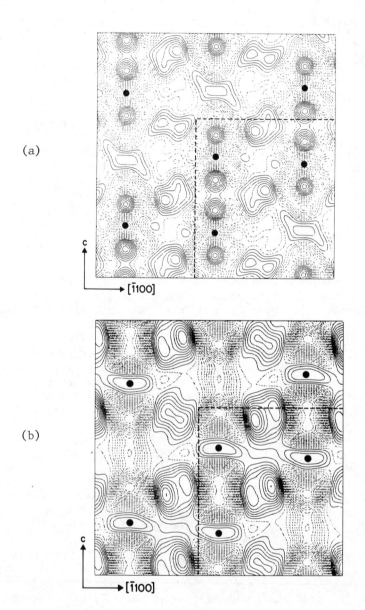

Fig. 2: Experimental deformation density of (a) Ti_2O_3 and (b) V_2O_3 in the (11$\bar{2}$0) section shown in Fig. 1. Contour intervals are at 0.02 eÅ^{-3}. Negative contours are represented as dotted lines starting at 0.0 eÅ^{-3}. The filled circles indicate the metal atom sites, and the square formed by broken lines the limits of the corresponding theoretical map.

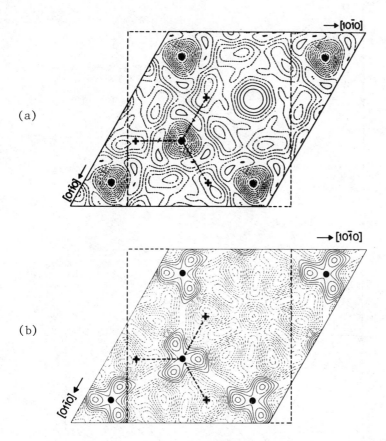

Fig.3 : Experimental deformation density of (a) Ti_2O_3 and (b) V_2O_3 in a section perpendicular to the c-axis bisecting the M' – M" line. The contour intervals are as in Fig.2. The filled circles and the crosses represent the projections of the nearest metal and oxygen atom sites respectively. The square formed by broken lines indicates the limits of the corresponding theoretical map.

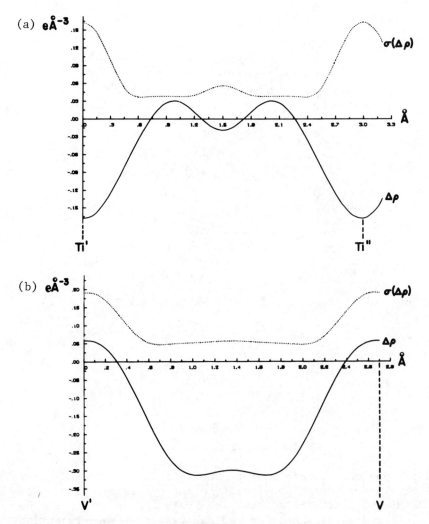

Fig.4 : The deformation density Δρ (solid line), and its error
σ(Δρ) (dotted line) along the M – M' and M' – M" lines in
(a) Ti$_2$O$_3$ and (b) V$_2$O$_3$.

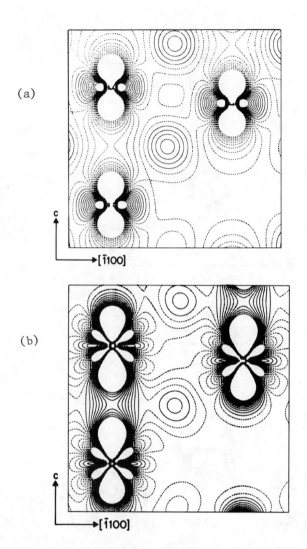

Fig.5 : Theoretical deformation density corresponding to the experimental map shown in fig.2, using the same contour intervals.

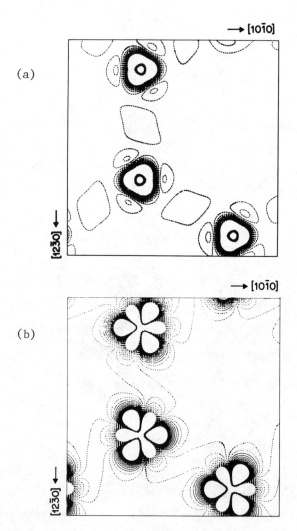

Fig.6 : Theoretical deformation density corresponding to the
experimental map shown in fig.3, using the same contour
intervals.

deformation densities and errors along the M – M' and M' – M" lines
are represented in Fig.4. The theoretical maps corresponding to
Figs. 2, 3 are shown in Figs. 5, 6 respectively.

The experimental and theoretical maps agree with each
other concerning the nature of the metal-metal bonds. The main
disagreements are close to the metal atom sites, where the
experimental errors are large (Fig.4), at regions around the
oxygen atoms, and in the interstitial regions, due to the
theoretical approximations used.

Examining Figs. 2, 5 for Ti$_2$O$_3$, it is clear that the
significant structure along the c-axis corresponds to metal-metal
σ-bonds along the c-direction, or to an occupied bonding a$_{1g}$ band
as found theoretically. For V$_2$O$_3$ we obtain positive deformation
around the V atoms perpendicular to the c-axis, and negative
deformation parallel to it.

Examining Figs. 3, 6 for V$_2$O$_3$, it turns out that the
positive deformations around the V atoms have a shape of a three-
leaf clover, pointing about 180° away from the projections of the
nearest oxygen atom sites. This is consistent with an occupied e$_\pi$
band, as found theoretically, and corresponds to metal-metal
π-bonds between orbitals pointing away from the nearest oxygen
atoms. For Ti$_2$O$_3$ we get in this section negative deformation
triangular wells around the Ti atom sites. There is a disagreement
between theory and experiment concerning the directions of the
triangle corners. In the theoretical map they are pointing towards
the projections of the directions of the closest oxygen atoms,
while in the experimental maps they are rotated, and point towards
some intermediary direction. The reason for this discrepancy may
be the theoretical assumption that the e$_\pi$ band orbitals are somewhat occupied, and that on the other hand, the spherical
approximation used ignores the fact that the 3d orbitals in the
covalent bond are mainly e$_\sigma$ (pointing towards the close oxygen
atoms).

Acknowledgement – The authors thank J.M. Honig for supplying
the single crystals used in this work.

REFERENCES

1. D.B. McWhan and J.P. Remeika, Phys. Rev. B 2, 3734 (1970).

2. G.V. Chandrashekhar, Q.W. Choi, J. Moyo and J.M. Honig,
   Mat. Res. Bull. 5, 999 (1970).

3. J. Dumas and C. Schlenker, J. Phys. (Paris) C 4, C4-41 (1976);
   J. Mag. & Mag. Mat. 7, 252 (1978).

4. N.F. Mott, "Metal Insulator transition" (London, Taylor &
   Francis Ltd., 1974).

5. J.B. Goodenough, Prog. Sol. St. Chem. 5, 145 (1971).

6. J. Ashkenazi and T. Chuchem, Phil. Mag. 32, 763 (1975).

7. J. Ashkenazi and M. Weger, J. Phys. (Paris) C4, C4-189 (1976).

8. M.G. Vincent, K. Yvon, A. Grüttner and J. Ashkenazi,
   Acta Cryst. A 36 (1980) in the press;
   M.G. Vincent, K. Yvon and J. Ashkenazi, Acta Cryst. A 36 (1980)
   in the press;
   J. Ashkenazi, M.G. Vincent, K. Yvon and J.M. Honig, submitted
   to J. Phys. C (1980).

* Present address : Biozentrum der Universität Basel
  Klingelbergstrasse 70 - CH-4056 Basel, Switzerland.

SCF-X$_\alpha$-SW CALCULATIONS OF ELECTRON STATES IN RARE EARTH

ION DOPED CRYSTALS

G.M. Copland and M. Mat Salleh[*]

Queen Elizabeth College

Campden Hill Road, London W8 7AH, U.K.

The SCF-X$_\alpha$-SW method has been applied to the calculations of
electronic states in rare earth ion doped crystals. The
conventional method is modified by use of a second Watson sphere.
For Yb^{3+} and Tm^{2+} in CaF$_2$, the crystal field parameters and
transferred hyperfine interaction parameters have been calculated
using this method. The agreement with experiment is reasonably
good for the hyperfine interactions, but less good for crystal
field interactions. This method is felt to be reasonably promising
for calculations of rare earth electronic states but further work
is necessary to refine the technique.

In recent years, the Self-Consistent Field X$_\alpha$ -Scattered
Wave Method (SCF-X$_\alpha$-SW) has been developed and applied to a variety
of molecular clusters. The basic method has been extensively
reviewed, see for example ref.[1] . This method has certain
computational advantages over other molecular orbital methods,
particularly in dealing with large clusters, and clusters containing
heavy metal ions. The work that we wish to report in this
communication is part of a study of the applicability of the
SCF-X$_\alpha$-SW method to rare earth ion complexes [2,3] .

We report specifically on the system Yb^{3+} in CaF$_2$ and
Tm^{2+} in CaF$_2$ where the rare-earth ion substitutes for the Ca^{2+} ion.
The systems were chosen because there exists extensive spectroscopic
data on these systems, particularly concerning transferred hyperfine interactions and covalency effects [4,5] . These data give a
measure of the degree of electron transfer between the central
rare earth ion and the fluorine ligands. Together with the crystal

265

field splittings these enable us to compare the predictions of the
SCF-X_α-SW calculations of electronic structures with experimental
data.

        The basic method of calculation rests upon a muffin-tin
approximation to the potential in the cluster, which is usually
surrounded by a "Watson sphere" to stabilise the charge of the
complex. The initial atomic potentials from which the muffin tin
potential is derived are calculated using the Herman and Skillman
Hartree Fock Slater computer program [6] with the α parameter for
each atomic species in the cluster taken from the work of Schwartz
[7,8] . The cluster used for the calculations reported here
consisted of the rare earth ion at the centre surrounded by a cube
with fluorine ions at each vortex. The radii of the muffin tin
spheres around each atomic centre were chosen such that spheres
around adjacent centres just touched. The intersphere region was
given a constant potential with an α value taken as the weighted
average of the atomic values for the cluster. The basic parameters
for the calculation are given table I.

        Table I : Parameters for SCF-X_α-SW Calculations

Ion	α	Sphere radius (a.u.)
Yb^{3+}	0.69317	1.8849
F^-	0.73732	2.5749
Intersphere	0.73241	
Outersphere	0.73732	7.0347

        As in the cases of Pr^{3+} in $LaCl_3$ [2] and the ytterbium
garnets [3] , it was found that this simple configuration did not
lead to satisfactory results. It is obvious that the basic cluster
considered above cannot be a satisfactory model of the real
system as it ignores all next nearest and further neighbouring
ions. The surrounding ions will have charge distributions which
significantly overlap the outersphere drawn around the basic
complex, thus modifying the potential in the intersphere region.
These effects can be largely accounted for by the simple device
of introducing a second "Watson Sphere" centred on the centre of
the cluster whose radius is chosen to just touch muffin tin spheres
drawn around the first nearest neighbour Ca ions. This inner
Watson sphere is thus chosen to have a radius of 3.2648 a.u. The
outer sphere charge is taken to be equal in magnitude but opposite
in sign to the overall charge of the cluster (5e for Yb F_8 and 6e
for Tm F_8) and the inner sphere charge is the opposite charge to
the total ionic charge from the nearest neighbour Ca ions which
falls within the outersphere. For the geometry considered here,

this is close to 12e, and thus the inner sphere charge was taken to be -12e.

Use of this extra sphere enabled the computation to find the appropriate number of rare-earth orbitals to be bound and provided a realistic picture of the electronic potential in the cluster. A similar approach has also been reported by Weber et al. [9].

Calculations were performed using spin-polarised and spin-unpolarised symmetrised orbitals on the rare-earth ion and self-consistency was assumed to have been reached when the relative change in potential between iteration cycles was less than 5×10^{-2} at any point in space. Frozen cores were used for the rare-earth ions, only 4f, 5s and 5p orbitals were taken to be "thawed". Energy levels and charge distributions were calculated for all orbitals of the rare earths and fluorine ions.

From these results crystal field parameters and hyperfine interaction parameters were calculated. A detailed discussion of the derivations of these parameters from the SCF-X$_\alpha$-SW calculations is not possible in a short communication of this nature and will be presented more fully elsewhere. We merely report the important results here. In Table II the predicted crystal field parameters are shown, together with other calculations using different techniques and experimental values.

Table II : Crystal Field Parameters for Tm^{2+} and Yb^{3+} in CaF_2 (in cm^{-1}).

	Tm^{2+}		Ref.	Yb^{3+}		Ref.
	b_4	b_6		b_4	b_6	
Experimental	45.8	5.1	[10]	44.7	6.9	[18]
M-O Calculation	47.3	5.7	[4]	68.0	8.5	[5]
Point Charge Model	14.4	0.8		16.1	0.6	
Previous X$_\alpha$ prediction	-31	-3.6	[11]	-	-	
This work	18.5	6.8		11.6	3.1	

The crystal field hamiltonian is defined to have the form :

$$V_c = 33b_4[C_0^{(4)} + \sqrt{\frac{5}{14}}(C_{-4}^{(4)} + C_4^{(4)})] - \frac{143}{60}b_6[C_0^{(6)} - \sqrt{\frac{7}{2}}(C_{-4}^{(6)} + C_4^{(6)})]$$

(13)

Only cubic sites have been considered here, non-cubic sites arising from charge compensation for the Yb^{3+} have not been considered.

       In Table III we present a summary of measured and calculated hyperfine interaction parameters for these systems. Definitions of the parameters may be found in [14].

Table III : Transferred Hyperfine Interaction Parameters (MHz)

	Tm^{2+}	Yb^{3+}	Ref.
A_s(calc)	-3.33	4.09	This work
A_s(expt)	2.58	1.67	[15]
A_p(calc)	1.42	8.91	This work
A_p(expt)	3.18	7.81	[16]

       From the results summarised in Tables II and III, we can draw some conclusions as to the validity of this method for investigating rare earth ion electronic states in crystals. Other work reported elsewhere on Eu^{2+} and Gd^{3+} in CaF_2 [17] and more detailed work on the systems described in this paper support the conclusions that the technique can produce reasonable values for transferred hyperfine interactions and covalent spin transfer parameters. The level of agreement between experimental and calculated values is felt to be satisfactory at this stage. The crystal field values however are less satisfactory. The calculations appear consistently to underestimate b_4 while giving better results for b_6. The work of Baker [4], [5] gives a better overall fit but this is based on a semi-empirical MO approach using covalency parameters. This does not perhaps give as basic a picture of the electronic structure as the SCF-X_α-SW model used here. It is perhaps hardly surprising that the crystal field parameters are less satisfactorily described by this method as their calculation is based on the small differences between large energies of 4f orbitals, which are very sensitive to the model parameters such as interionic spacing and muffin tin radii. The charge distributions and charge transfer between spheres is much less sensitive to these parameters and it is from these factors that the hyperfine interaction parameters were derived. No attempt was made in these exploratory calculations to adjust the interionic spacings to "optimise" the agreement between experiment and the calculation although it is recognised that crystallographic data cannot give accurate values for the spacings in doped crystals.

A more detailed account of this work will be published elsewhere but the authors feel that these results are sufficiently encouraging to warrant publication at this stage and to lead to further work in these systems with the SCF-X$_\alpha$-SW method.

One of us (MM) acknowledges the Universiti Kebangsaan Malaysia for financial support.

REFERENCES

*Present Address : Physics Department, National University of Malaysia, Kuala Lumpar, Malaysia.

1. N. Rösch, Electrons in finite and infinite structures. Eds. P. Phariseau and L. Scheire (Plenum : New York 1977).

2. M. Mat Salleh and G.M. Copland, Journal of Physics C : Solid State Physics 11 L777 (1978).

3. C.J.P. Graf and G.M. Copland, (Proceedings of 1980 Annual Conference of Condensed Matter Division of the European Physical Society).

4. J.M. Baker, Journal of Physics C : Solid State Physics 10 3323 (1977).

5. J.M. Baker et al., Journal of Physics C : Solid State Physics 11, 3071 (1978).

6. F. Herman and S. Skilman, Atomic Structure Calculations (Prentice Hall : Englewood Cliffs 1963).

7. K. Schwartz, Physical Review B5, 2466 (1972).

8. K. Schwartz, Theoretica Clinica Acta 34, 225 (1974).

9. J. Weber, H. Berthou and C.K. Jørgensen, Chemical Physics Letters 45, 1 (1977).

10. Z.J. Kiss, Physical Review 127, 718 (1962).

11. J.L. Alves and M.L. De Siqueira, International Journal of Quantum Chemistry 11S 75 (1977).

12. J.M. Baker, W.B.J. Blake and G.M. Copland, Proceedings of the Royal Society A309, 119 (1969).

13. J.M. O'Hare, J.A. Detrio and V.L. Donlan, Journal of Chemical Physics 51, 3937 (1969).

14. J.M. Baker, Journal of Physics C : Solid State Physics 1, 1670 (1968).

15. R.G. Bessant and W. Hayes, Proceedings of the Royal Society A235, 430 (1965).

16. U. Ranon and J.S. Hyde, Physical Review <u>141</u>, 259 (1966).

17. M. Mat Sallah, Ph.D. Thesis, University of London 1979.

VALENCE RELAXATION IN THE 2p CORE PHOTOIONIZATION OF

MIXED VALENCE TmSe CRYSTAL

A. Bianconi, S. Modesti
Istituto di Fisica, Università di Roma, Roma, Italy

M. Campagna, K. Fischer
Institut für Festkörperforschung der KFA
D-5170 Jülich, West Germany

S. Stizza
Istituto di Fisica, Università di Camerino
62032 Camerino, Italy

The L_3 X-ray absorption spectrum of TmSe has been measured using synchrotron radiation emitted by the storage ring Adone. Evidence of homogeneous mixed-valence of TmSe and of the core hole induced relaxation has been found.

One foundamental point in core level photoionization is the effect of relaxation induced by the core hole. These effects are expected to be important in core level spectroscopy of all rare earths especially of stable chalcogenides and of compounds in the homogeneous mixed valence state[1]. A recent paper[2] has shown that the valence change induced by the creation of a hole in a core level has large effects on the Auger spectrum of $YbAl_3$. We have studied the core transitions from the $2p_{3/2}$ core level of TmSe to study the effect of core relaxation in the total photo-ionization cross-section.

It is well known that TmSe is in a homogeneous mixed valence state where the Tm ions fluctuate between a $4f^{13}5d^0$ and $4f^{12}5d^1$ configuration[3]. From bulk measurements for stoichiometric samples the mixing-valence-ratio $r = Tm^{3+}/Tm^{2+} \simeq 3$ has been determined[4].

Following the photoionization of the $2p_{3/2}$ core level in the "adiabatic limit" a fully relaxed valence band is expected. The effect of the core hole on the valence 4f 5d states can be estimated by using the (Z+1) analogy, i.e. the valence orbitals will be the same as the following Z+1 atom having a Z+1 positive charge localized in the nucleus. For stable Tm^{3+} configuration ($4f^{12}$) the $4f^{13}$ level will drop below the Fermi level after switching on the core hole. The Tm atom starts in $4f^{12}5d^1$ configuration of the Tm^{3+} ion and in the fully relaxed state it changes to the $4f^{13}$ configuration of the Tm^{2+} ion. Therefore it will no longer be possible to clearly distinguish the two valence states of the initial homogeneous mixed valence state in core level photoionization measurements. In fact the $M_{4,5}VV$ Auger spectrum of $YbAl_3$ does not show any effects due to valence fluctuations. The fully relaxed state is reached if the characteristic response time of the valence electrons is shorter than the lifetime of the excited state (the measuring time).

         In the "sudden" regime the characteristic time of the measurement is so short that the valence electron has no time to relax. In XPS measurements of core levels where a fast photo-electron is excited the "sudden" approximation is expected to be a good approximation although evidence for partial relaxation in rare earth metals and Ce compounds especially has been found[5,6].

         The experiments have been performed at the synchrotron radiation facility PULS using the X-rays emitted by the storage ring Adone. The X-rays are monochromatized by a Si(220) single channel-cut crystal at 17 m from the source. The resolution was about 1.5 eV at the L_3 threshold of Tm at 8650 eV. The samples were single crystals of TmSe in thickness d = 20 μ.

         The L_2 spectrum has been measured and we have found that except for one factor it was the same as the L_3 spectrum. The incident photon flux Io and the transmitted flux I were measured and the photoabsorption coefficient α = 1/d ln Io/I is plotted in Fig.1 as a function of the photon energy. The zero of the energy scale has been taken at the first maximum of the absorption derivative.

         The L_3 spectrum of rare earths is dominated by a "giant resonance" near the threshold due to transitions to d states near the Fermi level[7]. This resonance is partially due to the high density of d-states but the large oscillator strength should be explained by the local character of the final states confined in the core region by the high centrifugal potential acting on the d-electrons. Such large atomic resonance covers all the other molecular or solid state effects which determine the X-ray Absorption Near Edge Structures. The Fano like-shape[8] well describes this transition to a quasi bound state superimposed on a continuum. The spectrum can be fitted by an asymmetric lorentzian and an arctan.

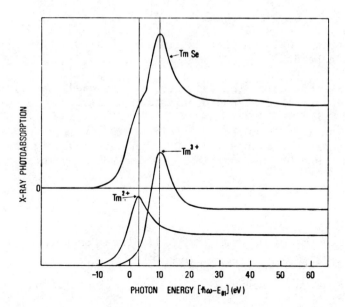

Fig.1 : The L$_3$ X-ray photoabsorption spectrum of single crystal
TmSe. In the lower part the individual contributions of
the Tm^{3+} and Tm^{2+} ions are plotted. The pre-edge continuum
background has been subtracted.

       The curve in figure 1 has been deconvoluted in two
curves due respectively to Tm^{3+} and Tm^{2+} ions. The intensity ratio
between the spectrum due to Tm^{3+} and Tm^{2+} is r = 1.6 to be compared
with the expected mixing valence ratio 3. The lifetime of the
excited state is estimated from the full width of the "giant
resonance" (Γ = 7.40 $\pm$ 0.20 eV) to be T = 0.9 × 10^{-16} sec.

       The evidence of transitions due to Tm^{3+} ions indicates
that we are not in a fully "adiabatic" regime also if valence band
relaxation is present as it is shown by the weak intensity of the
Tm^{3+} component. Such an intermediate regime between the "sudden"
and the "adiabatic" regime is expected if the characteristic time
of the photoexcitation is close to the response time of valence
electrons. Therefore we have an extimate for the relaxation time
in TmSe $\tau \gtrsim 0.9 \times 10^{-16}$ sec.

       The lineshape of the resonances is determined by the
autoionization channels of the excited states like the

$2p^5(4f^{13}5d^0) \, \varepsilon d^* \rightarrow 2p^6(4f^{12}5d^0)$ + photo electron which is the same final state in continuum photoexcitation of a 4f electron in the d-conduction band.

In conclusion the data show an intermediate relaxation regime in the core level photoionization of TmSe.

Thanks are due to the staff of the project for utilization of synchrotron light PULS and particularly to Dr. F. Comin and Dr. S. Mobilio for experimental help.

REFERENCES

1. G.K. Wertheim and M. Campagna, Sol. State Commun. 26, 553 (1978).

2. W.F. Egelhoff, G.G. Tibbetts, Phys. Rev. Lett. 44, 482 (1980).

3. M. Campagna, G.K. Wertheim, E. Bucher, Structure and Bonding 30, 99 (1976).

4. D. Botlogg, A. Schlegel, P. Wachter, Physica 86-88B, 229 (1977).

5. F. Crecelius, G.K. Wertheim, D.N.E. Bucham, Phys. Rev. B18, 6519 (1978).

6. J.C. Fuggle, R. Lässer, R. Beyss, M. Campagna and F. Hulliger, to be published.

7. S. Nakai, C. Sugiura, S. Kunii, T. Suzuki, Jpn. J. of Appl. Phys. 17, Suppl. 17-2, 197 (1978).

8. U. Fano, J.W. Cooper, Rev. Mod. Phys. 40, 441 (1968).

OCCUPATION POTENTIAL VERSUS COULOMB CORRELATION ENERGY

IN 4-f INSULATORS

F. López-Aguillar, J. Costa-Quintana, and J.S. Muñoz

Departamento de Electricidad y Electrónica
Universidad Autónoma de Barcelona

Bellaterra – Barcelona, Spain

Hufner and Werteim's estimate the correlation energy between $4f^n$ and $4f^{n+1}$ configurations by photoemission experiments to be of the order of 6 eV while others give a value of $\sim$10.5 eV. We have established a method based on the energy differences per unit cell with two configurations as a Coulomb correlation, from the band structure and an estimation of the collective effects by perturbative methods which by introducing the U correlation in the MT potential allows us to find the excited states showing some hybridation and deslocalization and a widening of 4f bands. This broadening could explain the discrepancy observed in the above experimental results.

In the last seven years there has been an increasing interest in the 4f excited states in metals as in semiconductor and insulator compounds, from a theoretical and experimental point of view. This intererst arises because many magnetic and conduction properties stem from the 4f level position.

Herbst et al.[1,2] have a method to find 4f states and the position of the occupied 4f levels in metals computing the energy difference per unit cell that is needed to change the crystal valence using the 5d 6s band as an intermediate stage between Δ_+ and Δ_-. The correlation in the solid is replaced by that of the renormalized atom since the 4f states are localized. This requires a knowledge of ionic states and the 5d 6s band structure.

Recently Hedén et al.[3,4] and Hufner and Wertheim[5] have
published estimates of the Coulomb correlation energy from X-Ray
photoemission measurements for 4f electronic states to be of the
order of 6 eV in metals and semiconductors.

An indirect method to determine Coulomb correlation
energy in semiconductors have been given by Batlogg et al.[6]. They
assume that U (Coulomb correlation) might be associated with the
difference between successive ionization potentials in the same
atom and that such potentials behave linearly respect to the number
of electrons. As this method deals with free ions crystal effects
as screening, correlations and Madelung potentials are neglected.

Lang et al.[7] have determined the position of the 4-f
states above and below the Fermi level by a combined procedure
of Bremsstrahlung isochromat spectroscopy and X-Ray Photoelectron
spectroscopy. There seems to be large differences (80 %) from
Hufner and Wertheim's results for some metals and yet a little
larger for semiconductors.

We have established a method which gives results in
agreement with Lang et al. and Herbst's but under some assumptions
could also explain Hufner and Wertheim's data. Our model differs
from Herbst's in that the band energies include the f band states.
We also have taken into account short range interactions in a
Hubbard multiband expansion. It seems too that it will be
applicable to intermetalic alloys as well as to 4f or 5f oxides
compounds.

In this method two many body hamiltonians[8] are defined
for two crystal configurations, one stable and the other with an
electronic transfer (one per unit cell) between 2N ions. Coulomb
correlations are determined by the energy differences of the
ground states which are estimated by means of perturbation theory :
the perturbative part being exact up to 2nd order and from 3rd order
by temporarily ordered diagrams, whose intermediate states include
only scattering processes among particles.

The crystal binding energy is given by[9]

$$E = <H_o> + <H_{int}> + \sum <\mu|V_{loc}|\mu><n_\mu>$$

where $<H_o>$ is the total energy from band structure,

$$<H_{int}> = \sum_{\lambda\mu} (R_{\lambda\mu\lambda\mu} - R_{\lambda\mu\mu\lambda}) <n_\lambda> <n_\mu>$$

R being a matrix represented in space

$$(|f_1\sigma> + |d_1\sigma'>) \otimes (|f_2\sigma''> + |d_2\sigma'''>)$$

satisfying $R = (I-GV)^{-1}V$ and matrix elements of V are

$$\langle ij|V|st\rangle = \int \psi_i^*(\vec{r}_1) \, \psi_j^*(\vec{r}_2) \, \frac{1}{|\vec{r}_1 - \vec{r}_2|} \, \psi_s(\vec{r}_1) \, \psi_t(\vec{r}_2) d\vec{r}_1 d\vec{r}_2$$

the ijst indexes stand for wave functions centered in the atoms with symmetry compatible with its coordinations respect to neighbouring ions.

Collective effects are introduced in G, a functional matrix whose elements depend on band structure, and it is of the form

$$G_{\lambda\mu}(\omega) = (1 - \langle n_\lambda \rangle)(1 - \langle n_\mu \rangle) \sum_{n=o}^{\infty} (-1)^n$$

$$\int \frac{N_\lambda(e)de}{(\omega + e)^{n+1}} \int N_\mu(e') \, e'^n \, de'$$

G is diagonal in 576 dimension space, and (e) stands for unoccupied states density[10].

Our results agree with Herbst's calculations. We believe, however, that with our model Hüfner and Wertheim's results could also be explained if a new choise of initial conditions is made.

In this paper two solutions are given : the first one is strictically collective and the second within the context of band structure. In both cases the most important point as the Coulomb correlation is concerned is the hybridation of the 4f states with more extended ones. This hybridation affects both valence as conduction bands giving a pattern similar to that stemmed from mixed valence compounds since to consider some hybridation may be interpreted as an ionic co-existence of several valence states. The f-d hybridation determines the ratio of occupation of the site lattice by different oxidation degree ions.

In our model the alteration of Coulomb correlation within this hybridation process is carried out by a change in the representation space of the V and R matrices and the corresponding modified G. The space representation may now be written

$$\{|(a_1^\lambda f_1^\lambda + b_1^\mu d_1^\mu)\sigma\rangle + |(a_1'^\nu f_1'^\nu + b_1'^\xi d_1'^\xi)\sigma'\rangle\}$$

$$\otimes \{|(a_2^\alpha f_2^\alpha + b_2^\beta d_2^\beta)\sigma''\rangle + |(a_2'^\gamma f_2'^\gamma + b_2'^\delta d_2'^\delta)\sigma'''\rangle\}$$

indexes 1,2 refer to particle states 1,2 while band indexes are

denoted by gree k. subindexes and the σ's are spin variables.
A matrix (A) to change the base from no-hybridized to hybridized
defines the process in such a way that

$$A_1 \otimes A_2 = B$$

$$R_h = (I - V_h G')^{-1} V_h = B^{-1} (V^{-1} - G')^{-1} B$$

G' being diagonal but now it corresponds to a new base of functions
and it has identical form to the no-hybridized ones. Matrix B
remains defined by hybridation coefficients of the new basis.

Results of the binding energy taking into account this
space are necessarily reduced depending on f-d hybridation
coefficients. Thus, shared occupation of $\langle n_f \rangle | f \rangle + (1 - \langle n_f \rangle) | d \rangle$
with $\langle n_f \rangle \approx 0.7$ cuts the Coulomb correlation down to 20 %. To
achieve a cut of 40 % which is the discrepancy observed in Eu
measurements by Hüfner and Lang a mixing degree of $\langle n_f \rangle \approx 0.5$
should be needed. In other words a change in the Coulomb correlation
produced by hybridation bring us near Hüfner and Wertheim and
Batlogg et al. results but we think that only hybridation, does
not explain the above discrepancies.

The second solution arises from the framework of band
structure including the effects of Coulomb correlation as an
additional potential that acts on the unoccupied f states, that
is, collective effects are included in the particle states. This
is accomplished adding a new potential term to crystal potential
models. This new term has the form

$$\Sigma_\ell | f_\ell > u_\ell < f_\ell | \tag{1}$$

where the ℓ subindex acts on the empty states of the stable
configuration and U_ℓ are the corresponding Coulomb correlation.
At this stage, as we need to describe the effects of such a term
we introduce the concept of occupational potential. We call
occupational potential the energy difference between an occupied
f state and the first unoccupied 4f, in other words this is the
difference between the correlation energy and the band width of
the unoccupied f states.

This potential (1) leads to a localization in energy of
the unoccupied f-state concomitant with the more spreaded 5d 6s
states resulting in a larger probability of f-d-s-hybridation
that implies a charge deslocalization. Thus, the relation between
maximum ($r \simeq 0.35$ a.u.) and minimum charge density points at the
ionic radii distance ($r = 2.7$ a.u.) changes from 2.7×10^4 in
occupied states to 7.5×10^2 in unoccupied ones. In addition the
radii of maximum charge for the unoccupied states being now

$r \simeq 0.65$ a.u. Therefore, we find widening (≈ 0.3 Ry) in the energy interval of the conduction band where the f orbitals have influence.

The above potential gives rise to an energy difference of $\sim U_\ell$ between the top of 4f occupied and unoccupied band, that in the case of Europium trioxide is of 0.8 Ry. The occupation potential ($U - \delta_{4f} = 0.5$ Ry) means a decrease of 40 % in the position of the unoccupied states, that could explain the observed discrepancies.

Finally, we may conclude that in order to identify quantitatively the Coulomb correlation energy and our occupational potential the separation between the top of 4f occupied band and the conduction band has to be larger or equal to the energy difference per unit cell needed to change the valence state in the whole crystal, or the collective effects due to hybridation be almost equal to the empty f band width.

Acknowledgement   -   One of us (J.C.Q.) wishes to thank the Ministry of Universities and Research for financial help.

REFERENCES

1. J.F. Herbst, R.E. Watson and J.W. Wilkins, Physical Review B, 17, 3089 (1978).

2. J.F. Herbst, Physical Review B, 21, 427 (1980).

3. P.O. Hedén, H. Löfgren and S.B.M. Hagström, Physical Review Letters, 26, 432 (1971).

4. ----------, Physica Status Solidi B, 49, 721 (1972).

5. S. Hüfner and G.K. Wertheim, Physical Review B, 7, 5086 (1973).

6. B. Batlogg, E. Kaldis, A. Schegel and P. Wachter, Physical Review B, 12, 3940 (1975).

7. J.K. Lang, Y. Baer and P.A. Cox, Physical Review Letters, 42, 74 (1979).

8. J. Costa-Quintana and F. López Aguilar (to be published).

9. V_{loc} is a local potential which has already been taking into account in the band structure calculation; is determined by means of :

$$\langle i|V_{loc}|j\rangle = \sum_\mu (\langle i\mu | \frac{1}{r_{1,2}} | j\mu\rangle - \langle i\mu | \frac{1}{r_{1,2}} | \mu j\rangle)\langle n_\mu\rangle$$

μ-index refers to occupied f/d states.

10. ω is twice the energy difference between empty and filled bands.

ELECTRONIC PROPERTIES OF THE INTERMETALLIC SEMICONDUCTOR CsAu

F. Meloni[*] and A. Baldereschi

Laboratoire de Physique Appliquée

EPF-Lausanne, Switzerland

The intermetallic compound CsAu is studied by the self-consistent local-density pseudopotential method based on hard-core Cs and Au pseudopotentials adjusted to atomic spectra and valence wavefunctions. CsAu is predicted to be a semiconductor with an indirect energy gap of 1.3 eV and with the lowest direct threshold at 3.0 eV. The analysis of the valence electron charge density shows that the crystal ground state corresponds to the ionic configuration $Cs^+ Au^-$.

The physical properties of the alkalinoble metal compounds have been investigated in recent years and show very interesting features. RbAu and CsAu have a special place among the compounds of this family since they are intermetallic semi-conductors where gold plays the role of the anion (1,2). The remaining alkaligold compounds LiAu, NaAu and KAu have metallic properties.

CsAu is by far the most studied among the alkali-noble metal compounds and has been investigated in both the liquid (3,4) and solid phases. It crystallizes in the cubic CsCl structure with a = 4.263 Å (1). According to optical (1) and photoemission data (2,5,6) CsAu is a semiconductor with relatively high ionicity. Its electronic energy bands have recently been studied by the APW (7) and the relativistic KKR (6) methods. Both calculations confirm the semiconducting behaviour of this compound but they differ quantitatively from each other even with respect to the value of the lowest energy gap. Here we report a self-consistent local-density pseudopotential calculation which allows a satisfactory interpretation of all the available experimental data.

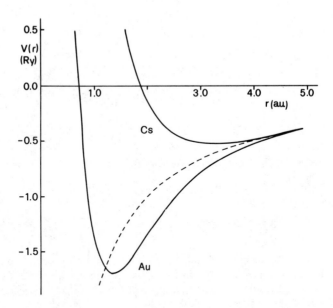

Fig.1 : Atomic hard-core model potentials of Cs and Au. The
broken line represents the Coulomb potential -2/r.

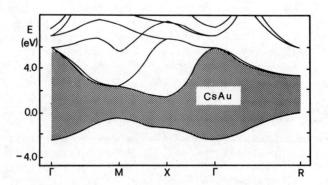

Fig.2 : Self-consistent local-density pseudopotential energy
bands of CsAu.

We first construct local hard-core atomic pseudopotentials
of Cs and Au which reproduce the lowest s and p experimental atomic
energy levels (8) and the corresponding calculated valence-electron
wavefunctions outside the core region (9). The analytic expression
of our model bare potentials is

$$V(r) = -\frac{2}{r}(1 - e^{-\alpha r}) - B\, e^{-\beta r} + A\, e^{-\alpha r}$$

where atomic units are used. The first term on the right side
represents the Coulomb potential energy of the valence electron
and is truncated in the core volume in order to avoid unnecessary
divergencies. The parameter α is chosen so that $1/\alpha$ is the radius
of the atomic core. The second term represents the additional
attraction due to the finite dimension of the core and contains
both electrostatic and exchange interactions between valence and
core electrons. The values of the parameter B and β which appear
in this term are obtained from a fit to atomic Hartree-Fock
potentials (9). Orthogonality to core states is accounted for by
the last term. The parameter A measures the strength of this
repulsion and is determined by fitting atomic energy levels and
the position of the maximum of 6s Hartree-Fock valence function
(10). The resulting atomic parameters expressed in atomic units
are $A = 45.0$, $B = 20.0$, $\alpha = 2.3$, $\beta = 1.6$ for Au and $A = 24.0$,
$B = 2.7$, $\alpha = 1.5$, $\beta = 1.0$ for Cs.

The atomic model potentials are represented in Fig.1.
In agreement with the electro-negativity values $\chi_{Cs} = 0.7$ and
$\chi_{Au} = 2.4$ (11) we find that the gold potential $V_{Au}(r)$ is
considerably more attractive than $V_{Cs}(r)$ and this explains why Au
plays the role of the anion in CsAu. From an electronic point of
view the difference between V_{Cs} and V_{Au} or between χ_{Cs} and χ_{Au}
is due to the presence of a filled $5\,d^{10}$ shell in the latter. A
valence s or p electron experiences a $\frac{-2}{r}$ Coulomb potential for
$r > 4$ au in both Cs and Au. At shorter distances from the nucleus
the electron experiences core effects. In the case of Cs, it is
strongly repelled by the 5s or 5p core electrons to which it must
be orthogonal. In the case of Au, it first overlaps the 5 d
electrons and feels an additional attraction due to incomplete
nuclear screening before being strongly repelled by the 5s or 5p
core electrons at distances $r < 2$ au. The atomic pseudopotential
difference shown in Fig.1 is directly related to the semiconducting properties of CsAu.

The energy bands of CsAu are then calculated by the self-
consistent local-density pseudopotential method. The bare crystal
potential is constructed from a linear combination of the atomic
model potentials described above. The electron-electron interactions

are taken into account by the local-density approximation based on Slater exchange with coefficient 1 and on Wigner correlation.

The resulting self-consistent energy bands are reported in Fig. 2 and show that CsAu is a semiconductor with a minimum indirect gap of 1.3 eV between the top of the valence band at $R = \frac{2\pi}{a} (\frac{1}{2}, \frac{1}{2}, \frac{1}{2})$ and the bottom of the conduction band at $X = \frac{2\pi}{a} (\frac{1}{2}, 0, 0)$. The energy bands give a minimum direct energy gap at 3.03 eV at $M = \frac{2\pi}{a} (\frac{1}{2}, \frac{1}{2}, 0)$. We notice, however, that there are several points in the Brillouin zone where the direct energy gap is only slightly larger. For example it is 3.10 eV at R, 3.11 eV at X and is smaller than 3.20 eV along the M-X axis. Since we have neglected the spin-orbit interaction which will modify somewhat the lower p-like conduction bands we cannot make a definite statement on the exact location of the lowest direct gap.

The optical data (1) indicate a strong absorption threshold of excitonic nature at 2.5 eV and a somewhat weaker absorption at lower energies. According to our calculation we interpret the low-energy "weak" absorption as due to indirect transitions whose oscillator strength is probably enhanced by the presence of defects in the sample and the structure at 2.5 eV as the lowest direct exciton. Our results provide an upper bound of 0.5 eV for the binding energy of the latter. In fact considering the spin-orbit splittings in the conduction bands the direct inter-band threshold will be less than 3.0 eV. Finally we mention that the calculated width of the CsAu s-like valence band is 2.7 eV which compares well with the value 2.6 eV obtained by photo-emission measurements (2,5).

The bonding type in CsAu is illustrated in Fig.3 by the contour plot of the electronic valence charge density in the diagonal (110) plane. The valence electrons are all concentrated around Au thus confirming the ionic $Cs^+ Au^-$ configuration suggested by the electro-negativity values and by the core shifts recently observed in photoemission measurements (2). Fig. 3 shows an important anion-anion contact along the cubic axis and from this we derive an ionic radius of Au^- of $1/2$ a = 2.13 Å. The presence of anion-anion contacts in CsAu is also indicated by the fact that the cation-anion distance in RbAu is only 0.14 Å smaller than in CsAu (1) while the Rb^+ radius is 0.21 Å smaller than that of Cs^+ (11).

We are at present extending our study of the electronic properties of alkali-noble metal compounds to RbAu. Since the electronegativity of Rb (χ_{Rb} = 0.8) is only slightly larger than

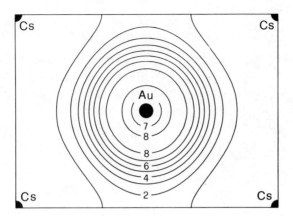

Fig.3 : Valence electron charge density of CsAu in the diagonal
        (110) plane. Units are electrons / unit cell.

that of Cs the electronic properties of RbAu are expected to be
similar to those of CsAu except for a slightly smaller ionicity.
Preliminary self-consistent pseudopotential energy bands of RbAu
confirm the above statement and show that RbAu is an indirect
semiconductor with an energy gap of 0.6 eV and corresponds to
the Rb^+Au^- ionic ground-state configuration.

REFERENCES

* Also GNSM-CNR, Istituto di Fisica, Università di Cagliari,
  Italy.

1. W.E. Spicer, A. Sommer and J.G. White, Phys. Rev. 115, 57 (1959).

2. G.K. Wertheim, C.W. Bates Jr. and D.N.E.Buchanan, Solid State
   Commun. 30, 473 (1979).

3. F. Hensel, Adv. Phys. 28, 555 (1979).

4. J. Robertson, Solid State Commun. this issue.

5. W.E. Spicer, Phys. Rev. 125, 1297 (1962).

6. H. Overhof, J. Knecht, R. Fisher and F. Hensel, J. Phys. F 8,
   1607 (1979).

7. A. Hasegawa and M. Watabe, J. Phys. F 7, 75 (1977).

8. C.E. Moore, Atomic Energy Levels (US Govt. Publishing Office,
   Washington, 1949).

9. C. Froese-Fisher, Atomic Data 4, 301 (1972) for Cs and J.P. Desclaux, Atomic Data 12, 311 (1973) for Au.

10. W. Andreoni, A. Baldereschi, F. Meloni and J.C. Phillips, Solid State Commun. 25, 245 (1978).

11. L. Pauling, The Nature of the Chemical Bond (Cornell Univ. Press, Ithaca, 1960).

THE PRESSURE DEPENDENCE OF THE DENSITY OF STATES AND

THE ENERGY GAP OF TRIGONAL Te

J. von Boehm and H. Isomäki

Electron Physics Laboratory and Department
of General Sciences, Helsinki University of Technology

SF-02150 Espoo 15, Finland

Using the recent pressure dependent self-consistent symmetrized
orthogonalized-plane-wave (SOPW) results (general non-muffin-tin
X_α potential with $\alpha = 1$, 235 SOPWs at each of symmetry points
Γ, A, K, H, K' and H') the densities of states (DOS) at the
pressures 0, 8 and 38 kbar are calculated for trigonal Te. The
quadratic Lagrangian interpolation of the energies evaluated in
the final self-consistent potential with 201 OPWs at 131 symmetry
independent k-points and the Gilat-Raubenheimer method are used.
Our DOS agrees closely with the DOS of Starkloff and Joannopoulos
at 0 and 8 kbar and with the X-ray photoemission results of
Schlüter et al. The pressure dependence of the transitions
associated with smallest energies is also investigated.

Trigonal Te crystallizes into parallel helical chains
arranged in a hexagonal array. This uncommon structure is quite
anisotropic; the intrachain bonding is strong whereas the inter-
chain bonding is relatively weak. The dispersion of the energy
bands is therefore somewhat weaker in the directions perpendicular
to the chains. Applied pressure forces the chains closer together
which increases the valence electron density between the chains
and broadens the bands. Te is known to undergo a semiconductor-
metal transition at a pressure of $\approx$ 40 kbar[1-7]. However, there
has been some uncertainty about the location of the minimum
optical gap under pressure[8] as well as about the driving mechanism
and the exact pressure of the transition[1-7,9]. To study the effects
caused by pressure we present the behaviour of the bulk of the
bands of Te up to 38 kbar by means of the density of states (DOS).
The question of the location of the minimum optical gap is studied
by comparing the pressure dependence of the smallest transition

287

energies with each other and with experiments.

As the starting point we used the recent self-consistent symmetrized orthogonalized-plane-wave (SCSOPW) results by Isomäki et al.[10]. The SCSOPW calculations were performed within the X_α scheme (α = 1 and 2/3) using the general non-muffin-tin potential and 235 basis functions at each of the high-symmetry points Γ, A, K, H, K' and H' (for a detailed description of the SCSOPW method see Ref.11). The eigenvalues for the DOS are calculated in the final SCSOPW α = 1 potential with 201 orthogonalized-plane-waves at 131 symmetry independent regularly spaced k-points. Starting from these eigenvalues energies at a finer grid are calculated with the quadratic Lagrangian interpolation. Finally the DOS is calculated with the Gilat-Raubenheimer method[12]. (For a more detailed description see Refs. 13 and 14).

The lattice constants used at the pressures 0, 8 and 38 kbar are given in Table 1. The calculated DOS diagrams at these pressures are shown in Fig.1 (full lines). The following contributions from the valence band (VB) and conduction band (CB) triplets can be separated along increasing energy : VB1 (two-peaked s-bonding), VB2 (two-peaked p-bonding), VB3 (p-non-bonding), CB1 (p-antibonding) and CB2 (already mixing strongly with upper CBs). The increasing pressure broadens the bands which lowers the peak heights (though with a few exceptions) and closes and fills the valleys between the triplets. The minimum SCSOPW α = 1 gap at H closes at $\approx$ 22 kbar[10]. The filling of the valley between VB3 and CB1 at 38 kbar (the lowest diagram in Fig.1) is almost invisible. This shows that the overlap of the VB and CB edges is limited to a small region around point H. (We are aware that a more accurate treatment of this type of local crossing with a finer grid in the Lagrangian interpolation would yield a small filling).

Table 1 : The lattice constants

	0 kbar	8 kbar	38 kbar
a(nm)	0.44572[a]	0.4378[b]	0.4241[b]
c(nm)	0.5929[a]	0.5948[b]	0.5972[b]
u(a)	0.2633[a]	0.267[c]	0.2705[c]

[a]Ref.15   [b]Ref.5   [c]Ref.16

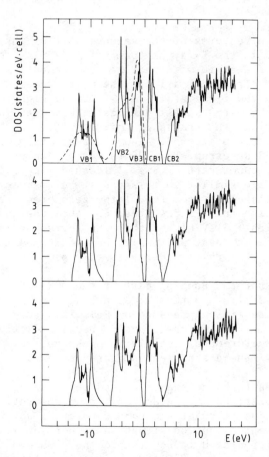

Fig.1 : The effect of pressure on the SCSOPW α = 1 DOS of Te.
The uppermost, intermediate and lowest diagram (full lines)
are calculated at the pressures 0, 8 and 38 kbar,
respectively. For lattice constants see Table 1. All
diagrams are given on the same scale. The 30 lowest
bands are used. The valence band maximum of each diagram
is placed at the origin of the energy axis. The XPS result
of Schlüter et al.[18] (dashed line) is given in arbitrary
units.

The two uppermost DOS diagrams in Fig. 1 are very similar to the corresponding self-consistent pseudopotential DOS[17] up to $\approx$ 5 eV. The similarity extends even to very fine structural details. This is a remarkable result because the two calculations are completely independent.

The main features of the valence part of the DOS are usually reflected in the X-ray photoemission (XPS) spectrum. However, it should be kept in mind that several essential assumptions are involved in a direct comparison. The matrix elements from the valence states to the final states at $\approx$ 1000 eV higher energies are assumed to be constants. All surface effects (the direct surface contribution, the transport to the surface and the escape from the surface) are neglected. Also the resolution of $\approx$ 0.5 eV rounds off the XPS spectrum. The XPS result of Schlüter et al.[18] is redrawn in the upper part of Fig. 1 (dashed line). The earlier XPS spectra[19,20] are similar to this result but reveal less structure. As it is clear from Fig. 1 the zero pressure valence DOS agrees well with the XPS result except that the upper peak of the VB1 DOS lies $\approx$ 1 eV higher than the corresponding XPS maximum. This may, at least partially, be attributed to the tendency of the X_α theory to underestimate gaps.

At 38 kbar the bonding part of the DOS (VB1 + VB2) of Te has become quite similar to that of Se at 140 kbar[21] (the semiconductor-metal transition pressure of Se). This seems to indicate that the bonding mechanisms of Te and Se become relatively similar near the semiconductor-metal transition pressure. However, in the upper parts of DOS there are clear differences.

It seems to be generally agreed that the absorption edge for light polarized $\perp$ c is due to (group theoretically) allowed direct transitions at the corner point H. However, for the light polarization $\parallel$ c the situation is not so clear because the direct transition at H is now forbidden. Although the majority of the authors seems to be in favour of the direct transition[22-24] there have also been speculations about the possibility of allowed indirect transitions[25]. The nature of the $\parallel$ c transition can be studied by comparing the pressure dependence of the experimental $\parallel$ c and $\perp$ c absorption edges with each other. Neuringer obtained for the first derivatives at zero pressure the values $-1.8 + 0.3$ and $-2.2 + 0.4$ $(10^{-2}$ eV/kbar) for $\parallel$ c and $\perp$ c respectively[26]. This would indicate a different transition for the two polarizations. However, the error bars are relatively large and in a later experiment Anzin et al.[27] obtained the same value $-2.034 + 0.004$ $(10^{-2}$ eV/kbar) for both polarizations. Kosichkin concluded[24] that, because also the magnetic absorption measurements[22,23] give the same gap value for both polarizations, the $\parallel$ c transition should be direct.

We can study this question theoretically by comparing
the pressure dependence of the transitions of smallest energies
with experiments. In doing that we assume that the neglect of the
relativistic effects and the underestimation of the gap region do
not essentially affect the conclusions. We consider from the SCSOPW
energies at symmetry points the three highest ones of the uppermost
VB and the three lowest ones of the lowest CB. For the comparison
an exponential function of the form $E(P) = A.exp(-\beta P) - B$ (P is
pressure, A, β and B are constants) is fitted to the transition
energies E at the pressures 0, 8 and 38 kbar (Table 2). From the
comparison of the gaps and their derivatives in Table 2 it is
obvious that the direct transition $H_3 \rightarrow H_1$ (for both values of α)
agrees best with the recent experimental results (three last rows
in Table 2). The other two possibilities are $H_3 \rightarrow A_1$ (A is the
middle point of the top face of the first Brillouin zone) and
$H_3 \rightarrow M_1$ (M is in the middle of adjacent corner points of the top
face). The energy of the transition $H_3 \rightarrow A_1$ is only ≈ 0.2 eV larger
than that associated with the transition $H_3 \rightarrow H_1$. The second
derivative of the transition $H_3 \rightarrow A_1$ agrees quite well with the
experimental values, especially for $\alpha = 1$. However, the magnitude
of the first derivative tends to be too small. On the other hand
the first and second derivatives of the transition $H_3 \rightarrow M_1$ agree
well with the experiments (though the magnitude of the first
derivative, especially for $\alpha = 2/3$, tends to be slightly too large).
However, the energy difference is already ≈ 0.5 eV larger than that
of the transition $H_3 \rightarrow H_1$. Thus, the above study of the pressure
dependence clearly favours the transition $H_3 \rightarrow H_1$ for both
polarizations of light, although the transitions $H_3 \rightarrow A_1$ and $H_3 \rightarrow M_1$
cannot be totally excluded.

Pine et al.[8] suggest that the minimum optical gap should
change from H → H to H → A before the semiconductor-metal
transition. Their empirical pseudopotential calculation yielded
the indirect gap H → A which closed at u $\approx$ 0.29 (compare with
Table 1). However, the difference $E(H_3 \rightarrow A_1) - E(H_3 \rightarrow H_1)$ from
Table 2 is a monotonously increasing function with respect to
pressure which is in a clear contradiction with the result of
Pine et al.

Acknowledgements   -   We wish to express our gratitude to Professors
T. Stubb and M.A. Ranta for their continuous support. We also want
to thank the staff of the computer center of Helsinki University of
Technology for their amicable co-operation. Earlier collaboration
with Dr. P. Krusius is also acknowledged. And last but not least we
wish to thank Mrs. T. Aalto for the typing of the manuscript.

Table 2 : The dependence of the transitions on pressure. The transitions $\Gamma_2 \to A_1$ and $M_2 \to A_1$ are not monotonously decreasing functions of pressure and a parabolic fit is used.

	A (eV)	β (1/kbar)	B (eV)	E(0) (eV)	$(\partial E/\partial P)_{P=0}$ (10^{-2} eV/kbar)	$(\partial^2 E/\partial P^2)_{P=0}$ (10^{-4} eV/kbar^2)
SCSOPW α=1						
$\Gamma_2 \to A_1$				1.1353	-0.0056	-0.1187
$\Gamma_2 \to H_1$	0.3586	0.01608	-0.5509	0.9095	-0.5766	0.9270
$\Gamma_2 \to M_1$	0.4492	0.02237	-0.9893	1.4386	-1.0050	2.2484
$H_3 \to A_1$	0.1588	0.08650	-0.3077	0.4665	-1.3740	11.8842
$H_3 \to H_1$ [a]	0.3539	0.05273	0.1132	0.2407	-1.8659	9.8379
$H_3 \to M_1$	0.4758	0.04816	-0.2940	0.7698	-2.2912	11.0336
$M_2 \to A_1$				0.5690	-0.0958	1.4231
$M_2 \to H_1$	0.09007	0.08812	-0.2531	0.3432	-0.7937	6.9943
$M_2 \to M_1$	0.2016	0.05905	-0.6707	0.8723	-1.1905	7.0301
SCSOPW α=2/3						
$\Gamma_2 \to A_1$				1.0696	0.0454	-0.2059
$\Gamma_2 \to H_1$	0.4194	0.01368	-0.5102	0.9296	-0.5737	0.7847
$\Gamma_2 \to M_1$	0.5205	0.02720	-0.8899	1.4104	-1.4159	3.8517
$H_3 \to A_1$	0.1198	0.1085	-0.05765	0.1774	-1.2993	14.0936
$H_3 \to H_1$ [a]	0.3346	0.05318	0.2972	0.03737	-1.7792	9.4616
$H_3 \to M_1$	0.5409	0.04861	0.02267	0.5182	-2.6291	12.7792
$M_2 \to A_1$				0.4029	0.2154	0.8133
$M_2 \to H_1$	0.03180	0.2265	-0.2311	0.2629	-0.7202	16.3104
$M_2 \to M_1$	0.2181	0.06167	-0.5256	0.7437	-1.3450	8.2938
Transmiss. [b]	0.3229	0.0630	0	0.3229	-2.034	12.82
Photocond. [b]	0.3296	0.061	0	0.3296	-2.01	12.26
Laser em. [c]	0.3335	0.0591	0	0.3335	-2.0	10.4

[a] Ref.10   [b] Ref.27   [c] Ref.8

REFERENCES

1. P.W. Bridgman, Proc. Am. Acad. Arts Sci. 74, 425 (1942).

2. P.W. Bridgman, Proc. Am. Acad. Arts Sci. 81, 165 (1952).

3. G.C. Kennedy & P.N. La Mori, J. Geophys. Res. 67, 851 (1962).

4. S.S. Kabalkina, L.F. Vereshchagin & B.M. Shulenin, Zh. Eksperim. Teor. Fiz. 45, 2073 (1963); Engl. transl. : Sov. Phys. JETP 18, 1422 (1964).

5. J.C. Jamieson & D.B. McWhan, J. Chem. Phys. 43, 1149 (1965).

6. F.A. Blum & B.C. Deaton, Phys. Rev. 137, A1410 (1965).

7. G.C. Vezzoli, Z. Kristallogr. 134, 305 (1971).

8. A.S. Pine, M. Menyuk & G. Dresselhaus, Solid State Commun. <u>31</u>, 187 (1979).

9. G. Doerre & J.D. Joannopoulos, Phys. Rev. Lett. <u>43</u>, 1040 (1979).

10. H. Isomäki, J. von Boehm, P. Krusius & T. Stubb, Phys. Rev. B <u>B22</u>, 2945 (1980).

11. J. von Boehm & P. Krusius, Int. J. Quantum Chem. <u>8</u>, 395 (1974).

12. G. Gilat & L. J. Raubenheimer, Phys. Rev. <u>144</u>, 390 (1966).

13. H. Isomäki and J. von Boehm, Physica <u>99B</u>, 255 (1980).

14. J. von Boehm and H. Isomäki, J. Phys. C: Solid St. Phys. <u>13</u>, 3181 (1980).

15. P. Unger & P. Cherin, in The Physics of Se and Te, edited by W.C. Cooper (Pergamon, Oxford, 1969), p.223.

16. R. Keller, W.B. Holzapfel & H. Schulz in Ref. 17 and in the paper by T. Starkloff & J.D. Joannopoulos, J. Chem. Phys. <u>68</u>, 579 (1978).

17. T. Starkloff & J.D. Joannopoulos, Phys. Rev. <u>B19</u>, 1077 (1979).

18. M. Schlüter, J.D. Joannopoulos, M.L. Cohen, L. Ley, S.P. Kowalczyk, R.A. Pollak & D.A. Shirley, Solid State Commun. <u>15</u>, 1007 (1974).

19. R.A. Pollak, S. Kowalczyk, L. Ley & D.A. Shirley, Phys. Rev. Lett. <u>29</u>, 274 (1972).

20. N.J. Shevchik, M. Cardona & J. Tejeda, Phys. Rev. <u>B8</u>, 2833 (1973).

21. J. von Boehm & H. Isomäki, J. Phys. C: Solid St. Phys. <u>13</u>, 4953 (1980).

22. P. Grosse & K. Winzer, Phys. Status Solidi <u>26</u>, 139 (1968).

23. H. Shinno, R. Yoshizaki, S. Tanaka, T. Doi & H. Kamimura, J. Phys. Soc. Jap. 35, 525 (1973).

24. Y.V. Kosichkin, in The Physics of Selenium and Tellurium, edited by E. Gerlach and P. Grosse (Springer, Berlin-Heidelberg-New York, 1979), Springer Series in Solid-State Sciences, Vol. 13, p.96.

25. S. Tutihasi, G.G. Roberts, R.C. Keezer & R.E. Drews, Phys. Rev. <u>177</u>, 1143 (1969).

26. L.J. Neuringer, Phys. Rev. <u>113</u>, 1495 (1959).

27. V.B. Anzin, M.I. Eremets, Y.V. Kosichkin, A.I. Nadezhdinskii & A.M. Shirokov, Phys. Status Solidi (a) <u>42</u>, 385 (1977).

ENERGY BAND STRUCTURES OF NiO BY AN LCAO METHOD

J. Hugel and C. Carabatos

Laboratoire de Physique des Milieux Condensés
Faculté des Sciences, Université de Metz

Ile du Saulcy, 57000 Metz, France

Valence bands of NiO based on an LCAO method have been computed
using 3d orbitals for nickel and 2s and 2p orbitals for oxygen.
These orbitals were obtained by solving a coupled Schrödinger
equation for the nickel and the oxygen ions with a local ionic
potential acting on the lattice cation and anion site. The Slater
Koster relations have been used for the development of the matrix
elements and the resulting overlap, potential and crystal field
integrals limited to the first and second neighbours have been
numerically integrated.
The 4s conduction band is determined by a parametrized LCAO
method. The radii of the ionic local crystal potentials as the
parameters of the conduction band were choosen in order to
reproduce the optical reflectivity results. For this the complex
relative dielectric constant $\varepsilon(\omega)$ has been computed considering
electric dipolar transitions between the occupied and the empty
bands. The refractive index, the reflectivity and absorption
curves were then deduced and the first absorption peak for NiO
has been attributed to 2p – 3d transitions.

1. INTRODUCTION

Several methods of band structure calculations have been
focussed on NiO in a non self-consistent[1,2,3] and in a self-
consistent manner[4,5]. In an earlier calculation[3] an energy gap
about 4 eV has been obtained whereas only recently a slight
overlap between the 2p and 3d bands has been found in agreement
with photoemission results[6]. Most of the mentionned band structure
calculations are limited to the valence bands; in the present work

we calculate the valence and conduction bands of NiO and compute to both the photoemissions results[6] and the optical curves[7].

In the second section we describe the general outline of the method. The numerical details are given in the third section. The results of the energy bands and the calculations of the relative dielectric constant are presented in the fourth section. The conclusions appear in the last section.

2. THEORETICAL SITUATION

The orbitals used in construction the Bloch functions are obtained by solving a Schrödinger equation with a local potential. Around a lattice site l with nuclear charge $Z^{(1)}$ and $N^{(1)}$ electrons the Schrödinger equation is written as

$$H^{(\ell)}(\underset{\sim}{r})\; \varphi_n^{(\ell)}(\underset{\sim}{r}) = E_n^{(\ell)}\; \varphi_n^{(\ell)}(\underset{\sim}{r})$$

and the local hamiltonian as :

$$H^{(\ell)}(\underset{\sim}{r}) = \frac{p^2}{2m} + V^{(\ell)}(\underset{\sim}{r})$$

where $\quad V^{(\ell)}(\underset{\sim}{r}) = -\frac{Z^{(\ell)} - N^{(\ell)}}{r} + V^{(\ell)}_{\text{Coul.}}(r) + V^{(\ell)}_{\text{exch.-corr.}}(r) + V^{(\ell)}_{\text{ionic}}(\underset{\sim}{r})$

within a sphere of finite radius

$$V^{(\ell)}(\underset{\sim}{r}) = 0 \quad \text{elsewhere}$$

The coulomb potential at a lattice site l is the short range coulomb part expressed in terms of the spherically averaged electron density at site l plus short range coulomb contributions from neighbouring sites.

$$V^{(\ell)}_{\text{coul.}}(r) = V^{(\ell)}_{\text{Sh}}(r) + \sum_{\ell'} V^{(\ell')}_{\text{Sh}}(r)$$

The neighbouring short range parts are expanded at site l using the α Löwdin[9] expansion. The short range coulomb part at site l is defined as :

$$V^{(\ell)}_{\text{Sh}}(r) = -\frac{1}{r} \int_r^\infty \rho^{(\ell)}(t) dt + \int_r^\infty \frac{\rho^{(\ell)}(t) dt}{t}$$

The exchange-correlation potential is taken as the local Slater[8] exchange corrected by a constant correlation factor α

$$V^{(\ell)}_{\text{exch.-corr.}}(r) = -3 \left(\frac{3}{32\pi^2} \frac{\rho^{(\ell)}_{\text{tot}}(r)}{r^2}\right)^{1/3} \alpha$$

The total density $\rho^{(\ell)}_{\text{tot}}(r)$ is computed from the local orbitals centered on the other sites.

$$\rho^{(\ell)}_{\text{tot}}(r) = \sum_n r^2 R^{(\ell)}_n(r) + \sum_{n\ell'} r^2 R^{(\ell')}_n(r)$$

where $R^{(\ell)}_n(r)$ is the radial part of the nth wave function. The second summation in the above relation is carried out again with the aid of the α-Löwdin[9] procedure. The ionic potential is the long range contribution due to the ionicity.

$$V^{(\ell)}_{\text{ionic}}(\underline{r}) = \sum_{\ell' \neq \ell} \frac{q_{\ell'}}{|\underline{r} - \underline{r}_{\ell'}|}$$

q_ℓ, gives value and sign of the ionic charge in units of the electronic charge.

The crystal eigenstates $E(k)$ are solutions of the secular equation.

$$\|H_{ij}(k) - E(k)\,S_{ij}(k)\| = 0$$

H_{ij} is the one electron crystal hamiltonian whose potential is the sum of the local potentials of finite radii over all lattice sites. The indices i and j refer to all different non orthogonalized Bloch functions constructed from the ionic orbitals. S_{ij} is the usual overlap matrix and is not diagonal. The hamiltonian matrix elements can be written as

$$H_{ij}(\underline{k}) = \sum_n \langle \varphi^{(o)}_i(\underline{r}) \,|\, \frac{p^2}{2m} + \sum_\ell V^{(\ell)}(\underline{r}) \,|\, e^{i k r_n} \varphi^{(n)}_j(\underline{r}) \rangle$$

By making use explicity of the eigenvalues E^1_n of the local hamiltonian H^1 the diagonal and off diagonal matrix elements become :

$$H_{ii}(\underline{k}) = E_i^{(o)} + \sum_{n\neq o} \langle\varphi_i^{(o)}(\underline{r})|V^{(n)}(r)|\varphi_i^{(o)}(\underline{r})\rangle$$

$$+ \sum_{\ell} e^{i\underline{k}\underline{r}}\{E_i^{(o)}\langle\varphi_i^{(o)}(\underline{r})|\varphi_i^{(\ell)}(\underline{r})\rangle + \langle\varphi_i^{(o)}(\underline{r})|V^{(\ell)}(r)|\varphi_i^{(\ell)}(\underline{r})\rangle\}$$

$$+ \sum_{\ell}\sum_{n\neq\ell} e^{i\underline{k}\underline{r}}\langle\varphi_i^{(o)}(\underline{r})|V^{(n)}(r)|\varphi_i^{(\ell)}(\underline{r})\rangle$$

$$H_{ij}(\underline{k}) = \sum_{\ell\neq o} e^{-i\underline{k}\underline{r}}\{E_j^{(\ell)}\langle\varphi_i^{(o)}(\underline{r})|\varphi_j^{(\ell)}(\underline{r})\rangle$$

$$+ \langle\varphi_i^{(o)}(\underline{r})|V^{(o)}(r)|\varphi_j^{(\ell)}(\underline{r})\rangle\}$$

$$+ \sum_{\ell}\sum_{n\neq\ell} e^{-i\underline{k}\underline{r}}\langle\varphi_i^{(o)}(\underline{r})|V^{(n)}(r)|\varphi_j^{(\ell)}(\underline{r})\rangle$$

3. NUMERICAL DETAILS

The total density on a lattice site is obtained by extending the α-Löwdin procedure to the third shell of neighbours. We adapt the classical Herman Skillman[10] computer program adjusted for an isolated ion in order to solve at the same time the Schrödinger equations for Ni^{2+} and O^{2-} with their own total densities until self-consistency is achieved. We get then the ionic orbitals with their corresponding energies and the local potentials. As the numerical resolution of the Schrödinger equation works only for a potential with spherical symmetry we include only the Madelung contribution of the long range Coulomb potential. The cubic symmetry term arising from the NaCl symmetry of NiO and which removes the degeneracy of some levels is treated in a perturbation theory. It acts only on the degenerate d levels and splits them into two levels with the well known 10 Dq splitting [11]. For the factor α we adopt the full Slater exchange[8] for the O^{2-} ion and retain the value 1.1 [12] for the Ni^{2+} ion. In order to determine the metal and oxygen potential radii Rm and Ro we impose first that the spherical potential volumes fill best the volume of the unit cell. Second we adjust the radii with the prescription to reproduce approximately the first reflectivity peak at the right energy. The results are Rm = 2.8 and Ro = 2.6 atomic units, the lattice parameter of NiO being 7.927 [13].

The ten basis Bloch sums are formed from the oxygen 2s, 2px, 2py and 2pz orbitals and from the transition metal 3dxy, 3dyx,

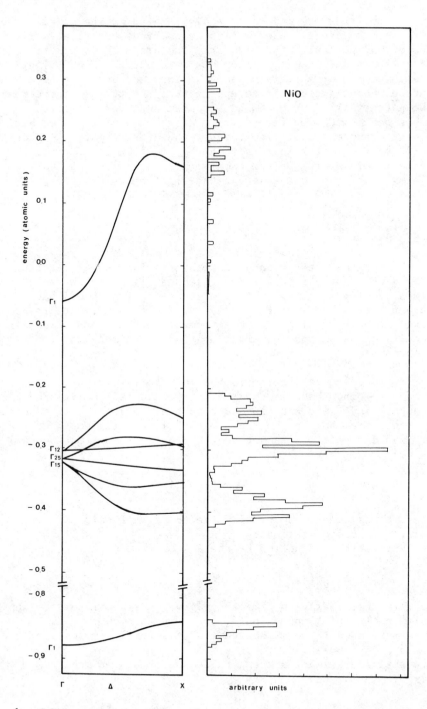

Fig.1 : LCAO energy-band for NiO along the Δ direction and density
of states.

3dzx, $3dx^2-y^2$, $3dz^2-r^2$ and 4s orbitals located at the origin. The matrix elements are expanded as linear combinations of a smaller number of independent integrals using the Slater Koster[14] relations. For the 2s, 2p and 3d matrix elements we neglect the contributions of the three center integrals and take into account only first and second neighbour two center interactions owing to the finite radii of the potentials. For the 4s hamiltonian and overlap matrix elements we include the five nearest neighbours of the same kind in order to avoid the overlap catastrophe. Furthermore we parametrise the two center integrals in the hamiltonian matrix elements involving the 4s function with the requirement that the remaining peaks come out at about the right energies.

4. RESULTS

The LCAO energy band results E(k) for the Δ direction and the density of states are presented in figure 1. The 2p and 3d bands overlap slightly and the width of the 2p and 3d bands are about 3 eV and 3.7 eV respectively. The 4s band width is about 11 eV. Our results are in agreement with the experimental photoemission data[6] and support comparison with more involved calculations[5].

To calculate the optical of NiO for every value of k we must know the interband momentum matrix elements $|eM_{vc}(k)|$ coupling the valence and the conduction band as the energy associated with each transition. For an isotropic material the imaginary part of the retive dielectric constant in the dipolar approximation is given by[15]

$$\varepsilon_2(\omega) = \frac{4\pi e^2}{m\omega^2} \sum_{v,c} \sum_{k} \frac{2}{(2\pi)^3} |eM_{vc}(\underset{\sim}{k})|^2 \delta(E_{vc}(\underset{\sim}{k}) - \hbar\omega)$$

In order to calculate numerically the above expression we make the following approximations.
i) We replace the δ function by the following expression

$$(h/\tau)/(E_{vc}(k) - \hbar\omega)^2 - (h/\tau)^2)$$

where τ is the relaxation time which gives a line width to each transition. In our computations τ is taken to 6.10^{-14} sec.
(ii) We make use of the effective mass theorem[16] already employed for other compounds[17,18].
Within this approximation the μ th component of the momentum matrix elements reads :

$$(e\, M_{ij}(\underset{\sim}{k}))^\mu = \frac{m}{\hbar} \sum_{\ell,n} V_{i\ell}^*(\underset{\sim}{k}) \frac{\partial H_{\ell n}(\underset{\sim}{k})}{\partial k_\mu} V_{nj}(\underset{\sim}{k})$$

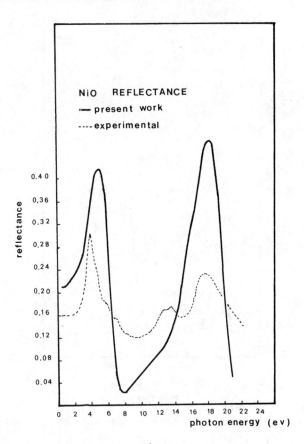

<u>Fig.2</u> : Reflectance spectrum of NiO.
       Present work and experimental results of Powell and Spicer.

where $V_{nj}(k)$ is the eigenvector associated with the j th crystal
state and satisfies

$$\sum_{\ell n} V_{i\ell}^*(\underset{\sim}{k})\, H_{\ell n}(\underset{\sim}{k})\, V_{nj}(\underset{\sim}{k}) = E_{ij}(\underset{\sim}{k})\, \delta_{ij}$$

iii) For the 3d band we discard the band model which permits two
electrons per state and safeguard Hund's rule. The considered band
to band transitions are the 2s and 2p bands to the remaining half-
filled 3d band, the 2s and 2p bands to the 4s band and the 3d band
to the 4s band.

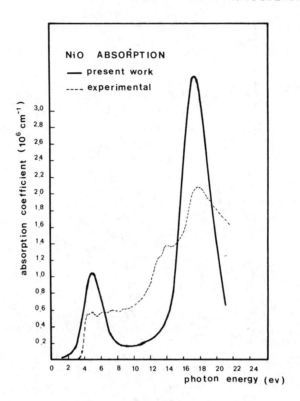

<u>Fig.3</u> : Absorption coefficient of NiO.
Present work and experimental results of Powell and Spicer.

        The use of the Kramers-Kronig technique to get the real
part of the relative dielectric constant allows for the reflectivity
and absorption curves to be easily deduced (fig.2 and 3). As we have
calculated separately the contributions of the above mentionned
band to band transitions in $\varepsilon_2(\omega)$ we attribute the first absorption
peak at 4.9 eV to 2p - 3d transitions. This assignment agree with
the conclusions of a SCF Xα method[19]. Our peak at 17.3 eV is due
to 2p - 4s transitions. The calculated dielectric constant is 7.15
and the refractive index 2.67 to be compared with the values 5.43
and 2.33[7].

5. CONCLUSION

        Although it is well admitted that the 2p and 4s ionic
levels spead out in bands and that the 3d electrons remain
localized about the lattice site a band picture derived from a
local potential taking into account the crystal surrounding is

still adapted to describe the essential optical properties of NiO. By considering the 3d levels as a band we cannot reproduce the fine structures of the optical spectra but we are able to reproduce the main features of the curves and to give their corresponding transitions.

Acknowledgement — The author wish to thank Professor F. Bassani and Dr. F. Casula for their stimulating discussions and helpful suggestions as for their constant interest in this work.

REFERENCES

1. J. Yamashita, J. Phys. Soc. Japan 18, 1010 (1963).

2. T. Wilson, Int. J. Quant. Chem. IIIS, 754 (1970).

3. L.F. Mattheis, Phys. Rev. B5, 290 (1972).

4. T.C. Collins, A.B. Kunz and J.L. Ivey, Int. J. Quant. Chem. Symp. 9, 519 (1975).

5. A.B. Kunz and G.T. Surrat, Solid State Commun. 25, 9 (1978).

6. D.E. Eastman and J.L. Freeouf, Phys. Rev. Lett., 34, 395 (1975).

7. R.J. Powell and W.E. Spicer, Phys. Rev. B2, 2182 (1970).

8. J.C. Slater, Phys. Rev. 81, 385 (1951).

9. P.O. Löwdin, Adv. Phys. 5, 1 (1956).

10. F. Herman and S. Skillman, Atomic structure calculations (Prentice Hall, Englewood cliffs, 1963).

11. F. Bassani, Theory of imperfect crystalline Solids, Trieste Lectures (IAEA, Vienna, 1971).

12. K. Schwartz, Phys. Rev. B5, 2466 (1972).

13. W.B. Pearson, A. Handbook of lattice spacings and structures of metals and alloys (Pergamon, New York, 1978).

14. J.C. Slater and G.F. Koster, Phys. Rev. 94, 1498 (1954).

15. F. Bassani and G. Pastori Parravicini, Electronic States and Optical Transitions in Solids (Pergamon, New York, 1975).

16. G. Dresselhaus and M.S. Dresselhaus, Phys. Rev. 160, 649 (1967).

17. D.D. Buss and V.E. Shirf, Electronic Density of States, NBS publication 223 (L.H. Bennet editor 1971).

18. Castet-Mejean and F.M. Michel-Calendini, J. Phys. C, Solid State Phys. 11, 2195 (1978).

19. K.H. Johnson, J.W. Messmer and J.W.D. Conolly, Solid State Commun. 12, 313 (1973).

CHEMICAL BONDING, RECIPROCAL FORM FACTORS, AND

WANNIER FUNCTIONS

Jean-Louis Calais[*]

Quantum Chemistry Group, University of Uppsala

Box 518, S-751 20 Uppsala, Sweden

Compton profile measurements of increasing accuracy have stimulated the interest in wave functions and densities in momentum space. The Fourier transform of the momentum density can be interpreted as a sum of autocorrelation functions in position space. This quantity - the reciprocal form factor - contains important information about the geometry and the chemical bonding in the system.

Wannier functions calculated directly via the Koster-Kohn variational principle provide theoretical reciprocal form factors to be compared with those obtained experimentally from Compton profiles. This gives a goodness criterion for the Wannier functions and opens up possibilities for new ways of describing chemical bonding in solids.

INTRODUCTION

The equivalence between "ordinary" space and momentum space is described in elementary textbooks in quantum mechanics. With rather few exceptions this equivalence is not exploited as much as would seem possible in the theory of the structure of matter.

The main exception in recent years is connected with the experimental and theoretical work on Compton profiles[1], which provides information about the momentum distribution in atoms, molecules, and solids. Important attempts are being made to connect the information obtained in a Compton scattering

305

experiment with other quantities of chemical and physical
importance.

Chemical bonding in solids is a field which has so far
received relatively little attention. Except for the basic
classification in metallic, ionic, covalent, and molecular
crystals there is very little about cohesive properties of
solids in most textbooks. The main reason for this situation
is very probably associated with the lack of suitable conceptual
tools[2].

It has been pointed out by Schülke[3] that the Fourier
transform of the momentum density contains information about the
lattice structure for crystals with filled bands. The properties
of this "reciprocal form factor" have been further investigated
both for atoms and molecules[4], and for solids[5]. This quantity
is thus a function of "ordinary space" containing information,
which is complementary to that obtained from the ordinary
electron density.

It seems very plausible that the reciprocal form
factor together with the electron density might provide an
interesting tool for the characterization of chemical bonding.
In a crystal the reciprocal form factor can be expressed in
terms of the Wannier functions, and this is of special interest
now when Wannier functions can be calculated directly via the
Koster-Kohn variational principle[6-8].

One of the basic difficulties in the description of a
chemical bond is the question as to what it is that is bonded.
The old question how to define an atom or an ion in a crystal
or a molecule has recently received what seems to be a
satisfactory answer by Bader and collaborators[9], in terms of
the topology of the electron density. With this characterization
of the entities that are bonded in a solid or a molecule we have
a firm basis on which a more detailed theory of chemical bonding
can be built.

In the present paper we derive certain relations for
the reciprocal form factor and related quantities, which should
be useful for extracting information about the chemical bonding
from these quantities.

BLOCH AND WANNIER FUNCTIONS IN MOMENTUM SPACE

In this section we recall some of the connections between Bloch and Wannier functions in ordinary space and momentum space. We want to stress the point that these results are not supposed to form the basis for actual calculations. For simplicity of notation we consider one single band which is doubly filled.

We consider a Born-Kármán (BK) region of a crystal with N unit cells, labelled by the lattice vectors $\bar{m} = \mu_1\bar{a}_1 + \mu_2\bar{a}_2 + \mu_3\bar{a}_3$, where the μ_i's are integers. The Bloch functions $(\bar{k},\bar{r})$ are labelled by the N wave vectors $\bar{k}$ of the first Brillouin zone (BZ), and are assumed to be normalized in BK, so that

$$\int_{BK} \psi^*(\bar{k}^1,\bar{r})\psi(\bar{k},\bar{r})dv = \delta(\bar{k}^1,\bar{k}) \tag{1}$$

The Wannier functions can then be written as

$$W(\bar{m},\bar{r}) = \frac{1}{\sqrt{N}} \sum_{k}^{BZ} \psi(\bar{k},\bar{r})e^{-i\bar{k}\cdot\bar{m}} =$$

$$= \frac{1}{\sqrt{N}} \cdot \frac{V}{8\pi^3} \int_{BZ} \psi(\bar{k},\bar{r})e^{-i\bar{k}\cdot\bar{m}} \, d\bar{k} \ , \tag{2}$$

where $V = NV_{oa}$ is the volume of the BK region and V_{oa} is the volume of a unit cell.

In momentum space the Bloch function is essentially the momentum eigenfunction[10],

$$\chi_B(\bar{k},\bar{p}^1) = (2\pi\hbar)^{-3/2} \int_{BK} \psi(\bar{k},\bar{r})e^{-\frac{i\bar{p}^1\cdot\bar{r}}{\hbar}} \, dv =$$

$$= \frac{V}{\sqrt{8\pi^3\hbar^3}} \sum_{\bar{K}} v(\bar{k}+\bar{K})\delta(\bar{k}+\bar{K},\frac{\bar{p}^1}{\hbar}) \ , \tag{3}$$

but it can also be written as the Fourier transform at $\bar{p}^1 - \hbar\bar{k}$ of the periodic part of the Bloch function in ordinary space.

The Wannier function in ordinary space is the Fourier transform of the momentum eigenfunction

$$W(\bar{m},\bar{r}) = \frac{V}{8\pi^3\sqrt{N}} \int_{all \ space} v(\bar{k})e^{i\bar{k}\cdot(\bar{r}-\bar{m})} \, d\bar{k} \tag{4}$$

In momentum space the Wannier function can be written as a product of a phase factor and the momentum space function corresponding to the Wannier function at the origin

$$\chi_w(\bar{m},\bar{p}^1) = e^{-\frac{i\bar{p}^1\cdot\bar{m}}{\hbar}} \chi_w(\bar{p}^1) \; ; \tag{5}$$

$$\chi_w(\bar{p}^1) = (2\pi\hbar)^{-3/2} \int_{BK} W(\bar{r})e^{-\frac{i\bar{p}^1\cdot\bar{r}}{\hbar}} dv \tag{6}$$

For a filled band the spatial part of the Fock-Dirac density matrix can be written either in terms of Bloch or of Wannier functions,

$$\rho(\bar{r},\bar{r}^1) = \sum_{\bar{k}}^{BZ} \psi(\bar{k},\bar{r})\psi^*(\bar{k},\bar{r}^1) = \sum_{m}^{BK} W(\bar{m},\bar{r})W^*(\bar{m},\bar{r}^1) \tag{7}$$

In momentum space we have similarly

$$\rho(\bar{p}^1,\bar{p}^{11}) = \sum_{\bar{k}}^{BZ} \chi_B(\bar{k},\bar{p}^1)\chi_B^*(\bar{k},\bar{p}^{11}) =$$

$$= \sum_{\bar{m}}^{BK} \chi_w(\bar{m},\bar{p}^1)\chi_w^*(\bar{m},\bar{p}^{11}) \; . \tag{8}$$

Combining (5) and (8) we have

$$\rho(\bar{p}^1,\bar{p}^{11}) = N\chi_w(\bar{p}^1)\chi_w^*(\bar{p}^{11})\delta(\bar{p}^{11} - \bar{p}^1,\frac{\bar{K}}{\hbar}), \tag{9}$$

where $\bar{K}$ is a reciprocal lattice vector.

AUTOCORRELATION FUNCTIONS

As pointed out by Benesch et al.[11] the "ordinary" form factor is proportional to the convolution of the momentum space density matrix. In terms of orbitals this means that it can be written as a sum of overlap integrals in momentum space, which can also be interpreted as autocorrelation functions.

The reciprocal form factor[3,4,5] is defined as the Fourier transform of the density in momentum space

$$B(\bar{r}) = \int \rho(\bar{p})e^{-\frac{i\bar{p}\cdot\bar{r}}{\hbar}} d\bar{p} \; . \tag{10}$$

In the case of a crystal with filled bands we get with (8), (9) and (10)

$$B(\bar{r}) = N\int\chi_w(\bar{p})\chi_w^*(\bar{p})e^{-\frac{i\bar{p}.\bar{r}}{\hbar}}\,d\bar{p} \,. \tag{11}$$

It is, however, more interesting to express $B(\bar{r})$ in terms of functions in position space. Writing the momentum density as a double Fourier transform of the position density matrix and using (7) we have

$$
\begin{aligned}
B(\bar{r}) &= \int\rho(\bar{r}^1,\bar{r}^1+\bar{r})dv^1 = \\[4pt]
&= \sum_k^{BZ} \int\psi(\bar{k},\bar{r}^1)\psi^*(\bar{k},\bar{r}^1+\bar{r})dv^1 = \\[4pt]
&= \sum_m^{BK} \int W(\bar{m},\bar{r}^1)W^*(\bar{m},\bar{r}^1+\bar{r})dv^1
\end{aligned} \tag{12}
$$

Since the Wannier functions are periodic this can be written as

$$B(\bar{r}) = N\int W(\bar{r}^1)W^*(\bar{r}^1+\bar{r})dv^1 \,, \tag{13}$$

which could also have been obtained directly from (6) and (11). This implies together with the orthogonality of the Wannier functions centred in different unit cells that the reciprocal form factor vanishes at direct lattice vectors.

From the practical computational point of view it is very important that $B(\bar{r})$ can be calculated from one Wannier function, although it contains information corresponding to all the Bloch functions in the band.

DISCUSSION

Already a superficial comparison of $\rho(\bar{r}) = \rho(\bar{r},\bar{r})$, (7), and (13) gives an indication of how $B(\bar{r})$ complements $\rho(\bar{r})$. The electron density is the sum of the squares of the Wannier functions centered in the different unit cells. The reciprocal form factor among other things tells us something about the localization of the Wannier functions. That is one of the crucial properties of these functions[7,8], and it seems plausible that a study of $B(\bar{r})$ from that point of view might be fruitful.

Insofar as the function $B(Z) = B(0,0,Z)$ is the one-dimensional Fourier transform of a directional Compton profile[5] the extensive work on chemical binding and Compton scattering[1] certainly has a bearing on the reciprocal form factor too. It

would be interesting to pursue that connection in greater
detail. Certain results obtained from an explicit study of the
reciprocal form factor have been published for metals[12],
insulators[13], and semiconductors[14].

It is clear that the experimentally available quantity
$B(\bar{r})$ is of great interest also to theory. First of all direct
overall comparisons between experimental and theoretical $B(\bar{r})$'s
should be made. Deviations between these two quantities (which
are much more interesting than agreement) should be analyzed with
respect to the approximations made to obtain the theoretical
function. This is of special interest with regard to the direct
calculation of Wannier functions, as the $B(\bar{r})$ comparison provides
an extra criterion of goodness. In that connection one must be
aware of possible correlation effects.

The use of the zeros of $B(\bar{r})$ as a check on the ortho-
normalization procedure[5,14] deserve some comments. For ionic
crystals approximate symmetrically orthogonalized free ion orbitals
seem to provide rather good approximations of the Wannier functions.
Within that approximation one can proceed in different ways. A
purely theoretical calculation of the cohesive energy as a function
of internuclear separation yields a certain equilibrium distance,
and for the lattice vectors corresponding to that distance the
theoretical $B(\bar{r})$ vanishes. One can also calculate $B(\bar{r})$ for the
experimental lattice parameter. Assuming that the orbitals occupied
in the free ions suffice to give a reasonable approximation of the
Wannier functions one can then use the zeros of $B(\bar{r})$ as a criterion
for the accuracy of the orthogonalization procedure used.

It would, however, seem more satisfactory to use purely
mathematical criteria for that accuracy, in particular since
considerably improved orthogonalization procedures now exist[15].
A Wannier function approximated by a linear combination of
completely orthogonal atomic orbitals[8] will then lead to zeros
in $B(\bar{r})$ for the lattice parameter considered, and other features
of that $B(\bar{r})$ could be compared with its experimental counterpart.

Acknowledgement - The author would like to thank Dr. W.
Weyrich for many valuable discussions about the topic of this
paper.

REFERENCES

* Supported by the Swedish Natural Sciences Research Council.

1. B. Williams (ed.), "Compton Scattering"; Mc Graw Hill, Inc.,
   New York 1977.

2. J.L. Calais and K. Schwarz, Israel Journal of Chemistry 19, 88 (1980).

3. W. Schülke, Phys. Stat. Sol. (b) 82, 229 (1977).

4. W. Weyrich, D. Pattison and B.G. Williams, Chemical Physics 41, 271 (1979).

5. P. Pattison and W. Weyrich, Journal of Physics and Chemistry of Solids 40, 213 (1979).

6. G.F. Koster, Physical Review 89, 67 (1953).

7. W. Kohn, Physical Review B 7, 4388 (1973).

8. J. von Boehm and J.L. Calais, Journal of Physics C 12, 3661 (1979).

9. R.F.W. Bader, T. Tung Nguyen-Dang and Y. Tal, Journal of Chemical Physics 70, 4316 (1979) and references therein.

10. J.C. Slater, "Quantum Theory of Molecules and Solids", vol.2; Mc Graw Hill Book Co., New York 1965.

11. R. Benesch, S.R. Singh and V.H. Smith Jr., Chemical Physics Letters 10, 151 (1971).

12. P. Pattison and B. Williams, Solid State Communications 20, 585 (1976).

13. P. Pattison, W. Weyrich and B. Williams, Solid State Communications 21, 967 (1977).

14. B. Kramer, P. Krusius, W. Schröder and W. Schülke, Physical Review Letters 38, 1227 (1977).

15. Ø. Ra, International Journal of Quantum Chemistry 10, 5, 57 (1976).

THE ELECTRONIC STRUCTURE OF THE RARE-EARTH IRON GARNETS AND

RARE-EARTH GALLIUM GARNETS AS CALCULATED BY THE SC-Xα-SW METHOD

C.J.F. Graf and G.M. Copland

Queen Elizabeth College

Campden Hill Road, London W8 7AH, U.K.

Using the SCF-Xα-SW method calculations of the electronic structure of the complexes $(Yb\ O_8Fe_2)^{7-}$ and $(Yb\ O_8Ga_2)^{7-}$ have been carried out. Results show that there is preferential transfer of charge from the rare-earth and ferric ions onto those oxygen ligands which are found experimentally to form the dominant superexchange paths between the ions.

In this paper we report recent calculations of the electronic structure of rare-earth iron garnet and rare-earth gallium garnet by the self-consistent-field Xα scattered-wave (SCF-Xα-SW) method. The SCF-Xα-SW method has been in past years successfully applied to many multiatomic systems. Its procedure and underlying assumptions are described elsewhere [1-5]. One advantage of the SCF-Xα-SW method over other molecular orbital methods is that it may be easily applied to systems composed of heavy atoms (e.g. rare-earth ions). The simplicity of this method is partly due to the use of the local exchange Xα approximation where a local exchange of the Slater type $(\rho 1/3)$ is multiplied by a parameter α[5,6] and to the possibility of avoiding the use of basis sets.

Here we have applied the method to the clusters $(Yb\ O_8Fe_2)^{7-}$ and $(Yb\ O_8Ga_2)^{7-}$. The Yb^{3+} ion in the rare-earth garnets is found at a (c) site[*] with local point symmetry D_2 with eight coordinated O^{2-} ions. The latter are situated at the

---

[*] The conventions for labelling ionic sites are those defined in ref.[7].

corners of a distorted cube centered on the Yb^{3+} ion at two distinct distances (see Fig.1). The angular coordinates of these groups of oxygen ions as obtained from X-ray data [8] of an yttrium gallium garnet host crystal are given by :

	r	θ	φ
Group A	2.3030 Å	56.15°	−167.49°
Group B	2.4323 Å	59.07°	98.76°

There are three sets of near neighbouring Fe^{3+} ions to an Yb^{3+} site in the rare-earth iron garnets. The two ferric ions considered in this work are those situated on the nearest neighbour (d) site at 3.08 Å from the Yb^{3+} ion, with two intermediate type (A) oxygen ions placed symmetrically about the (c)–(d) direction for each ion pair (Fig.1). It is now well established that the dominant Re–Fe superexchange interaction in the rare earth iron garnets arises between the rare earth ions and these two nearest neighbours (d) ferric ions [9,10] .

In addition to the one-electron energies, the other quantities available from the SCF-Xα-SW calculations are the charge associated with each "muffin-tin sphere" and the percent of atomic component for each molecular orbital. The SCF-Xα-SW parameters used in the calculations are given in Table I. The atomic spheres were chosen to be overlapping. The exchange parameters are

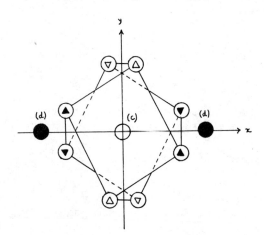

Fig.1 : The geometry of the cluster $(Yb\ O_8Fe_2)^{7-}$. Oxygen ions of
the type (A) are shown with a black triangle. Oxygen of
the type (B) with a white triangle. Δ means above the
xy plane and ∇ below. (c) and (d) are respectively the
Yb^{3+} and the Fe^{3+} sites.

those derived by Schwartz [11,12]. The values for the "outer-sphere" and "interatomic" regions were chosen as a weighted average. All calculations were done using the spin-restricted frozen core method. The stabilizing effects of the crystal environment were taken into account by two uniformly charged "Watson spheres" [13]. This technique avoided the pushing up into continuum of some energy levels.

Table II presents the results obtained for the ground state electron structure of the valence levels of $(Yb\ O_8Fe_2)^{7-}$. The Table gives the orbital energies and the electron density distribution in the different atomic spheres. All orbitals for the valence electrons of Yb^{3+} are shown. Also presented are those molecular orbitals which were not completely localized on the oxygen spheres or on the iron spheres. Analysis of the charge distribution of these molecular orbitals indicate an appreciable admixture between the wavefunctions of ytherbium, iron and type (A) oxygen ions. Thus they can contribute to the Re-Fe super-

Table I: Parameters used in the SCF-Xα-SW calculations of the clusters $(Yb\ O_8Fe_2)^{7-}$ and $(Yb\ O_8Ga_2)^{7-}$

Xα-SW parameters	Sphere radii (in a.u.)
Yb = 0.69317	$R_{Yb^{3+}}$ = 1.9508
O = 0.74447	$R_{O^{2-}}$ = 2.6456
Ga = 0.70690	$R_{Fe^{3+}} = R_{Ga^{3+}}$ = 1.2094
$(Yb\ O_8Fe_2)^{7-}$:	$R_{outersphere}$ = 7.2420
$\alpha_{outersphere} = \alpha_{interatomic}$	= $R_{outer\ Watson\ sphere}$
= 0.73376	$R_{inner\ Watson\ sphere}$ = 5.2820
$(Yb\ O_8Ga_2)^{6-}$:	
$\alpha_{outersphere} = \alpha_{interatomic}$	
= 0.73293	

Table II : Energy levels and charge distribution for $(Yb\ O_8Fe_2)^{7-}$

MO type*	Symmetry	Energy (in Rydbergs)	Charge distribution**			
			Yb sphere	O(A) sphere	O(B) sphere	Fe sphere
Yb 5s	A1	−4.2471	97.01	0.29	0.03	0.02
O(A) 2s	B3	−3.0769	3.34	23.80	0.01	0.47
O(A) 2s	B1	−3.0654	1.27	24.19	0.02	0.79
Yb 5p	B3	−2.6241	82.12	3.14	−	0.05
Yb 5p	B1	−2.5715	90.75	1.47	0.21	0.04
Yb 5p	B2	−2.5281	95.10	0.24	0.32	0.01
O(A) 2p	B2	−2.1484	0.39	15.42	1.18	16.58
O(A) 2p	B1	−2.1473	0.71	15.23	1.26	16.63
O(A) 2p	A1	−2.0202	0.14	20.71	0.89	6.52
O(A) 2p	B3	−2.0174	0.25	20.25	1.11	6.95
O(A) 2p	A1	−1.9721	1.24	23.38	0.63	1.03
O(A) 2p	B2	−1.9421	1.39	24.41	0.01	0.18
O(A) 2p	B3	−1.9268	7.35	22.65	0.07	0.55
O(A) 2p	B1	−1.9232	5.63	23.40	0.03	0.06
Yb 4f	B1	−1.630888	95.31	0.90	0.01	0.39
Yb 4f	B1	−1.627939	99.25	0.07	0.02	0.03
Yb 4f	B2	−1.627660	99.22	0.09	0.01	0.03
Yb 4f	A1	−1.627568	99.17	0.13	−	0.02
Yb 4f	B3	−1.627231	99.77	0.25	−	0.03
Yb 4f	B2	−1.626974	99.48	0.05	−	0.03
Yb 4f	B3	−1.626679	99.67	0.02	−	−
Fe 3d	B2	−1.5444	0.39	11.62	−	26.20
Fe 3d	B1	−1.5412	0.20	11.40	−	25.82
Fe 3d	A1	−1.4548	0.09	5.19	−	38.37
Fe 3d	B3	−1.4533	0.22	5.13	−	38.41

* Main atomic component of the corresponding MO

** In % of one electron.

Table III : Energy levels of Yb^{3+} in $(Yb\ O_8Ga_2)^{7-}$

Orbital	Symmetry	Energy (in Rydbergs)
5s(2)	A_1	-3.783891
5p(2)	B_3	-2.138069
5p(2)	B_1	-2.204097
5p(2)	B_2	-2.067658
4f(2)	B_1	-1.158838
4f(2)	B_2	-1.157686
4f(2)	B_3	-1.156680
4f(1)	B_2	-1.156437
4f(2)	A_1	-1.156408
4f(2)	B_3	-1.155739
4f(2)	B_1	-1.14152

Occupation number in parentheses

exchange paths which produce an exchange field at the Yb site.
The orbitals belonging to the type (B) oxygens were almost
completely localized on the oxygen (B) atomic spheres. From
Table II we see that the predominant charge transfer from the

Yb site to the O(A) sites arises through the 5p shell rather than
from the 4f shells.

Table III presents the orbitals for the valence electrons
of Yb^{3+} in the $(Yb\ O_8Ga_2)^{7-}$ system. All the orbital energies are
on average scaled 0.4 to 0.5 Rydberg higher than those corresponding
to the $(Yb\ O_8Fe_2)^{7-}$ case. This occurred due to different Watson
sphere charges being used in both cases.

Comparisons between the energy levels of the 4f orbitals
of Yb^{3+} of both systems $(Yb\ O_8Fe_2)^{7-}$ and $(Yb\ O_8Ga_2)^{7-}$ show the
difficulty of explicitly calculating exchange splittings from these
calculations. It is obvious that covalency effects are important
in discussing crystal field effects and it is not possible to
simply assume that the crystal field effects appropriate to rare
earth ions in iron garnets can be obtained by using those

calculated in the structurally similar diamagnetic garnets. These preliminary results also show that the exchange interaction between the rare-earth ion and ferric ions cannot be treated as a small perturbation on the crystal field states [14].

Acknowledgement - This work was supported in part by Brazilian Institutions : CNPq and Universidade Federal do Parana.

REFERENCES

[1]   K.H. Johnson, Journal of Chemical Physics 45, 3085 (1966).

[2]   K.H. Johnson and F.C. Smith, Physical Review B5, 831 (1972).

[3]   K.H. Johnson, Advances in Quantum Chemistry 7, 143 (1973).

[4]   J.C. Slater and K.H. Johnson, Physical Review B5, 844 (1973).

[5]   J.C. Slater, Advances in Quantum Chemistry 6, 1 (1972).

[6]   J.C. Slater, The Self Consistent Field for Molecules and Solids, Vol.4, Quantum Theory of Molecules and Solids (McGraw Hill, New York, 1974).

[7]   International Tables of X-ray Crystallography (Kynoch Press : Birmingham 1952).

[8]   F. Euler and D. Bruce, Acta Crystallographic 19, 971 (1965).

[9]   I. Nowick and S. Ofer, Physical Review 153, 409 (1967).

[10] G.M. Copland, Chemical Physics Letters 7, 175 (1970).

[11] K. Schwartz, Physical Review B5, 2466 (1972).

[12] K. Schwartz, Theoretica Chimica Acta 34, 225 (1974).

[13] R.E. Watson, Physical Review 111, 1108 (1958).

[14] Velichy and Veltrichy.

DIRECTED COMPTON PROFILES AND DIAMAGNETIC SUSCEPTIBILITIES

FOR RUBIDIUM HALIDE CRYSTALS

K. Mansikka and O. Aikala

Department of Physical Sciences
University of Turku

SF-20500 Turku 50, Finland

Directed Compton profiles and magnetic susceptibilities of RbF, RbCl, and RbBr crystals have been calculated using the LCAO-approximation for the Bloch states of the electrons. The ortho-gonalization effects involved are treated by Löwdin's symmetrical method to the second nearest neighbour approximation. The results obtained show significant solid state effects and they are in a general agreement with experimental data (when available).

In previous investigations, a model based on the LCAO-approximation for the calculation of X-ray scattering factors, Compton profiles, and diamagnetic susceptibilities for ionic crystals has been introduced[1,2,3]. The numerical applications to a series of ionic solids show that this model predicts solid state effects on the ions of a crystal in good agreement with experiments.

By definition the Compton profile is the projection of the electron momentum density onto the photon scattering vector, and it is determined by relations

$$J(p_x) = \iint \rho(\underline{p}) \, dp_x dp_y \,, \qquad (1)$$

$$\rho(\underline{p}) = (2\Pi)^{-3} \iint d\underline{r}d\underline{r}' e^{i\underline{p} \cdot (\underline{r}-\underline{r}')} \rho(\underline{r},\underline{r}') \,, \qquad (2)$$

where $\rho(\underline{p})$ is the total momentum density and $\rho(\underline{r},\underline{r}')$ denotes the one-matrix of the electronic system in the initial state. The variables $\underline{p}$ and $\underline{r}$ have, of course, their usual meanings.

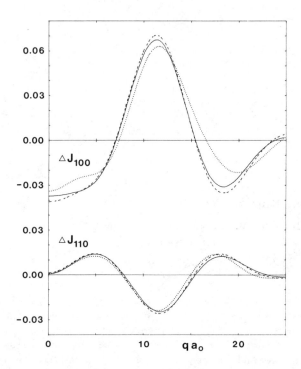

<u>Fig.1</u> : The convoluted theoretical anisotropies ΔJ_{100} and ΔJ_{110}
for RbF (dotted lines), RbCl (dashed lines), and RbBr
(solid lines vs. a reduced variable qa_o (a_o = lattice
parameter of a crystal). $\Delta J_{hkl} = J_i - J_{hkl}$, where J_i is
the isotropic profile, q denotes the momentum variable.

<u>Table 1</u> : Theoretical and experimental values of the molar
susceptibility for RbF, RbCl, and RbBr (cgs units $\times 10^6$)

| | Present work | | Powder crystal |
	$X_{free-ion}$	$X_{crystal}$	X_{exp}^5
RbF	−37.8	−35.4	−31.9
RbCl	−55.5	−51.2	−46.0
RbBr	−67.4	−62.1	−56.4

The theoretical results indicate[4] that the Compton profile is a sensitive quantity to changes in the one-electron wave functions, especially at the low values of the momentum variable. On the other hand, a dominant contribution to the total momentum density at the low momentum values is resulting from the diffuse valence states. The Compton profile is therefore a useful physical quantity for studying crystalline effects on the atoms and for testing electronic wave functions.

In the case of diamagnetic solids, a comparable quantity to the Compton scattering intensity is the magnetic susceptibility, because both of them are determined by the square of the vector potential of the external electromagnetic field. In Compton scattering, the vector potential is a time-varying quantized field, whereas it is usually assumed as a static potential of the uniform magnetic field when considering diamagnetic effects. In this sense, diamagnetism can be interpreted as a macroscopic Compton effect of the uniform magnetic field. The susceptibility of ionic solids within the first-order perturbation theory is given by

$$\chi = - \frac{1}{B} \frac{\partial <H'>}{\partial B} \tag{3}$$

where B is the external magnetic field and H' denotes the magnetic hamiltonian.

Diamagnetic susceptibility is also rather sensitive to reveal solid state effects on the atoms of crystals for it is proportional to the expectation value of r^2 over the crystal ion. This value is primarily determined by the diffuse overlapping valence states of the atoms, which undergo most remarkable changes during a crystal formation from the corresponding free atoms.

In this work, we have calculated the isotropic and directed Compton profiles and diamagnetic susceptibilities of RbF, RbCl, and RbBr crystals using the above LCAO-approximation. The Compton profiles show significant solid state effects both in the isotropic and directional cases (Figure 1). The same holds true for the susceptibilities, which are also in a general agreement with the experimental values[5] whose accurate remeasurements are very much desired. This would be important because in the heavy ionic crystals the relativistic effects may be important.

REFERENCES

1. O. Aikala and K. Mansikka : Physik der Kondensierten Materie
   11, 243 (1970).

2. O. Aikala, W. Jokela, and K. Mansikka : Journal of Physics C <u>6</u>, 1116 (1973).

3. K. Mansikka and S. Mikkola : Journal of Physics C <u>7</u>, 3737 (1974).

4. B. Williams (editor) : Compton Scattering, McGraw-Hill, Great Britain (1977).

5. Handbook of Chemistry and Physics (1973-74).

OSCILLATOR STRENGTH SUMS AND EFFECTIVE ELECTRON NUMBER

IN OPTICAL SPECTROSCOPY OF SOLIDS

D.Y. Smith
Argonne National Laboratory
Argonne, Illinois 69439, USA
and
Physikalisches Institut (Teil 2) der
Universität Stuttgart, 7 Stuttgart 80, Germany

E. Shiles*
Virginia Commonwealth University
Richmond, Virginia 23284, USA

Recently developed composite optical data for metallic aluminum are employed to illustrate various definitions of n_{eff}, the number of electrons effective in an absorptive process. Two effects are demonstrated : the "screening" of valence-electron processes by the polarizable ion cores and the redistribution of oscillator strength between valence and core electrons.

    Composite sets of optical data based on a self-consistent iterative Kramers-Kronig analysis of a wide variety of experimental measurements have recently become available for a number of simple substances[1,2]. The various definitions of $n_{eff}(\omega)$, the number of electrons effective in optical processes up to energy $\hbar\omega$,[3] have been illustrated with this data and two effects are apparent :
1. the "screening" of real valence- or conduction-electron processes by virtual processes in the core states, and 2. the "transfer" of oscillator strength between core and valence electrons.

OSCILLATOR STRENGTH SUMS

The f sum rule is a major guide to analysis of optical data. However, in principle the f sum rule holds only for the infinite range which includes all absorptions of the system. Consequently, care must be employed in interpreting finite-energy f sums. A number of definitions for partial oscillator strength sums are possible because the f sum rule takes different forms for the various optical constants. The three most common are for the imaginary parts of the dielectric function, $\varepsilon = \varepsilon_1 + i\varepsilon_2$, the refractive index, $N = n + i\kappa$, and the energy-loss function, ε^{-1} [4]:

$$\int_0^\infty \omega\varepsilon_2(\omega)d\omega = \frac{\pi}{2}\omega_p^2 , \tag{1}$$

$$\int_0^\infty \omega\kappa(\omega)d\omega = \frac{\pi}{4}\omega_p^2 , \tag{2}$$

and

$$\int_0^\infty \omega Im[\varepsilon(\omega)^{-1}] d\omega = -\frac{\pi}{2}\omega_p^2 , \tag{3}$$

where $\omega_p = (4\pi ne^2/m)^{1/2}$ is the plasma frequency.

These rules all have the same physical content [5]. They arise from causality and the equations of motion. These physical requirements translate into the mathematical conditions that the optical properties are analytic and have high-frequency behaviour depending only on the electron density, n, charge, e, and mass, m.

For example, the complex dielectric function $\varepsilon(\omega)$, is analytic in the upper half plane, except for a pole at the origin in the case of conductors, and has an asymptotic behavior

$$\varepsilon(\omega) = 1 - \omega_p^2/\omega^2 + \dots , \omega \to \infty \tag{4}$$

Related analytic properties are shared by the complex refractive index, $N = \varepsilon^{1/2}$, and the inverse of the dielectric function. Further, all these functions have asymptotic high-frequency behaviours involving only ω_p^2/ω^2 even though the functions are vastly different at finite frequencies. Combining the dispersion relations for these quantities with this asymptotic behaviour yields [6] the several f sum rules.

In analogy with these rules it has become common to define the effective number of electrons contributing to optical processes up to energy $\hbar\omega$ by [3,7]:

$$n_{eff}(\omega)\big|_{\varepsilon} = \frac{m}{2\pi^2 e^2} \int_0^{\omega} \zeta\varepsilon_2(\zeta)d\zeta , \qquad (5)$$

$$n_{eff}(\omega)\big|_{\kappa} = \frac{m}{\pi^2 e^2} \int_0^{\omega} \zeta\kappa(\zeta)d\zeta , \qquad (6)$$

and

$$n_{eff}(\omega)\big|_{\varepsilon^{-1}} = -\frac{m}{2\pi^2 e^2} \int_0^{\omega} \zeta Im[\varepsilon^{-1}(\zeta)] d\zeta . \qquad (7)$$

Here it is necessary to distinguish the particular $n_{eff}(\omega)$ by the subscript ε, κ, and ε^{-1}. In the limit of $\omega \to \infty$ all three functions approach the electron density. However, at intermediate frequencies the functions are quite different and cannot be interpreted as physical electron densities.

An example of the three forms of $n_{eff}(\omega)$ is given in Figure 1 for recently compiled values of the optical properties of metallic aluminum[2]. The three quantities are strikingly different below the onset of core transitions at the $L_{II,III}$ edge, but merge at higher energies.

From the formulation[8] of the dielectric function in terms of dipole oscillator strengths, it follows that $n_{eff}(\omega)\big|_{\varepsilon}$ may be consistently interpreted as the integrated dipole oscillator strength. In general $n_{eff}(\omega)\big|_{\kappa}$ and $n_{eff}(\omega)\big|_{\varepsilon^{-1}}$ do not have as simple an interpretation, but for isolated absorptions they are simply related to $n_{eff}(\omega)\big|_{\varepsilon}$. This may be seen[7] by considering a model system consisting of an absorber imbedded in a transparant dielectric medium with a real, constant dielectric function ε_b.

This model would apply, for example, to the absorption spectrum of valence or conduction electrons in a solid in which the absorption of the valence and core electrons are well separated. For energies at which there is significant valence-electron absorption the principal effect of the core states is simply to contribute a dispersionless, real term to the dielectric function.

At energies intermediate between the "core" and "valence" absorptions of the model the dielectric function has the limiting value

$$\varepsilon(\omega) = \varepsilon_b - \omega_{p,v}^2/\omega^2 + \dots , \qquad \omega \to \hat{\omega}. \qquad (8)$$

Here $\hbar\hat{\omega}$ is an energy large compared with the transition energy of the valence electrons, but less than the lowest core excitation.

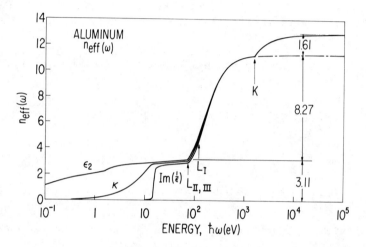

<u>Fig.1</u> : The effective number of electrons effective in absorptive
processes up to energy $\hbar\omega$, $n_{eff}(\omega)$, for metallic aluminum
based on the composite optical data of Shiles et al.,
Ref.2.

The quantity $\omega_{p,v}$ is the apparent plasma frequency of the valence electrons,

$$\omega_{p,v}^2 = \frac{4\pi e^2}{m} \, \Sigma \, f_v \, , \tag{9}$$

where $\Sigma \, f_v$ is the total oscillator strength of the valence electrons. The corresponding limits of the refractive index and ε^{-1} are

$$N(\omega) = \varepsilon_b^{1/2} - \omega_{p,v}^2 / 2\varepsilon_b^{1/2}\omega^2 + \ldots \, , \quad \omega \to \hat{\omega}, \tag{10}$$

and

$$\varepsilon^{-1}(\omega) = \varepsilon_b^{-1} + \omega_{p,v}^2 / \omega_b^2 \omega^2 + \ldots \, , \quad \omega \to \hat{\omega} \, . \tag{11}$$

Sum rules then follows[7] for the finite-energy interval covering the valence-electron absorption in the approximation that $\omega_{p,v} \ll \hat{\omega}$:

$$\int_0^{\hat{\omega}} \zeta \varepsilon_2(\zeta) d\zeta \approx \frac{\pi}{2} \, \omega_{p,v}^2 \, , \tag{12}$$

$$\int_0^{\hat{\omega}} \zeta \kappa(\zeta) \, d\zeta \approx \frac{\pi}{4} \, \omega_{p,v}^2 / \varepsilon_b^{1/2} \, , \tag{13}$$

and

$$\int_0^{\hat{\omega}} \zeta \text{Im}[\varepsilon^{-1}(\zeta)] d\zeta \approx - \frac{\pi}{2} \, \omega_{p,v}^2 / \varepsilon_b^2 \, . \tag{14}$$

In agreement with the interpretation that $n_{eff}(\omega)|_\varepsilon$ gives the total dipole oscillator strength up to energy $\hbar\omega$, Eq. 12 is independent of core processes as may be seen by the absence of terms in ε_b. However, both Eqs. 13 and 14 and, hence, $n_{eff}(\hat{\omega})|_\kappa$ and $n_{eff}(\hat{\omega})|_{\varepsilon^{-1}}$ depend on ε_b. That is, they are affected by virtual processes in the core states.

This behaviour is demonstrated for aluminum in figure 1 by the $n_{eff}(\omega)$ values in the region between the bulk plasmon resonance at 15 eV and the onset of core transitions at 72.7 eV. In aluminum the conduction- and core-electron absorptions are well separated. The oscillator strength of the three conduction electrons is virtually exhausted above the plasmon resonance so that the $n_{eff}(\omega)$ values rise only slightly as the $L_{II,III}$ edge is approached. To within computational error the ratios between the various $n_{eff}(\omega)$ values are consistent with Eqs. 12–14 for values of ε_b between 1.03_3 and 1.04_0. The corresponding value of ε_b calculated by a Kramers-Kronig transformation of the core electron absorption is 1.03_6, in good agreement with these values.

EXCHANGE OF OSCILLATOR STRENGTH

The f sum rule for a set of n particles states that

$$\sum_{\ell} f_{j+\ell} = n \ , \tag{15}$$

where the sum is taken over all final states ℓ for transition from a given initial state j. Here the dipole oscillator strength is given by[9]

$$f_{j+\ell} = \frac{2m}{3\hbar^2} (E_\ell - E_j) \left| <\Psi_j | \vec{R} | \Psi_\ell > \right|^2 \ , \tag{16}$$

where E_k is the energy of state Ψ_k, and $\vec{R} = \sum_s \vec{r}_s$ is the dipole operator for the system. Occasionally this rule is interpreted as implying that each electron makes a contribution of unity to the f sum. This is incorrect. Generally, electrons in the higher-lying states of an interacting system make contributions per electron greater than unity while the contribution of electrons in lower energy states is less than unity. This is illustrated in Table I. The experimental values shown are based on the composite data of Shiles et al.[2] for aluminum metal. They show a total oscillator strength sum for the two K-shell electrons of only 1.6_1, while the three conduction electrons exhibit an oscillator strength of 3.1_1.

Table I : The oscillator strengths for the energy levels of aluminum

Shell	Occupation	Metal	Atomic theory
Conduction or valence	3	3.1_1	3.12
$L_{II,III}$	6	8.2_7	$\left.\begin{matrix} 6.98 \\ 1.35 \end{matrix}\right\}8.33$
L_I	2		
K	2	1.6_1	1.55
TOTAL	13	12.9_9	13.00

The "transfer" of oscillator strength from core to valence electrons arises from the Pauli principle and is well known in atomic and X-ray physics[10,11], but appears to have been observed only for d-band materials in solids[12-14]. The essential elements of the effect may be understood within the one-electron approximation. In this model each electron is pictured as moving in a fixed potential arising from the other charges in the system. Each electron has an f sum of unity provided transitions to all other one-electron states are included. However, in the actual many-electron system transitions to states occupied by other electrons are, of course, forbidden.

In this approximation the f sum for a single electron may be divided into a sum over physically-allowed transitions to vacant states and one over the Pauli-principle-forbidden transitions to occupied states. This yields rules for the individual electrons of the form

$$\sum_{\text{vacant}} \tilde{f}_{j \to \ell} = 1 - \sum_{\text{occupied}} \tilde{f}_{j \to \ell} \, , \qquad (17)$$

where the one-electron oscillator strengths $\tilde{f}_{j \to \ell}$ are given by the analog of Eq.16, but using the one-electron energies and wave functions. The sign of the various one-electron oscillator strengths depends on the difference in the energies of the states and is positive for absorptions and negative for emissions. Thus, for a single electron the oscillator strength sum for transitions to vacant states is greater than unity if the electron lies at the top of the Fermi distribution so that only emissions forbidden by the Pauli principle occur on the right-hand side of Eq.17. Similarly, the sum is less than unity for an electron at the bottom of the distribution since then only forbidden absorptions contribute to the right-hand side. Notice that a transition which is forbidden in emission for one electron is forbidden in absorption for another. Hence, a sum over all the individual one-electron oscillator strengths yields just the f sum rule for the entire system, Eq.15, the sums over occupied states in Eq.17 all cancelling in pairs.

In the case of aluminum the Pauli principle forbidden transitions are 1s↔2p, 1s↔3p, 2s↔2p, 2s↔3p, 3s↔2p, and 3s↔3p. Numerical estimates of the relevant energies and matrix elements were made in the Hartree-Fock-Slater approximation using Herman and Skillman's atomic wave function programs[15]. The results are given in Table I by the column labled atomic theory. The agreement between the observed oscillator strength distribution in the solid and that predicted for the atom is remarkably close. This might at first seem surprising since the spectra of the outermost electrons is quite different for the two, but it may be remarked that the f sum is an integral over the entire spectrum. Further, the core states are almost identical in both the atom and the metal.

An interesting point brought out by the calculation is that the 2s level suffers a large loss of oscillator strength to the 2p level. This leads to an oscillator strength of 1.35 for the L_I shell compared with 6.98 for the $L_{II,III}$ shell. As a result, the L_I edge is difficult to distinguish from the $L_{II,III}$ background in the absorption data of metallic aluminum [2].

Acknowledgements - The authors would like to thank K. Jakobi for computer system aid. One of us (D.Y.S.) wishes to express gratitude to Prof. Dr. H. Pick for his encouragement and hospitality. Support for this research was provided by the U.S. Department of Energy and the Deutsche Forschungsgemeinschaft.

REFERENCES

* Present address : ECAC, IITRI, Annapolis, Maryland 21402, USA.

1. E. Shiles and D.Y. Smith, Bulletin of the American Physical Society, 22, 92 (1977); 23, 226 (1978); and 24, 335 (1979).

2. E. Shiles, T. Sasaki, M. Inokuti and D.Y. Smith, to be published.

3. See, for example, H. Ehrenreich, in The Optical Properties of Solids, edited by J. Tauc (Academic, New York, 1966), Sec.9, p.146.

4. See, for example, F. Stern, in Solid State Physics, edited by F. Seitz and D. Turnbull, (Academic, New York, 1963), Vol.15.

5. M. Altarelli and D.Y. Smith, Physical Review B 9, 1290 (1974).

6. M. Altarelli, D.L. Dexter, H.M. Nussenzveig and D.Y. Smith, Physical Review B 6, 4502 (1972).

7. D.Y. Smith and E. Shiles, Physical Review B 17, 4689 (1978).

8. See, for example, D. Pines, Elementary Excitations in Solids (Benjamin, N.Y., 1963).

9. H.A. Bethe and E.E. Salpeter, Quantum Mechanics of One- and Two-Electron Atoms (Springer Verlag, Berlin, 1957).

10. Y. Sugiura, Philosophical Magazine 4, 495 (1927).

11. R. de L. Kronig and H.A. Kramers, Zeitschrift für Physik 48, 174 (1928).

12. H.R. Philipp and H. Ehrenreich, Physical Review 129, 1550 (1963).

13. T.E. Faber, Advances in Physics 15, 547 (1966).

14. N.V. Smith, Advances in Physics 16, 629 (1967).

15. F. Herman and S. Skillman, Atomic Structure Calculations (Prentice-Hall, Engelwood-Cliffs, N.J., 1963).

A DIFFUSION MODEL FOR 1/f NOISE FOR SAMPLES OF FINITE DIMENSIONS

L. Bliek

Physikalisch-Technische Bundesanstalt

D-3300 Braunschweig, FRG

Noise generated by fluctuations of a parameter which is subject
to a diffusion equation is shown to have a 1/f power spectrum
over a wide range of frequencies, provided that the thickness
of the sample is small compared to its other dimensions.

1. INTRODUCTION

1/f noise is a quite general phenomenon turning up in
many different fields[1-4]. A generally accepted explanation has
not yet been given and it is sometimes even considered to be
difficult to construct mathematical models for its origin[4]. In
the following a model is presented which is based on the
assumption that 1/f noise is caused by fluctuations of a parameter u, for which a continuity equation

$$\frac{\partial u}{\partial t} = D \, \Delta u \qquad (1)$$

holds. This differential equation is the well known diffusion or
heat conductivity equation. Although it appears to be generally
accepted that such a diffusion like model cannot describe 1/f
noise[5,6], one finds that for systems with finite dimensions the
power spectrum may have a 1/f region covering many frequency
decades. Further regions with simple, but different dependencies
on frequency are found below and above the 1/f part of the
spectrum. The results of the model calculation indicate that for
diffusion-like systems 1/f noise should be typical of planar
structures, where the sample thickness is small compared to the
other two dimensions.

2. CALCULATION OF THE NOISE SPECTRUM

Through an infinite medium, for which equation 1 holds, a fluctuation u_j occuring at a time t_j and having the initial shape $\phi(\vec{r}-\vec{r}_j)$ will spread according to

$$u_j(\vec{r}-\vec{r}_j,t-t_j) = \int\!\!\int\!\!\int_{-\infty}^{+\infty} \frac{\phi(\vec{r}'-\vec{r}_j)}{8\pi D(t-t_j)^{3/2}} \exp \frac{(\vec{r}-\vec{r}')^2}{4D(t-t_j)} \, d\vec{r}'. \quad (2)$$

If positive and negative, but otherwise equal fluctuations occur independently of each other and randomly in space and time with a constant probability $\rho\nu$ which is the same for either sign, the resulting power spectrum can readily be calculated using Carson's theorem. One finds

$$S_u(f,\vec{k}) = \frac{2\,\rho\nu}{k^4 D^2 + 4\pi^2 f^2} \left\{ \left[\int\!\!\int\!\!\int_{-\infty}^{+\infty} \phi(\vec{r})\cos(\vec{k}\vec{r})d\vec{r} \right]^2 \right.$$

$$\left. + \left[\int\!\!\int\!\!\int_{-\infty}^{+\infty}\phi(\vec{r})\sin(\vec{k}\vec{r})d\vec{r} \right]^2 \right\}. \quad (3)$$

For fluctuations that start as δ-functions in space, i.e. for

$$\phi(\vec{r}) = K\delta(\vec{r}) \quad (4)$$

this method to obtain the spectrum is equivalent to the introduction of a Langevin source $L(\vec{r},t)$ in eq.1, with

$$\langle L(\vec{r},t)\cdot L(\vec{r}+\vec{R},t+\nu)\rangle = 2\rho\nu v^2 K^2 \delta(\vec{R})\,\delta(\nu). \quad (5)$$

$S_u(f,\vec{k})$ is the Fourier transform of the autocorrelation function $\Phi_u(\nu,\vec{R})$ of the fluctuations in u. If one is only interested in the variation with time of u at an arbitrary, but fixed point in the system the quantity to be considered is $\Phi_u(\nu,0)$. Its Fourier transform with respect to time is given by

$$S_u(f) = \int\!\!\int\!\!\int_{-\infty}^{+\infty} S_u(f,\vec{k})d\vec{k} \quad (6)$$

Choosing $\phi(\vec{r})$ as given by eq.4 one finds

$$S_u(f) = \frac{\rho\nu\ K^2}{4(\pi D)^{3/2}} f^{-1/2} \quad (7)$$

For samples of finite size one cannot use Carson's theorem as for infinite systems and also different boundary conditions have to be applied. For bar-shaped samples with dimensions L_x, L_y and L_z both problems can be dealt with by extending the sample to infinity and replacing $u_j(\vec{r}-\vec{r}_j, t-t_j)$ with

$$u'_j(\vec{r}-\vec{r}_j, t-t_j) = \lim_{N_x, N_y, N_z \to \infty} \frac{1}{(2N_x+1)(2N_y+1)(2N_z+1)}$$

(8)

$$\sum_{i=y,x,z} \sum_{n_i=-N_i}^{N_i} u_j(\vec{r}-\vec{r}_j, -\vec{L}_n) + u_j(\vec{r}+\vec{r}_j, -\vec{L}_n) \}$$

where

$$\vec{L}_n = (n_x L_x, n_y L_y, n_z L_z) .$$

(9)

u'_j produces no flow through the boundaries of the original sample. The corresponding noise spectrum is still as given by eq.3 but $\vec{k}$ is now restricted to values for which

$$\vec{k}^2 = \frac{4\pi^2 m_x^2}{L_x^2} + \frac{4\pi^2 m_y^2}{L_y^2} + \frac{4\pi^2 m_z^2}{L_z^2}$$

(10)

holds with integer values for m_x, m_y and m_z. Choosing once more $\Phi(\vec{r})$ as given by equation 4 one finds for $S_u(f)$:

$$S_u(f) = (\rho v K^2/V) \sum_{m_x=-\infty}^{+\infty} \sum_{m_y=-\infty}^{+\infty} \sum_{m_z=-\infty}^{+\infty} \frac{1}{8\pi^4 (m_x^2/L_x^2 + m_y^2/L_y^2 + m_z^2/L_z^2)^2 D^2 + 2\pi^2 f^2}$$

(11)

Any of the sums in eq.11 can be replaced by an integral as long as the corresponding sample dimension is large compared to $(2\pi D/f)^{1/2}$. This produces up to three frequency regions in which the noise power spectrum depends on f in different ways. Calculating the integrals one finds if $L_x \geqslant L_y \geqslant L_z$:
For $f \gg 2\pi D/L_x^2$, $2\pi D/L_y^2$, $2\pi D/L_z^2$

$$S_u(f) \approx \frac{\rho v K^2}{4(\pi D)^{3/2}} f^{-1/2}$$

(12)

which holds exactly for infinite samples (eq.7).
For $f \gg 2\pi D/L_x^2$, $2\pi D/L_y^2$ but $f \leqslant 2\pi D/L_z^2$

$$S_u(f) \approx \frac{\rho v K^2}{8\pi D L_z} f^{-1} .$$

(13)

Here the noise spectrum is the same as for an infinite two-dimensional sample. The region extends from $f \approx 4D/\pi L_y^2$ to $f \approx \pi D/4L_z^2$ and may, for typical sample dimensions, cover many frequency decades. It should be missing, however, for wirelike samples with $L_y \approx L_z$.

For $f \gg 2\pi D/L_x^2$ but $f \ll 2 D/L_y^2$, $2\pi D/L_z^2$

$$S_u(f) \approx \frac{\rho \nu K^2}{4\pi^{3/2} D^{1/2} L_y L_z} f^{-3/2} \tag{14}$$

as for one-dimensional infinite samples.

For $f \ll 2\pi D/L_x^2$, $2\pi D/L_y^2$, $2\pi D/L_z^2$

$$S_u(f) \approx \frac{\rho \nu K^2}{2\pi^2 L_x L_y L_z} f^{-2} \ , \tag{15}$$

since here the only appreciable contribution to $S_u(f)$ comes from the term with $m_x = m_y = m_z = 0$ in eq.11.

For uncorrelated fluctuations eq.15 should be valid down to $f = 0$ where it obviously diverges. The divergence can be removed, if a suitable correlation is introduced. This produces an additional factor

$$C = \frac{4 \pi^2 f^2 \tau^2}{1 + 4\pi^2 f^2 \tau^2} \tag{16}$$

on the left hand side of eq.3 and 11. Hence the spectrum becomes flat for $f \ll 1/\tau$ and simultaniously $f \ll 2\pi D/L_x^2$, $2\pi D/L_y^2$, $2\pi D/L_z^2$. In this part of the spectrum

$$S_u(f) \approx 2\rho \nu K^2 \tau^2/(L_x L_y L_z) \ . \tag{17}$$

At very high frequencies, comparable to $2\pi D/a^2$, where a is the lattice constant, the general restriction of meaningful k values for periodic lattices to $-2\pi/a < k < 2\pi/a$ becomes effective and the limits in the sums of eq.11 have to be replaced by L_x/a, L_y/a and L_z/a. In the high-frequency limit eq.11 reduces to

$$S_u(f) \approx \frac{4 \rho \nu K^2}{\pi^2 a^3} f^{-2} \tag{18}$$

Upon increasing f, $S_u(f)$ decreases now more rapidly than according to eq.12. This leads to a finite value for

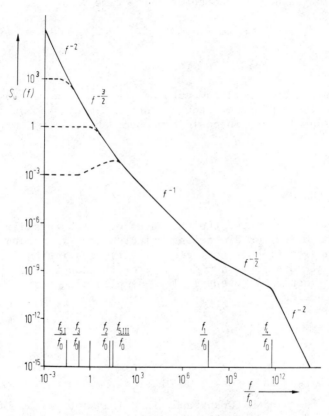

<u>Fig.1</u> : Calculated noise power spectrum $S_u(f)$ is in units of
$$\rho v K^2 L_x^3/(8\pi^4 D^2 L_y L_z) = \rho v K^2/(2\pi^2 f_o^2 L_x L_y L_z)$$
$$f_o = 2\pi D/L_x^2, \quad f_1 = \pi D/(4L_z^2), \quad f_2 = 4D/(\pi L_y^2),$$
$$f_3 = 4D/(\pi L_x^2), \quad f_4/f_o = (2^5/\pi^4)^{1/3} L_x^2/a^2 \approx 0,688 \cdot L_x^2/a^2,$$
$$f_{5,I} = 1/(2\pi\tau_I) \quad \text{and} \quad f_{5,III} = 1/(2\pi\tau_{III})$$

$$\int_{\infty}^{\infty} S_u(f)df = \Phi_u(0,0) = <u^2> \ .$$

An example of a calculated spectrum is shown in the figure for a sample with $L_x/L_y = 10$, $L_x/L_z = 10000$, $a = 0,01\ L_z$ and three different values of the correlation time :

$$\tau_I = 10^{3/2}.L_x^2/(4\pi^2 D)\ ,\ \tau_{II} = L_x^2/(4\pi^2 D) = 1/(2\pi f_o),$$

$$\tau_{III} = 10^{-3/2}L_x^2/(4\pi^2 D)\ .$$

For the figure, the parameters of the sample were chosen such as to make the different frequency dependencies show up clearly. A much wider 1/f region could have been obtained by increasing L_x/L_y.

3. CONCLUSION

In samples, the thickness of which is small compared with the other two dimensions, fluctuations in a parameter for which a diffusion or heat conductivity equation holds, can cause 1/f noise over a wide range of frequencies. On the other hand it should be absent in long samples with a small quadratic cross section.

REFERENCES

1. J. Clarke and G. Hawkins, Physical Review B 14, 2826 (1976).

2. B. Guinot and M. Granveaud, IEEE Transactions on Instrumentation and Measurement IM-21, 396 (1972).

3. J. Clarke and R.F. Voss, Journal of the Acoustical Society of America 63 (1), 258 (1978).

4. F.N. Hooge, Physica 83B, 14 (1976).

5. K.M. Van Vliet and A. Van der Ziel, Physica XXXIV, 415 (1958).

6. T.G.M. Kleinpenning, Physica 84B, 353 (1976).

NONLINEAR OPTICAL PROPERTIES AND FLUORESCENCE QUENCHING

IN NDGD-BORATE LASER CRYSTALS

F. Lutz

Institut für Angewandte Physik

Hamburg 36, Jungiusstrasse 11, F.R.G.

Recently grown rare earth borate laser crystals (e.g. $NdAl_3(BO_3)_4$) show second harmonic generation (SHG) comparable to that of $LiNbO_3$. The SHG efficiency is reduced at higher Nd^{3+} concentrations In $Nd_xGd_{1-x}Al_3(BO_3)_4$ crystals. Variation of the Nd^{3+} concentration results in a change of the fluorescence lifetime τ for $Nd_xGd_{1-x}Al_3(BO_3)_4$ between $\tau = 48 \pm 3$ µsec ($x = 0,01$) and $\tau = 19 \pm 1$ µsec ($x = 1$) due to concentration quenching. These nonlinear optical materials are able to form a SHG laser.

The $NdX_3(BO_3)_4$ crystals ($X_3 = Al^{3+}$, Ga^{3+}, Cr^{3+}) are lasers at $\lambda_1 = 1,06$ µm and $\lambda_2 = 1,3$ µm, having the highest optical gain observed so far [1].

The crystals were grown by a flux method with $Li_2O-B_2O_3$ as a flux. $NdAl_3(BO_3)_4$ crystallizes polymorphic in the space groups C2 and $C2/_c$ [2]. The space group R32 which has been reported earlier [3,4] is not realized in our crystals.

Lacking inversion symmetry C2 allows nonlinear optical properties ($C2/_c$ does not) which were investigated by second harmonic generation (SHG). The SHG efficiencies of crystalline powders (10 µm in diameter) of RE borates and $LiNbO_3$ were measured following the method described by Kurtz and Perry [5]. A Nd-YAG

339

laser was used (λ = 1,06 μm) as the pump source. The results are given in figure 1.

Crystals of $Gd_{.59}La_{.41}Ga_3(BO_3)_4$ and $Nd_{.01}Gd_{.64}La_{.35}Al_3(BO_3)_4$ show the highest SHG efficiency of all RE borates [6].

The decrease of SHG efficiency with increasing Nd^{3+} concentration as to be seen from the lower part of figure 1 can be explained by two effects :
1) The frequency doubled radiation (λ = 5300 Å) created in the SHG process can be absorbed by Nd^{3+} ions.
2) Crystals with high Nd^{3+} concentration ($x \stackrel{>}{=} 0,5$) show a higher contribution of the $C2/_c$ phase, which has no nonlinear properties.

The concentration quenching of Nd^{3+} in $Nd_xGd_{1-x}Al_3(BO_3)_4$ was determined by measuring the fluorescence lifetime τ of the metastable laser level $^4F_{3/2}$ of Nd^{3+}. Figure 2 shows the inverse lifetime $\tau^{-1}(x)$ as function of Nd^{3+} concentration in $Nd_xGd_{1-x}Al_3(BO_3)_4$.

The observed decay rate τ^{-1} in figure 2 can be written as

$$\tau(x)^{-1} = \tau_o^{-1} + q(x) ,$$

where τ_o is the radiative lifetime and $q(x)$ is the quenching rate, which depends on the Nd^{3+} concentration [7,8]. Extrapolating the measured data of figure 2 for $x \to 0$ yields a radiative lifetime of τ_o = 50 μsec. The observed linear concentration dependence of the decay rate τ^{-1} in $Nd_xGd_{1-x}Al_3(BO_3)_4$ is typical for high concentrated Nd^{3+} laser materials which exhibit weak concentration quenching. Even for full Nd^{3+} concentration the fluorescence quenching is weak. The quenching process is probably a phonon assisted cross relaxation [9].

Laser action in crystals of $Nd_xGd_{1-x}Al_s(BO_3)_4$ is possible within the full concentration range without severe losses due to nonradiative processes. Crystals with a Nd^{3+} concentration of x = 0.1 show the highest SHG efficiencies. The performance of a laser and a modulator within the same Nd^{3+} borate crystals is realizable.

Acknowledgement — The author would like to thank Miss B.Buchhorn for growing crystals and T.Härig for measuring the fluorescence lifetime.

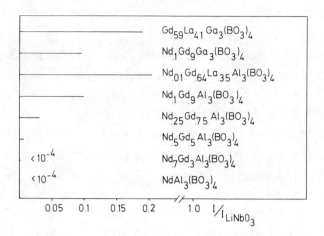

Fig.1 : Measured SHG intensities of RE borates (principal the
system $Nd_xGd_{1-x}Al_3(BO_3)_4$) compared with the SHG intensity
of $LiNbO_3$.

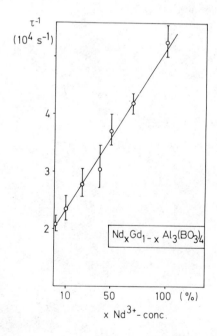

Fig.2 : The decay rate τ^{-1} vs. Nd^{3+} concentration in
$Nd_xGd_{1-x}Al_3(BO_3)_4$.

REFERENCES

1. G. Huber, Neodymium Miniature Lasers in Current Topics in Materials Science, ed. E. Kaldis, North-Holland Amsterdam (1980) 1.

2. O. Jarchow, F. Lutz and K.H. Klaska, Z.f. Krist. 149 (1979) 162.

3. A.A. Ballmann, Am. Min. 47 (1962) 1380.

4. H.Y.-P. Hong and K. Dwight, Mat. Res. Bull. 9 (1974) 1661.

5. S.K. Kurtz and T.T. Perry, J. Appl. Phys. 39 (1968) 3798.

6. F. Lutz, R. Orlowski, T. Bosselmann and G. Huber, to be published.

7. D. Fay, G. Huber and W. Lenth, Opt. Commun. 28 (1979) 117.

8. A. Lempicki, Opt. Commun. 23, 3 (1977) 376.

9. F. Lutz, B. Buchhorn and T. Härig, to be published.

INTRINSIC BIREFRINGENCE IN CuBr CRYSTALS DUE TO

SPATIAL DISPERSION

J.L. Deiss
Laboratoire de Spectroscopie et d'Optique du
Corps Solide, Associé au C.N.R.S. n° 232
Université Louis Pasteur
5, rue de l'Université – 67000 Strasbourg, France

O. Gogolin and E. Tsitsishvili
Institute of Cybernetics,
Tbilissi URSS

C. Klingshirn
Institut für Angewandte Physik
7500 Karlsruhe, Germany

Intrinsic birefringence in single crystals of CuBr has been measured in the vicinity of the absorption edge at 300 K. The strong enhancement of the birefringence in the direction $\vec{q}$ // [110] near the gap is caused by the spatial dispersion of the dielectric function. An attempt is made to deduce band-parameters from the experimental data.

INTRODUCTION

In the electric dipole approximation, cubic crystals are optically isotropic; the dielectric constant is a scalar and no optical birefringence should be observed according to classical crystal optics. However, birefringence in cubic crystals like Si, Ge, GaAs, ZnSe ... has been reported when light propagating along the [110] direction and has been attributed to spatial dispersion[1,2,3].

In the presence of spatial dispersion, the dielectric constant is q-dependent ($\vec{q}$ being the wavevector of light) and is no longer a scalar. If we expand the dielectric tensor $\varepsilon_{ij}(\omega,\vec{q})$ as a function of $\vec{q}$, we obtain :

$$\varepsilon_{ij}(\omega,\vec{q}) = \varepsilon_{ij}(\omega) + i\gamma_{ijk}(\omega)q_k + \alpha_{ijkl}(\omega)q_k q_1 \qquad (1)$$

In crystals with symmetry O_h and T_d (as CuBr), the tensor γ_{ijk} vanishes. For $\vec{q} /\!/ [110]$, there are two different values of the transverse dielectric constant $\varepsilon_T(1\bar{1}0)$ and $\varepsilon_T(001)$, resulting in an optical birefringence given by[4] :

$$\Delta\varepsilon = 2n\Delta n = \varepsilon(1\bar{1}0) - \varepsilon(001) = 1/2[(\varepsilon_L - \varepsilon_T)_{001} - (\varepsilon_L - \varepsilon_T)_{111}]$$

$$(2)$$

where ε_L is the longitudinal dielectric constant.

A strong dispersion of this birefringence is observed in the vicinity of dispersion centers, like energy gaps or excitons, because α_{ijkl} is a function of ω. This effect has been explained by a band structure model taking into account the band warping[2]. In this band model, due to the finiteness of the $\vec{q}$ vector, the degenerate valence bands (VB) split up away from $\vec{k} = 0$, giving rise to partially polarized transitions. According to equation (2), a strong dispersion of the birefringence is observed only if the VBsplittings along the directions [001] and [111] are different (band warping) and is expressed by :

$$\Delta n = D(\hbar\omega)^2 + c_0 g(\omega/\omega_0)(\hbar\omega)^2 \qquad (3)$$

EXPERIMENTAL RESULTS

The spatial dispersion birefringence was measured at 300 K using single oriented CuBr crystals grown by vapor phase. The samples are parallelepipeds of about $6 \times 8 \times 4$ mm with oriented [001], [110], [1$\bar{1}$0] faces cut from crystals blocks. They were polished and chemically etched to eliminate strains and surface depolarizing effects resulting from cutting and polishing. The transmitted light through the crystals of thickness e = 3 to 4 mm was measured between crossed $I^{\perp}$ and parallel $I^{/\!/}$ polarizers for $\vec{q} /\!/ [110]$. The polarization vector being at 45° of the [001] axis of the crystal, we have :

$$I^{\perp} = I^{/\!/} \sin^2\Delta/2, \text{ where } \Delta = 2\pi(n_{/\!/} - n_{\perp})e/\lambda$$

Δ is the phase difference, from which we calculate Δn. The samples used in these preliminary measurements had more or less internal strains, and were selected to present an as weak as possible accidental birefringence. The measured birefringence should be corrected to eliminate the built-in stress, however this correction

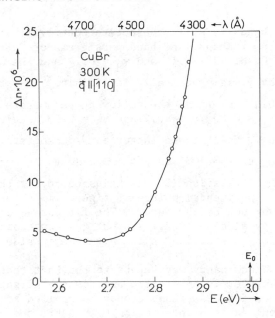

<u>Fig.1</u> : The measured birefringence Δn due to spatial dispersion
versus photon energy in CuBr at 300 K.

does practically not alter the strong dispersion on the high energy
side near E_0.

The observed birefringence Δn is shown in figure 1.
As we approach the dispersion center E_0 (here the 1S exciton of
CuBr with E_0 = 2.9974 eV), the birefringence is strongly increased.
The experimental curve (solid line) gives the intrinsic
birefringence due to spatial dispersion.

DISCUSSION

To explain the enhancement of the birefringence in CuBr,
we tried the band model of Higginbotham and Cardona[4,5] which
explained well the observed birefringence in GaAs, ZnSe ... This
band model taking into account the VB warping, in absence of
exciton interaction, gives the birefringence expressed by the
relation (3). In this expression, the second term represents the
birefringence due to the VB warping, and the parameter c_0 is
related to the Luttinger band parameters γ_1, γ_2 and γ_3. In CuBr,
the direct transitions at the fundamental edge take place between
a degenerate Γ_8 VB and a Γ_6 CB at $\vec{k}$ = 0. If we shift by an
amount $\vec{q}$ in k-space, as a result of a finite wavevector, the Γ_8
VB splits in a heavy and light hole bands. The amount of warping

of these split bands is still subject to controversy. From the
band calculation of Khan[6], the band parameters of CuBr are
derived: γ_1 = 0.78, γ_2 = 0.19 and γ_3 = 0.17 and the warping
characterized by γ_3 is practically negligible. In a recent paper,
Mattausch and Uihlein[7] found the following parameters to explain
their results γ_1 = 0.83, γ_2 = 0.2 and γ_3 = 0.53, indicating a
stronger warping. Using these latter band parameters, we obtained
for CuBr a value c_0 = 3.10^{-7} (eV)$^{-2}$ compared to 10^{-6} for GaAs.

The parameter D of relation (3) was adjusted to fit the theoretical
curve with the experimental points of figure 1. A satisfactory
fit could however not be obtained.   The fit was acceptable only
on the low energy side below 2.7 eV, but the expression (3) cannot
explain the strong dispersion on the high energy side near E_0.

          This discrepancy may be due to the fact that in CuBr,
the dispersion center is an exciton and not a band gap. The
theory elaborated for GaAs, ZnSe cannot be applied here as it
neglects a strong exciton interaction. In CuBr, the excitonic
Rydberg is very large (R $\simeq$ 100 meV) in comparison to the usual
semiconductors (for example R $\simeq$ 4 meV in GaAs). As a result,
the 1S exciton created below the band gap is still present at
300 K - its energy position is at E_0 = 2.9974 eV - and acts as
a dispersion center[8]. Work is actually in progress to explain
more satisfactorily this strong enhancement of the intrinsic
birefringence in CuBr in the excitonic polariton model.

REFERENCES

1. J. Pastrnak and K. Vedam, Physical Review, B3, 2567 (1971).

2. P.Y. Yu and M. Cardona, Solid State Communications, 9, 1421
   (1971).

3. M. Bettini and M. Cardona, Proceedings of the 11th International
   Conference on the Physics of Semiconductors, Warsaw (1972),
   1072.

4. V.M. Agranovich and V.L. Ginzburg, Spatial Dispersion in
   Crystal Optics and the Theory of Excitons, Wiley, New-York
   (1966).

5. C.W. Higginbotham, M. Cardona and F.H. Pollack, Physical
   Review, 184, 821 (1969).

6. M.A. Khan, Physica Status Solidi (b), 60, 641 (1973).

7. H.J. Mattausch and C. Uihlein, Solid State Communications,
   25, 447 (1978).

8. J. Ringeissen, J.G. Gross, S. Lewonczuk, to be published.

INFRARED ABSORPTION SPECTRUM OF IONIC POWDERS

J.M. Gerardy* and M. Ausloos

Institut de Physique, B5
Université de Liège

B-4000 Sart-Tilman/Liège 1, Belgium

The infrared absorption spectrum of powders has already been
discussed elsewhere, and examined along different lines of
approximation. We have solved exactly Maxwell's equations for
an arbitrary distribution of spherules to all 2^t polar orders.
We present here the dipolar and quadrupolar effects on the
infrared absorption spectrum of ionic (NiO) simple systems,
in the long wavelength limit approximation, when the spherule
dielectric constant has a simple theoretical form. Square and
tetrahedron clusters are used for illustration.

I. INTRODUCTION

The infrared spectrum of powder made from dielectric
particles embedded in a dielectric matrix can be investigated
by solving the Laplace equation, which is strictly valid in the
static case[1]. Much work, as those of Maxwell-Garnett[2], or of
Clippe et al.[3], have been developed in order to take into
account the polarization field due to the presence of the
particles, but only dipolar-dipolar interactions were considered.
More recently, Mc Phedran and Mc Kenzie[4,5] have calculated the
conductivity of cubic lattices of spheres in the static case,
including poles of order 2. Because of the high symmetry of the
lattice, such a calculation can be made within a reasonable
computer time. Ausloos et al.[6] have already extended Clippe et
al. theory[7] to more asymmetrical clusters.

347

Our aim is to calculate the infrared spectrum of simple clusters of spheres, for which the symmetry simplifications do not occur. In order to do so, let us consider an arbitrary cluster of N spheres embedded in a dielectric matrix (dielectric constant ε_e). Laplace equation can be solved if we use a potential V in the form :

$$V = \sum_{j=1}^{N} V_j \tag{1}$$

where

$$V_j(\vec{r}) = \sum_{n=1}^{\infty} \sum_{m=-n}^{+n} b_{nm}(j) Y_{nm}(\theta_j, \varphi_j)/r_j^{n+1} \tag{2}$$

defined for a point $\vec{r}$ situated outside each sphere. The usual spherical harmonics $Y_{nm}(\theta_j, \varphi_j)$ are expressed in a reference frame centered on the j-sphere. Then r_j, θ_j, φ_j are the spherical coordinates of the point $\vec{r}$ in this j frame.

We have already shown[7] that the $b_{nm}(j)$'s, which describe the electrical 2^n polar moment of the j-sphere of radius R_j, obey the relation :

$$b_{nm}(j) + \Delta_n(j) \sum_{k \neq j} \sum_{q=1}^{\infty} \sum_{p=-q}^{+q} h_{qpnm}(k/j) b_{qp}(k) = -\Delta_n(j) d_{nm} \tag{3}$$

where

$$b_{qpnm}(k/j) = r_j^{n+1} \int_0^{\pi} \int_0^{2\pi} \frac{Y_{qp}(\theta_k, \varphi_k)}{r_k^{q+1}} Y_{nm}^{\star}(\theta_j, \varphi_j) \sin\theta_j . d\theta_j . d\varphi_j \tag{4}$$

calculated on the surface $(r_j = R_j)$ of the j-sphere. $\Delta_n(j)$ is the 2^n polar susceptibility defined as a function of the dielectric constant $\varepsilon(j)$ of the j-sphere by

$$\Delta_n(j) = \frac{n[\varepsilon(j) - \varepsilon_e]}{n\varepsilon(j) + (n+1)\varepsilon_e} R_j^{2n+1} \tag{5}$$

The d_{nm}'s are the coefficients of the expansion (2) of the incident uniform field $\vec{E}_0$ in terms of the spherical harmonics.

In order to obtain the $b_{nm}(j)$'s, one must stop the q sum in eq.(3) at a finite value of q say q = t. In this paper, two approximations are used : t = 1 (dipolar interactions only) and t = 2 (dipolar and quadrupolar interactions). Although, as stated

above, our aim is to describe arbitrary clusters, it is useful to
test the theory and the influence of approximations on simple
clusters containing a few spheres (N = 4, below), and possessing
some symmetrical shape.

II. RESULTS

In order to present cluster spectra, we have calculated
the power dissipated within the spheres, rapported to their total
volume V and to the incident energy :

$$W(\omega) = \text{Im} \left[\frac{1}{V} \sum_{j=1}^{N} \varepsilon(j) \int_{v_j} |E(j)|^2 dv\right] \qquad (6)$$

where $\vec{E}(j)$ is the field within the j sphere and v_j is the j sphere
volume. We have taken for $\varepsilon(j)$ the form

$$\varepsilon(j) = \varepsilon_\infty + \frac{\varepsilon_0 - \varepsilon_\infty}{1 - f(f + i\gamma)} \qquad (7a)$$

with

$$f = \omega/\omega_T \qquad (7b)$$

where ε_∞ (ε_0) is the high (low) frequency dielectric constant,
γ a damping factor, and ω_T the transverse optical phases frequency.
For NiO spheres in a vacuum matrix ($\varepsilon_e = 1$), we have $\varepsilon_0 = 12.0$,
$\varepsilon_\infty = 5.4$ and $\omega_T = 401$ cm^{-1}.

We have first investigated a symmetrical tetrahedron
cluster. The four identical NiO spheres have the same radius
R = 75 Å and are closely packed. The direction of the incident
field was chosen a) perpendicular to a face of the tetrahedron,
b) parallel to an edge, c) parallel to the bisecting line of one
angle of one basis triangle vertex. Fig. 1 and 2 show W as a
function of f for the tetrahedron cluster in the respective
t = 1, t = 2 case. The principal effect was that the three
incident fields give the same curve for W(ω). This is due to
the high symmetry of the cluster. In figure 1, i.e. for the
t = 1 approximation, both peaks are similar to those of a two
sphere cluster. On figure 2 (t = 2 approximation) drastic effects
due to the quadrupolar interactions in a compact cluster are seen.
New peaks and structure appear and the two purely dipolar peaks
of figure 1 are displaced. The sharpness of shoulders is only
due to the finite mesh used for plotting the curves.

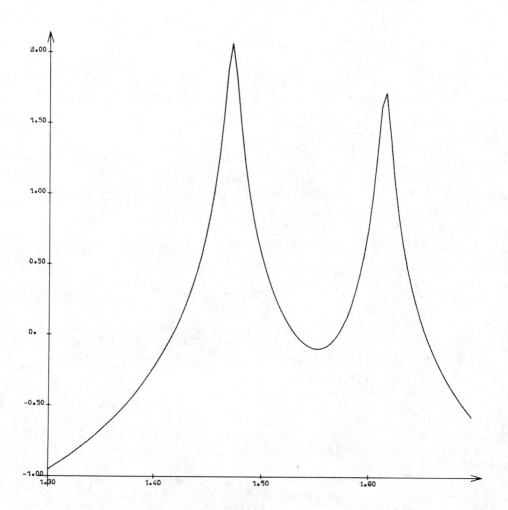

<u>Fig.1</u> : Logarithm of the dissipated power W as a function of
       $f = \omega/\omega_T$ for a packed tetrahedron of NiO spheres ($\gamma = 0.02$).

       Approximation $t = 1$ is used. A single curve is obtained
       for different field incidence (see text for the geometry).

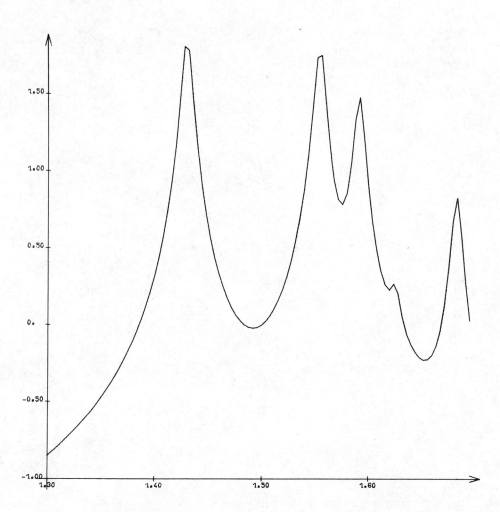

<u>Fig.2</u> : Log.W as a function of f for a packed tetrahedron of NiO
spheres (γ = 0.02) in the t = 2 approximation. For three
different geometries, the same power is dissipated into
the spheres.

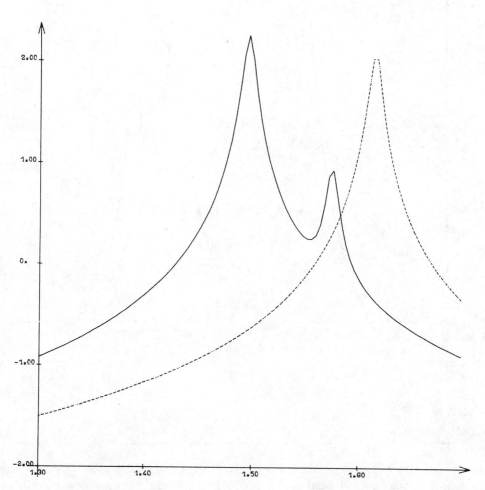

Fig.3 : Log.W as a function of f for a packed square of NiO spheres
        (γ = 0.02). Approximation t = 1 is used.
        Full line : geometry a) and b); dash line : geometry c)
        (see text).

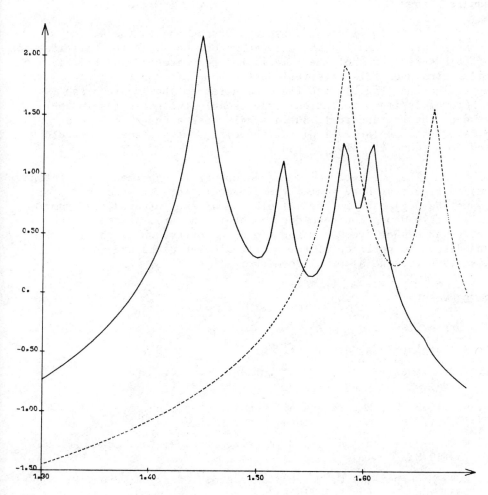

Fig.4 : Log.W as a function of f for a packed square of NiO spheres
($\gamma = 0.02$). Approximation $t = 2$ is used. The notations are
the same as for figure 3.

On figure 3 and 4, we have represented W as a function
of f, respectively for the t = 1 and t = 2 approximation, for a
cluster of four identical touching spheres (radius 75 Å), situated
on the corners of a square. The incident field is taken a) parallel
to an edge, b) parallel to a diagonal, c) perpendicular to the
square plane.

On both figures, curves corresponding to the a) and b)
geometry are identical while it seems obvious that the polarizations
of each sphere are not the same in the two cases a) and b), the
dissipated energy is nevertheless the same. In the t = 1
approximation (fig.3), the field parallel to the square plane
(whatever is its orientation) induces only two peaks (as in the
case of the tetrahedron). But a single peak appears (due to a
"transversal" polarization) for the c) geometry : such a structure
had been found in ref.3 already.

The influence of the quadrupolar interactions is pointing
out on figure 4. Here again, the effect is very drastic, and
cannot be neglected in the calculation of the infrared spectrum
of a highly dense ionic powder (compact cluster). A second peak
occurs at higher frequency for the c) geometry, while the splitting
and displaced position of the purely dipolar peaks are quite
important for the a) or b) geometry.

REFERENCES

* Boursier IRSIA.

1. R. Ruppin, Phys. Stat. Sol. (b) 87, 619 (1978).

2. J.C. Maxwell-Garnett, Philos. Trans. R. Soc. London 203, 385
   (1904).

3. P. Clippe, R. Evrard and A.A. Lucas, Phys. Rev. B, 14(4),
   1715 (1976).

4. R.C. Mc Phedran and D.R. Mc Kenzie, Proc. R. Soc. Lond. A 359,
   45 (1978).

5. D.R. Mc Kenzie, R.C. Mc Phedran and G.H. Derrick, Proc. R.
   Soc. Lond. A 362, 211 (1978).

6. M. Ausloos, P. Clippe and A.A. Lucas, Phys. Rev. B 18, 7176
   (1978).

7. J.M. Gerardy and M. Ausloos, Phys. Rev. (to be published)

HOLE TRAPPING ENERGIES AS EVIDENCE FOR THE EXISTENCE OF FREE SMALL POLARONS IN OXIDE CRYSTALS

R.T. Cox[*]

Département de Recherche Fondamentale
Section de Résonance Magnétique
CEN-Grenoble, 85X
38041 Grenoble-Cedex, France

The energy needed to liberate a hole trapped by an Mg^{2+} ion in Al_2O_3 (sapphire) was deduced from the temperature dependence (T = 220-300 K) of the recombination kinetics following UV excitation of compensated crystals. The result obtained, 0.7 eV, is much smaller than the value deduced from the activation energy of the p-type conductivity observed for uncompensated $Mg:Al_2O_3$ crystals at 1600-1900 K by Wang and Kröger. A possible interpretation is that free holes form small polarons in Al_2O_3, with site-to-site jumping energy about 2.4 eV at high T. Theoretical calculations by Mackrodt et al support this hypothesis. It is suggested that free holes may also form small polarons in MgO and CaO.

1. INTRODUCTION

In this communication, we discuss the properties of free holes in insulating, diamagnetic, metal oxides like Al_2O_3, MgO, etc (we exclude materials like NiO with unfilled d shells, which constitue a separate problem).It has generally been supposed that free holes behave as "large polarons" in these oxides, that is that they occupy high-mobility, delocalised states derived from the valence band edge. We will examine recent evidence that suggests that this is incorrect, that in at least some oxides the holes behave as "small" polarons. A small polaron is a hole concentrated onto one or two lattice atoms, the hole being localised by the potential well resulting from the lattice

distortion it creates. At low temperature, a small polaron is
self-trapped : it has low mobility except at temperatures high
enough to induce site-to-site jumping.

It is well known that free holes form small polarons
in the ionic halides (NaCl, CaF_2, etc.). This was discovered in
studies of the effects of ionising irradiation at liquid nitrogen
temperature[1]. The irradiation creates free electrons and free
holes; the electrons (large polarons) migrate instantly to traps
whereas the holes remain virtually immobile at 77 K. Thus the holes
(called "V_k centres") can be studied by ESR or by optical
spectroscopy after irradiation.

By contrast, in similar studies on metal oxides no centre
which could be a self-trapped hole has been reported, although
holes trapped by defects or impurities are frequently observed.
This has been taken to mean that holes do not self-trap in oxides,
a reasonable conclusion given the considerable width of the valence
band in oxides (of order 5-10 eV compared to 1 eV for halides).
A small polaron will exist only if the distortion around a
localised hole can reduce the potential energy enough to offset
the kinetic energy increase inherent in localisation; since a
localised state is equivalent to a linear combination of states
from the whole valence band, this kinetic energy increase is of
order half the valence band width[2].

Nevertheless, recent results suggest that small polarons
may exist in the oxides. We consider first the case of Al_2O_3
(sapphire).

2. THERMAL ANNEALING OF TRAPPED HOLES IN Al_2O_3

The results discussed here and in section 3 came from
a study of the physical and chemical properties of Al_2O_3 crystals
doped with magnesium impurity (this study was conducted in
cooperation with H.A. Wang and F.A. Kröger, Department of
Materials Science, University of Southern California, Los Angeles).

In the Al_2O_3 lattice, Mg^{2+} ions occupy Al^{3+} sites and
thus can trap a hole to form a neutral centre which has a
characteristic ESR spectrum[3] and a strong optical absorption
peaking at 490 nm. We represent this centre by the symbol Mg_{Al}^{o}.
This is the notation commonly used in semiconductor physics
and it helps emphasize that the trapped hole centre is a neutral
acceptor : thermal emission of the trapped hole ionises the
acceptor negatively :

$$Mg^o_{Al} \rightarrow Mg^-_{Al} + h^+ \qquad (1)$$

where Mg^-_{Al} represents an Mg^{2+} ion without a trapped hole.

One main aim of our study was to measure E_i the energy difference in the ionisation process (1). To do this, we used UV irradiation to create metastable Mg^o_{Al} centres, then studied the kinetics of their decay as a function of temperature.

These experiments were done with "compensated" samples, crystals of Al_2O_3 that had been double-doped with Mg and a transition metal impurity, either Cr or Fe. In thermal equilibrium of the electron distribution, such crystals contain Mg^{2+} ions which are charge-compensated by Cr^{4+} or Fe^{4+} ions. That is, in semiconductor language, negatively ionised acceptors Mg^-_{Al} are compensated by positively ionised donors Cr^+_{Al} or Fe^+_{Al}.

The samples were irradiated in situ at various temperatures in the cavity of an ESR spectrometer; the irradiation source was an argon UV laser ($h\nu$ = 3.4-3.7 eV). This creates the trapped hole centre Mg^o_{Al} as well as Cr^{3+} or Fe^{3+} ions (that is Cr^o_{Al} or Fe^o_{Al}), all of which can be detected conveniently by ESR.

The photoexcitation is equivalent to :

$$Mg^-_{Al} + D^+_{Al} + h\nu \rightarrow Mg^o_{Al} + D^o_{Al} \qquad (2)$$

where D = Cr or Fe, but the true nature of this process is uncertain and does not concern us here.

When the laser is switched off, the neutral acceptors and donors disappear as the system returns to thermal equilibrium. We have studied their decay rates from -55 to +30 C; over this range the rates vary by more than four orders of magnitude. For a given sample, the decay curves (concentrations against time) that we get at different temperatures differ only by a time-scaling factor (expansion or contraction along the time axis). This factor varies as $\exp(-E_{anneal}/kT)$ with, for both the Mg + Cr and Mg + Fe doped samples :

$$E_{anneal} = 0.7 \text{ eV} \qquad (3)$$

Since the result is independent of whether the donor impurity is Cr or Fe and since Mg^o_{Al} is known to anneal at similar temperatures for samples in which the donor is titanium[3] or manganese[4], we attribute the annealing process to hole release followed by recombination of free holes with trapped electrons. Here, we are assuming that free-hole diffusion is not rate-limiting, see section 4. Then E_{anneal} = 0.7 eV gives us a first estimate for the acceptor ionisation energy E_i.

3. INFORMATION FROM CONDUCTIVITY STUDIES

For semiconductor materials, the ionisation energy of acceptor centres is often determined from the temperature-dependence of the thermally excited free-hole concentration as deduced from electrical conductivity measurements. This requires uncompensated crystals, samples in which the acceptors are neutral in thermal equilibrium due to the absence of charge-compensating defects.

To obtain uncompensated $Mg:Al_2O_3$ samples, one must first avoid introducing charge-compensating impurities (Cr, Fe, Ti, ...); the uncompensated state is then produced by oxidising the samples in air at high temperature, which removes charge-compensating intrinsic defects (anion vacancies V_O^{++} or interstitial cations Al_i^{3+})[5]. ESR and optical studies show large concentrations of Mg_{Al}^o centres after oxidation. These centres exist in thermodynamic equilibrium, in contrast to the centres created metastably by irradiation of compensated samples. (A similar uncompensated state can be produced for lithium acceptors in MgO[6]).

Wang and Kröger have measured the electrical conductivity in air at 1600-1900 K for high purity $Mg:Al_2O_3$ crystals grown at LETI-Cristaltec, Grenoble[7] (this follows earlier studies by Mohapatra and Kröger[8] of crystals with lower Mg content). They observe the expected p-type conductivity, increasing exponentially with temperature.

Conductivity is proportional to the product of free-hole concentration and mobility $N_h \cdot \mu_h$. Since N_h depends on the acceptor ionisation energy E_i, the value of E_i can be extracted from the conductivity data, provided that μ_h is not thermally activated, that is provided free holes behave as large polarons. With this assumption, and after correction for the complicating effect of an increase in the concentration of charge-compensating defects with temperature, the data obtained by Wang and Kröger can be fitted by :

$$E_i = 3.05 \text{ eV} \ldots \tag{4}$$

(Here, we neglect the energy distribution of free-polaron states; taking standard, spherical band statistics would correct this figure to 2.85 eV.)

This result is much higher than our own estimate $E_i = 0.7$ eV, taken from the annealing data.

4. THE SMALL POLARON HYPOTHESIS

One can reconcile the results of the two sets of experiments by proposing that holes in Al_2O_3 behave as small polarons, a suggestion originally due to Kröger.

Making this hypothesis, we replace (4) by :

$$E_i + E_{jump} = 3.05 \text{ eV} \tag{5}$$

where E_{jump} is the activation energy of the polaron motion. If E_{anneal} is a good estimate of E_i, then $E_{jump} = 2.4$ eV.

If the activation energy for hole jumping is indeed this high, we must reexamine a hypothesis made earlier (section 2); there we assumed that the hole diffusion is not the rate-limiting step in the sequence hole-release → hole-diffusion → recombination occurring in our annealing kinetics studies at T = 220-300 K. We recall therefore some properties of the motion of free polarons from site to site.

A potential barrier of height $E_{barrier}$ separates the equi-energy configurations of the polaron system. Classical jumping over the barrier dominates the site-to-site motion only when $T > h\nu_c/k$, where ν_c is a frequency characteristic of the lattice vibrations. For Al_2O_3 the optical phonon frequencies run from 300 to 900 cm^{-1} [10,11] so we expect $h\nu_c/k$ to be between 430 and 1300 K. The 1600-1900 K range of the conductivity experiments appears high enough for polarons to do near-classical jumping. That is $E_{jump} \simeq 2.4$ eV would be a good estimate of $E_{barrier}$ (> 90 % of the true height, see Fig.2, ref.9).

Although this classical motion would be frozen at 220-300 K, a small polaron would still diffuse rapidly by phonon-assisted tunneling through the barrier at these lower temperatures. This process has an effective activation energy E_{eff} much less than the barrier height. From Fig.2 of ref.9, E_{eff} < 0.7 eV is possible at 220-300 K if it is mainly the higher frequency modes of Al_2O_3 that are active.

We believe that hole diffusion is not the rate-limiting step in our annealing experiments. If it were, that is if the $10^3 - 10^{-1}$ sec. characteristic times for annealing at 220-300 K corresponded to polaron lifetimes, we believe we would have identified ESR spectra for small polarons in Al_2O_3 in numerous irradiation studies we have done at much lower temperatures (77 K); Since we have seen no such spectra, we identify E_{anneal} with E_i.

5. INFORMATION FROM THEORETICAL CALCULATIONS

The small polaron hypothesis is supported by preliminary results from a theoretical study of the energies of various defect centres in Al_2O_3 that is being done by Mackrodt and co-workers[12]. They use the Harwell "HADES" procedure for calculating the effect of electronic polarisation and lattice distortion around a defect, assuming the crystal to be an assembly of Al^{3+} and O^{2-} ions.

This ionic crystal model evidently takes no account of the true electronic structure of the crystal (band structure) and therefore cannot readily be used to study the properties of delocalised holes (large polarons). However, it is naturally suited for calculating the properties of a hypothetical "minimally small" polaron because, in an ionic model, this is just an O^- ion on an O^{2-} site.

Mackrodt et al have calculated the difference between the energy of a free small polaron having this structure and the energy of an O^- ion next to an Mg^{2+} ion (that is a trapped small polaron). We emphasize that the ESRstudies[3] show that the latter structure is a good description of the centre Mg^o_{Al}. The cation site in Al_2O_3 has six anion neighbours : three at distance 1.86 Å and three at 1.97 Å. The ESR spectrum can be attributed to an O^- ion on the site at 1.97 Å.

Only preliminary calculations have been done so far but the results are thought to be within 0.2 eV of those a more extended calculation would give. For the site at 1.97 Å, these approximate calculations give E_i = 0.8 eV.

This is encouragingly close to our experimental value for hole emission, 0.7 eV, increasing our confidence in the small polaron interpretation of the experiments.

6. OTHER OXIDES

Calculations of the same type have been done for trapped hole centres in MgO and CaO during recent theoretical. studies of the energies of defects in these oxides[13-15]. In Table I, we list cases where theoretical and experimental estimates of the hole-trapping energies are both available.

We see that the calculated energies for trapping of holes at cation vacancies and at Li^+ ions in MgO agree quite well with the experimental E_i; for V_{Ca}^- in CaO, the two theoretical estimates differ by 0.7 eV but the experimental value lies between them.

Table 1 : Experimental release-energies for a hole trapped by the cation vacancy in MgO, by Li^+ in MgO, by the cation vacancy in CaO and by Mg^{2+} in Al_2O_3; theoretical energies for binding a small polaron to the same defects. The superscripts are reference numbers.

Crystal	Defect	Hole-trapping energy (eV)	
		(Experiment)	(Theory)
MgO	V_{Mg}^-	1.6 [21]	1.32 [15]
			1.41 [13]
MgO	Li_{Mg}^o	0.6 [23,24]	0.71 [15]
CaO	V_{Ca}^-	1.1 [22]	0.84 [15]
			1.53 [14]
Al_2O_3	Mg_{Al}^o	0.7 this work	≈0.8 [12]

We note that the use of the "O^- ion on an O^{2-} site" model for the trapped hole centres is justified by ESR studies of the structure of these centres. The authors of the calculations cited in Table I also took this model for the free hole, not because they considered this to be a valid model for the free hole but as a convenient, first step towards estimating the energy for emission of a large polaron.

Nevertheless, if the agreements in Table I are not to be attributed to pure accident, one must conclude either
(i) if the experiments do measure the energy of emission of large polarons, the large-polaron and small-polaron energies are very similar (a suggestion made by Mackrodt and Stewart[15]) or
(ii) free holes do indeed behave as small polarons in these oxides.

There is one other oxide for which evidence exists that free holes form small polarons (although this is not a metal oxide), namely silica glass SiO_2 (see[16,17] and references therein). This has been considered to be a rather special case, related to the peculiar electronic structure of silica : in this material a localised state can form entirely from a narrow "non-bonding" band that is separated from the rest of the valence band. Also, the disorder of the glass structure may contribute to stabilising the small polaron.

The results we have discussed in this paper suggest that silica is not necessarily a special case but that small polaron formation could conceivably be a rather general phenomenon in oxides.

We mention that Hughes[18] has studied Al_2O_3 using the same techniques (carrier transit-time measurements) that gave such useful information for silica but unfortunately could not deduce the Al_2O_3 free hole mobility.

7. CONCLUSION

Mott and Stoneham[16,19] have proposed that the small polaron has a molecular structure in silica, analogous to the "V_k centre" structure of the polaron in halides. In an ionic description, their model would be an $(O_2)^{3-}$ molecular ion on a pair of O^{2-} sites. Whether or not this model is valid for silica, we feel that if small polarons exist in metal oxides like Al_2O_3, MgO, CaO, they would be localised on a single anion. In ESR studies of trapped hole centres in these oxides (and indeed in a very wide variety of octahedrally and tetradedrally coordinated oxides) the hole is always found to localise on a single anion next to the trapping defect. Since there is no apparent tendency to form a molecular structure in the trapped polarons, we suggest it is also improbable in free polarons.

Finally, we suggest that the lack of any spectroscopic evidence (ESR or optical spectra) for self-trapped holes in oxides could mean that the polaron tunneling rate is very much higher in oxides than in halides, that is that the polarons migrate rapidly to traps. Some indication of this comes from measurements of the rate at which a hole trapped by a cation vacancy or by an Li^+ ion in MgO reorientates around the octahedron of anions containing the trap[20]. These trapped polarons tunnel from site to site $10-10^3$ times per second at 4 K.

REFERENCES

* Chercheur CNRS.

1. T.G. Castner and W. Känzig, J.Phys.Chem.Solids 3, 178 (1957).

2. A.M. Stoneham, Theory of Defects in Solids, 653, Clarendon Press, Oxford, (1975).

3. R.T. Cox, Solid State Commun. 9, 1989 (1971).

4. S. Geschwind, P. Kisliuk, M.P. Klein, J.P. Remeika and D.L. Wood, Phys. Rev. 126, 1684 (1962).

5. R.T. Cox, J. de Physique 34, suppl. C9, 333 (1973).

6. J.B. Lacy, M.M. Abraham, J.L. Boldu O., Y. Chen, J. Narayan and H.T. Tohver, Phys. Rev. B 18, 4136 (1978).

7. H.A. Wang, F.A. Kröger and R.T. Cox, to be published.

8. S.K. Mohapatra and F.A. Kröger, J. American Ceramic Society 60, 141 (1977).

9. M.J. Norgett and A.M. Stoneham, J. Phys. C : Solid State Phys. 6, 238 (1973).

10. H. Bialas and H.J. Stolz, Z. Physik B 21, 319 (1975).

11. W. Kappus, Z. Physik B 21, 325 (1975).

12. W.C. Mackrodt, private communication.

13. M.J. Norgett, A.M. Stoneham and A.P. Pathak, J. Phys. C : Solid State Phys. 10, 555 (1977).

14. J.H. Harding, J. Phys. C : Solid State Phys. 12, 3931 (1979).

15. W.C. Mackrodt and R.F. Stewart, J. Phys. C : Solid State Phys. 12, 5015 (1979).

16. N.F. Mott, Advances in Physics 26, 363 (1977).

17. R.C. Hughes and D. Emin, The Physics of Silica and its
     Interfaces, 14, Pergamon Press, New York (1978) (Ed.:
     S. Pantelides).

18. R.C. Hughes, Phys. Rev. B 19, 5318 (1979).

19. N.F. Mott and A.M. Stoneham, J. Phys. C: Solid State Phys.
     10, 3391 (1977).

20. G. Rius, A. Hervé, R. Picard and C. Santier, J. de Physique
     37, 129 (1976).

21. A.J. Tench and M.J. Duck, J. Phys. C: Solid State Phys. 6,
     1134 (1973).

22. A.J. Tench and M.J. Duck, J. Phys. C: Solid State Phys. 8,
     257 (1975).

23. D.J. Eisenberg, L.S. Cain, K.H. Lee and J.H. Crawford Jr.,
     Appl. Phys. Lett. 33, 479 (1978).

24. K.H. Lee and D.J. Eisenberg, unpublished thermoluminescence
     measurements.

ASYMPTOTICS OF POLARON ENERGY VIA BROWNIAN MOTION THEORY

Bernd Gerlach
Institut für Physik, Universität Dortmund
D-4600 Dortmund 50, West Germany

and

Hajo Leschke
Institut für Theoretische Physik
Universität Düsseldorf
D-4000 Düsseldorf 1, West Germany

Making use of the functional (Wiener) integral representation, we determine non-perturbatively the exact asymptotic behaviour of the free energy and the ground-state energy of Fröhlich's polaron Hamiltonian in the weak and the strong-coupling limit. In particular, methods of modern Brownian-motion theory prove to be a powerful tool to control the strong-coupling limit.

1. INTRODUCTION AND SUMMARY OF RESULTS

Most of the theoretical work on polarons in ionic crystals and polar semiconductors[1] is based on Fröhlich's polaron Hamiltonian

$$H := \frac{\vec{p}^2}{2m} + \hbar\omega \sum_{\vec{k}} a_{\vec{k}}^{+} a_{\vec{k}} + \sum_{\vec{k}} \{ g_{\vec{k}} a_{\vec{k}} e^{i\vec{k}.\vec{r}} + h.c. \} \tag{1a}$$

with

$$g_{\vec{k}} := -\frac{i\hbar\omega}{|\vec{k}|} \left(\frac{\hbar}{2m\omega}\right)^{1/4} \left(\frac{4\pi\alpha}{V}\right)^{1/2} \tag{1b}$$

H models the interaction of an electron (position $\vec{r}$, momentum $\vec{p}$, band mass m) with the scalar phonon field (quantization volume V, Bose operators $a_{\vec{k}}$, $a_{\vec{k}}^{+}$) associated to a branch of longitudinal optical lattice vibrations (wave vector $\vec{k}$, frequency ω). The interaction strength is characterized by α, the dimensionless electron-phonon coupling constant.

Since the value of α for different realistic materials ranges from 0.01 to roughly 6 or even larger, there have been considerable efforts of solid-state theorists to predict on the basis of (1) the "self energy", the "effective mass" and other observable quantities of the polaron as a function of α.

Among the various mathematical methods which have been used in this connection the beautiful method of path or functional integration initiated by Feynman[2] takes an exceptional position in that it allows one to eliminate exactly the phonon degrees of freedom. As a consequence, it is possible to obtain results very economically which appear to be difficult or even impossible to get within conventional formalisms.

An example is given by Feynman's[2] Gaussian two-parameters variational upper bound on the ground-state energy of (1) which contains enough flexibility to provide a smooth interpolation between the weak and the strong-coupling regime and (therefore) is believed to be rather accurate for intermediate coupling too.

From a somewhat exacting point of view however Feynman's bound has one defect. While it coincides for sufficiently small α (weak coupling) with the result obtained by ordinary perturbation theory, for large α it only comes pretty close to the strong-coupling-limit result suggested by existing strong-coupling (adiabatic perturbation) theories[3].

Although the generalization of Feynman's approach to an arbitrary Gaussian approximation[4] gives a slight improvement for intermediate coupling, it suffers from the same defect in the strong-coupling limit.

Not surprisingly, it is the Gaussian ansatz that is wrong, not the functional integral. Recently, borrowing methods from the modern theory of Brownian motion, it became clear[5] how to extract the correct strong-coupling limit from the polaron functional integral.

To a large extent the motivation for this contribution comes from (these) ideas centered around the strong-coupling limit. But for the sake of completeness, in order to "round the picture", and, last but not least, to demonstrate the effectiveness of the functional method also in this case, we have included also the less involved and perhaps more known weak-coupling limit. Both limits are considered for arbitrary temperatures. More specificly, we will use the functional integral representation to control non-perturbatively the exact asymptotic behaviour of the polaron free energy in the weak and the strong-coupling limit.

Denoting by

$$F(\alpha,\beta) := -\beta^{-1}\ln \text{ trace } e^{-\beta H(\alpha)} \tag{2}$$

the free energy of (1) as a function of α and the inverse temperature β we begin the next sections by proving

$$F(\alpha,\beta) \leqslant F(o,\beta) - \alpha\hbar\omega\frac{(\pi\beta\hbar\omega)^{1/2}}{e^{\beta\hbar\omega}-1} M(\tfrac{1}{2},1,\beta\hbar\omega) \tag{3}$$

$$F(\alpha,\beta) \leqslant -\gamma\alpha^2\hbar\omega \tag{4}$$

Here $M(a,b,z)$ denotes the Kummer function[6] and the constant

$$\gamma := -\inf_{\psi} \{\tfrac{1}{2}\int d^3x |\vec{\nabla}\psi(\vec{x})|^2 - \frac{1}{\sqrt{2}}\int d^3x \int d^3x' \ \frac{\psi^2(\vec{x})\psi^2(\vec{x}')}{|\vec{x}-\vec{x}'|} \} \tag{5}$$

is defined as the negative infimum of Pekar's Hartree-type energy functional[7] taken over all dimensionless and normalized real wave functions $\psi(\vec{x})$ of a dimensionless real vector $\vec{x}$. Recently it has been proved[8] that there exists a minimizing ψ which is unique except for translations. The numerical value of γ is 0.108513 according to[9].

The inequalities (3) and (4) may be viewed as non-zero temperature generalizations of previously derived upper bounds on the ground-state energy $E_o(\alpha) = \lim_{\beta\to\infty} F(\alpha,\beta)$.

In fact, the function of $\varepsilon := 1/\beta\hbar\omega$ multiplying $\alpha\hbar\omega$ in (3) goes as $1 + (\varepsilon/4) + 0(\varepsilon^2)$ for small ε, leading to the well known[2,10] result $E_o(\alpha) \leqslant -\alpha\hbar\omega$. Since γ is independent of β, the $\beta\to\infty$ limit of (4) simply is Pekar's inequality[7] $E_o(\alpha) \leqslant -\gamma\alpha^2\hbar\omega$.

Although both inequalities (3) and (4) hold as they stand for all α (and β), (3) is sharp only for small and (4) only for large α.

In fact, by constructing appropriate lower bounds on F we will prove that the right-hand sides of (3) and (4) are the leading terms in an asymptotic evaluation of F for small and large α, respectively.

Although these exact results on the asymptotic behaviour of F in the weak and the strong-coupling limit confirm the results suggested (for $\beta \to \infty$) either by ordinary (second order) or adiabatic perturbation expansions, we want to stress that we will obtain them in a non-perturbative fashion. Of course, this circumstance is of particular importance for the strong-coupling limit.

2. WEAK-COUPLING LIMIT

In the following we use units in which $\hbar$, m and ω are unity.

We start from the representation

$$F(\alpha,\beta) = F(o,\beta) - \beta^{-1} \ln <e^{\alpha\Phi}>_{\beta} \qquad (6)$$

which is implicit in ^{2}. Here $<.>_{\beta}$ denotes functional (Wiener) integration over (conditional) Brownian-motion paths $\vec{R}(\tau)$ in 3-dimensional Euclidean space which start ($\tau = 0$) and end ($\tau = \beta$) at the origin. The functional $\Phi \equiv \Phi[\vec{R}]$ is defined as

$$\Phi[\vec{R}] := \frac{1}{\sqrt{2}} \int_{o}^{\beta} d\tau \int_{o}^{\beta} d\tau' \; \frac{G(\tau - \tau')}{|\vec{R}(\tau) - \vec{R}(\tau')|} \qquad (7)$$

where

$$G(\tau) := \frac{\cosh((\beta/2) - |\tau|)}{2\sinh(\beta/2)} \qquad (8)$$

is the harmonic oscillator temperature Green's function.

Using Jensen's inequality $<\exp \alpha\Phi>_{\beta} \geqslant \exp <\alpha\Phi>_{\beta}$ we get from (6)

$$F(\alpha,\beta) \leqslant F(o,\beta) - \alpha\beta^{-1} <\Phi>_{\beta} \qquad (9)$$

from which (3) follows by an explicit calculation of $<\Phi>_{\beta}$.

The following remarks may be helpful. Due to the Gaussian nature of Brownian motion one has[4]

$$<|\vec{R}(\tau) - \vec{R}(\tau')|^{-1}>_{\beta} = (\frac{\pi}{6}<|\vec{R}(\tau) - \vec{R}(\tau')|^2>_{\beta})^{-1/2} \qquad (10)$$

The combination of (7) and (8) with (10) and the mean square increment of (conditional) Brownian motion

$$<|\vec{R}(\tau) - \vec{R}(\tau')|^2>_{\beta} = \frac{3}{\beta}|\tau - \tau'|(\beta - |\tau - \tau'|) \qquad (11)$$

finally leads to an integral representation of the Kummer function.

By applying the inequality $\ln x \leqslant x - 1$ to (6) we get a lower bound on F

$$F(\alpha,\beta) \geqslant F(o,\beta) - \beta^{-1} <e^{\alpha\Phi} - 1>_\beta \qquad (12)$$

Contrasting (9) with (12) we find

$$\lim_{\alpha \downarrow 0} (F(\alpha,\beta) - F(o,\beta))/\alpha = -\beta^{-1}<\Phi>_\beta \qquad (13)$$

Hence we have proved that the right-hand side of (3) provides the leading terms in an asymptotic expansion of F for small α and that these terms, as it should be, coincide with those obtained by ordinary perturbation theory.

3. STRONG-COUPLING LIMIT

As has been remarked by Pekar[7], the $\beta \to \infty$ version of (4) can be derived as follows. Use a product ansatz made out of a normalized electron state $|\psi>$ times a normalized phonon state $|\phi>$ in the Rayleigh-Ritz principle for (1). The partial expectation value $H_\psi := <\psi|H|\psi>$ can be viewed as a Hamiltonian of non-interacting harmonic oscillators being "displaced" in a manner which depends on $|\psi>$. Hence the ground-state energy of H_ψ, i.e. the minimum of $<\phi|H_\psi|\phi>$ with respect to $|\phi>$, is explicitly known. In fact, after performing the $\vec{k}$-integration (for $V \to \infty$) and after scaling it is seen to be given by Pekar's functional times α^2. The remaining minimization with respect to $|\psi>$ determines γ.

The generalization (4) of Pekar's inequality to arbitrary β follows from the observation that F as a quantum free energy is an increasing function of β.

In order to give a non-perturbative proof that $-\gamma\alpha^2$ is the leading term of $F(\alpha,\beta)$ for large α, we look for an appropriate lower bound to contrast with (4).

To begin with we make use of $G(\tau) \leqslant G(o)$ which yields via (6)-(8)

$$F(\alpha,\beta) \geqslant F(o,\beta) - \beta^{-1} \ln <\exp\{\alpha G(o)\Theta_\beta\}>_\beta \qquad (14)$$

where

$$\Theta_\beta[\vec{R}] := \frac{1}{\sqrt{2}} \int_o^\beta d\tau \int_o^\beta d\tau' \; |\vec{R}(\tau) - \vec{R}(\tau')|^{-1} \qquad (15)$$

From the stochastic scaling relation $\vec{R}(\tau) \stackrel{\circ}{=} \vec{R}(\lambda\tau)/\sqrt{\lambda}$ of Brownian motion we have for all $\lambda > 0$ the identity

$$<\exp\{\alpha G(o)\Theta_\beta\}>_\beta = <\exp\{\frac{\alpha G(o)}{\lambda^{3/2}} \Theta_{\lambda\beta}\}>_{\lambda\beta} \qquad (16)$$

In particular, chosing

$$\lambda = (\alpha\beta G(o))^2 \qquad (17)$$

we have $\alpha G(o)\lambda^{-3/2} = (\lambda\beta)^{-1}$ and, with β fixed, $\lambda\beta \to \infty$ as $\alpha \to \infty$. Hence we can take advantage of the following implication

$$\overline{\lim_{T\to\infty}} \frac{1}{T} \ln <\exp\{\frac{1}{T}\Theta_T\}>_T \leqslant \gamma \qquad (18)$$

of Donsker's and Varadhan's beautiful work[11] on the asymptotic evaluation of certain Brownian-motion expectations for large "time" $T = \lambda\beta$.

In fact, after dividing (14) by λ, eqs. (16)-(18) give

$$\lim_{\alpha\to\infty} \frac{F(\alpha,\beta)}{\alpha^2} \geqslant - (\frac{\beta}{2} \coth \frac{\beta}{2})^2 \gamma \qquad (19)$$

where the explicit form of $G(o)$ according to (8) has been inserted. As we will see immediately, (19) is just sufficient for our purpose.

Abbreviating the lower limit on the left-hand side of (19) as $\underline{f}(\beta)$ and the corresponding upper limit af $\overline{f}(\beta)$, a comparison of (19) with (4) gives in the limit $\beta \to 0$

$$\underline{f}(o) := \lim_{\beta\downarrow 0} \underline{f}(\beta) = \lim_{\beta\downarrow 0} \overline{f}(\beta) = -\gamma \qquad (20)$$

Making again use of the monotonicity of F as a function of β we have $\underline{f}(\beta) \geqslant \underline{f}(o)$. This inequality in conjunction with (4) and (20) finally leads to the result

$$\lim_{\alpha\to\infty} \frac{F(\alpha,\beta)}{\alpha^2} = -\gamma \qquad (21)$$

stated in sec.1.

In addition, the above proof shows that the $\beta \to \infty$ version of (21) is

$$\lim_{\alpha\to\infty} \frac{E_o(\alpha)}{\alpha^2} = -\gamma \qquad (22)$$

As to the nymerical value of γ we want to recall that $\gamma > 1/(3\pi) = 0.106103$ by roughly 2 %. The value $1/(3\pi)$ is characteristic of Gaussian approximations to the strong-coupling limit. It arises for instance by choosing a Gaussian trial function in (5) or a quadratic trial action in Feynman's variational principle[2,4].

Acknowledgement - We thank Heta Röhrig for carefully typing the manuscript.

REFERENCES

1. For reviews see e.g.
   Polarons and Excitons (C.C. Kuper and G.D. Whitfield, eds.),
   Oliver and Boyd, Edinburgh and London 1963; Polarons in
   Ionic Crystals and Polar Semiconductors (J.T. Devreese, ed.),
   North Holland, Amsterdam and London 1972.

2. R.P. Feynman, Physical Review 97, 660 (1955);
   R.P. Feynman and A.R. Hibbs, Quantum Mechanics and Path
   Integrals, McGraw-Hill, New York 1965;
   T.D. Schulz in Polarons and Excitons, see ref.1.

3. For a recent strong-coupling theory and clarifying discussion
   of the older ones see E.P. Gross, Annals of Physics 99, 1
   (1976).

4. J. Adamowski, B. Gerlach and H. Leschke in Proceedings of the
   Workshop on Functional Integration, Theory and Applications
   Louvain-la-Neuve, 1979, to be published by Plenum, New York.

5. J. Adamowski, B. Gerlach and H. Leschke, Strong-Coupling
   Limit of Polaron Energy, Revisited, preprint, Universities
   of Dortmund and Düsseldorf, January 1980.

6. Handbook of Mathematical Functions (M. Abramowitz and
   I.A. Stegun, eds.), seventh printing, Dover, New York 1970.

7. S. Pekar, Untersuchungen über die Elektronentheorie der
   Kristalle, Akademie Verlag, Berlin 1954; englisch translation:
   Research on Electron Theory of Crystals, Translation series
   No. AEC-tr-5575, Physics, AEC, Division of Technical
   Information, Washington, D.C. 1963.

8. E.H. Lieb, Studies in Applied Mathematics 57, 93 (1977).

9. S.J. Miyake, Journal of the Physical Society of Japan 38,
   181 (1975).

10. J.D. Lee, F.E. Low and D. Pines, Physical Review 90, 297 (1953).

11. M.D. Donsker and S.R.S. Varadhan, Communications in Pure and
    Applied Mathematics 28, 1, 279, 525 (1975), 29, 389 (1976);
    M.D. Donsker and S.R.S. Varadhan in Functional Integration
    and its Applications (A.M. Arthurs, ed.), Clarendon, Oxford,
    1975.

PATH-INTEGRAL CALCULATION OF THE GROUND STATE ENERGY

OF A POLARON IN A MAGNETIC FIELD[*]

F.M. Peeters[o] and J.T. Devreese[+]

Departement Natuurkunde
Universitaire Instelling Antwerpen

Universiteitsplein 1, 2610 Wilrijk, Belgium

We report a path-integral calculation of the ground state energy of a polaron in a magnetic field. The present calculation is expected to be valid for arbitrary magnetic field strength and arbitrary electron-phonon coupling constant. The effective interaction of the electron with the longitudinal-optical phonons is simulated by an anisotropic Feynman model. For physical values of the coupling constant, energy values have been obtained which are lower than those reported in the literature up to now.

1. INTRODUCTION

An electron moving in a polar crystal under the influence of a magnetic field is described by the Hamiltonian[1] :

$$H = \frac{1}{2m} (\vec{p} + \frac{e}{c} \vec{A})^2 + \sum_{\vec{k}} \hbar \omega_{\vec{k}} (a_{\vec{k}}^+ a_{\vec{k}} + \frac{1}{2}) + \sum_{\vec{k}} (V_{\vec{k}} a_{\vec{k}} e^{i.\vec{k}.\vec{r}}$$

$$+ V_{\vec{k}}^* a_{\vec{k}}^+ e^{-i.\vec{k}.\vec{r}}) \qquad (1)$$

where use has been made of the conventional polaron notations. For convenience the magnetic field $\vec{B}$ = rot $\vec{A}$ = (0,0,B) is chosen along the z-axis and for the vector potential we take the symmetrical gauge $\vec{A} = (-\frac{B}{2}y, \frac{B}{2}x, 0)$.

In the present paper we are merely interested in the ground state energy :

$$E = \lim_{\beta \to \infty} F(\beta)$$

with the free energy :

$$F(\beta) = -\frac{1}{\beta} \ln Z$$

where the partition function Z can be written as a path integral 2,3,4 :

$$Z = \mathrm{Tr}e^{-\beta H} = \int d\vec{r} \int\!\!\int D\vec{r}(t) \, \exp(S[\vec{r}(t)]) \delta(\vec{r}(\beta)-\vec{r})\delta(\vec{r}(o)-\vec{r}).$$

(2)

The appropriate functional $S[\vec{r}(t)]$ (called action), in (2), can be split up into two parts[5] :

$$S = S_e + S_I \tag{3}$$

with :

$$S_e = \frac{m}{2\hbar^2} \int_o^\beta dt \, [\dot{\vec{r}}(t)]^2 + i\hbar\omega_c (x(t)\dot{y}(t) - y(t)\dot{x}(t))] \tag{4}$$

$$S_I = \frac{1}{2} \sum_{\vec{k}} |V_k|^2 \int_o^\beta dt \int_o^\beta ds \,[(1 + n(\omega_{\vec{k}}))e^{-\hbar\omega_{\vec{k}}|t-s|}$$

$$+ \, n(\omega_{\vec{k}})e^{\hbar\omega_{\vec{k}}|t-s|} \,] \exp\{i\vec{k}.[\vec{r}(t) - \vec{r}(s)]\} \tag{5}$$

where we defined $\omega_c = \dfrac{eB}{mc}$ and $n(\omega) = \{e^{\beta\hbar\omega}-1\}^{-1}$.

2. APPROXIMATION AND RESULT

Up to this date nobody has solved the path-integral (2) exactly. Therefore one is forced to make an approximation. Following Feynman, consider the upper bound to the exact free energy[2] :

$$F \leqslant F_{ph} + F_m - \frac{1}{\beta} <(S-S_m)>_{S_m} \tag{6}$$

with F_{ph} the free energy of the free phonons, S_m a certain trial action with partition function

$$Z_m = e^{-\beta F_m} = \int d\vec{r} \iint D\vec{r}(t) \exp(S_m[\vec{r}(t)]) \delta(\vec{r}(\beta)-\vec{r}) \delta(\vec{r}(o)-\vec{r}) \quad (7a)$$

and a corresponding expectation value:

$$<A>_{S_m} = \frac{1}{Z_m} \int d\vec{r} \iint D\vec{r}(t) \exp(S_m[\vec{r}(t)]) A[\vec{r}(t)] \delta(\vec{r}(\beta)-\vec{r}) \delta(\vec{r}(o)-\vec{r}).$$

$$(7b)$$

Although inequality (6) has been proved for real actions, we think that the above inequality is also valid for actions S and S_m which can be derived from a hermitian Hamiltonian. This suggestion is strengthened by the observation that the above inequality is in fact a path-integral formulation of the Bogolubov inequality which is valid for hermitian Hamiltonians.

In choosing the trial action S_m we start from the observation that only the self-interaction term S_I in (2) prevents us to evaluate the path-integral (2) exactly. Thus only this term has to be replaced. This term describes the electron phonon interaction. As is well-known[2,3,4,6] replacing the self-interaction by a quadratic one gives very good results at zero magnetic field. For non-zero magnetic field the electron will move predominantly in a plane perpendicular to the magnetic field, this will result in a direction dependent self-interaction. Therefore it is natural to extend Feynman's trial action to one with an anisotropic self-interaction :

$$S_m = S_e + S_{m,I} \quad (8)$$

with S_e given by Eq.(4) and :

$$S_{m,I} = \frac{1}{2} \sum_{i=1}^{3} C_i \int_o^\beta dt \int_o^\beta ds [(1+n(w_i))e^{-\hbar w_i|t-s|} + n(w_i)e^{\hbar w_i|t-s|}] \cdot$$
$$(r_i(t) - r_i(s))^2 \quad (9)$$

Because the magnetic field is along the z-axis one has $C_1 = C_2 = C_\perp$, $C_3 = C_\parallel$ and $w_1 = w_2 = w_\perp$, $w_3 = w_\parallel$. One can show that this action (8) can be derived from a hermitian Hamiltonian which describes a Feynman polaron model in a magnetic field with an anisotropic effective interaction and in which the virtual particle has an anisotropic mass.

With this trial action one obtains the following approximation to the polaron ground state energy at zero temperature :

$$\frac{E}{\hbar\omega} = \frac{1}{2} \left(\sum_{i=1}^{3} s_i + v_\parallel - 2w_\perp - w_\parallel \right) - \frac{C_\perp}{mv_\perp w_\perp} \sum_{i=1}^{3} \frac{\partial s_i}{\partial v} - \frac{C_\parallel}{mv_\parallel w_\parallel}$$

$$- \frac{\alpha}{2\sqrt{\pi}} (\hbar\omega)^2 \cdot \sqrt{\frac{\hbar}{2m\omega}} \int_0^\infty dt \frac{e^{-\hbar\omega t}}{\sqrt{H(t)}} \ln \left\{ \frac{\sqrt{D(t)} + \sqrt{H(t)}}{\sqrt{D(t)} - \sqrt{H(t)}} \right\} \quad (10)$$

with $v_i^2 = w_i^2 + \frac{4C_i}{m\hbar w_i}$; the three constants s_i are determined by

the equation $-s_i^2(s_i^2 - v_\perp^2) + \omega_c^2(s_i^2 - w_\perp^2) = 0$. The functions $H(t)$ and

$D(t)$ are given by :

$$H(t) = D(t) - D_H(t)$$

$$D(t) = \frac{t}{2m} \cdot \frac{w_\parallel^2}{v_\parallel^2} + d_o^2(1 - e^{-\hbar s_o t}) \quad (11)$$

$$D_H(t) = \sum_{i=1}^{3} d_i^2(1 - e^{-\hbar s_i t})$$

with

$$d_o^2 = \frac{\hbar}{2mv_\parallel} \cdot \frac{v_\parallel^2 - w_\parallel^2}{v_\parallel^2}$$

$$d_i^2 = \frac{1}{2s_i} \cdot \frac{s_i^2 - w_\perp^2}{3s_i^2 + 2(-)^i \omega_c s_i - v_\perp^2} \quad ; \quad i = 1,2,3 \quad (12)$$

The energy expression (10) is shifted by an infinite
constant contribution[3,4] resulting from the zero point energy of
the phonons and from a contribution resulting from the fact that
the electron is placed in a crystal with an infinite volume.
Taking the limit of small/strong magnetic field and small/strong
electron-phonon coupling, energy values are obtained which are
equal to the lowest one reported in the literature[7,8,9,10]. At
zero magnetic field the result (10) reduces to the well-known result
of Feynman[2].

Expression (10) contains four parameters $(v_\perp, v_\parallel, w_\perp, w_\parallel)$
which are chosen such that the energy is minimal. Such a variational
calculation has been performed numerically for $\alpha = 1$ and $\alpha = 4$.
The present results E are shown in table 1 together with the
perturbation result of Larsen[7] E_{LA} ($\alpha=1$), the adiabatic approximation
result of Kartheuser and Negrete[11] E_{KN} ($\alpha=4$) and the Green's function
result of Lepine and Matz[10] E_{LM}. The proportional lowering of the

present result in comparison with the Lepine and Matz result,
defined as $\Delta = \dfrac{E - E_{LM}}{E_{LM}} \cdot 100$, is given in the last column.

Table I : Numerical results for the ground state energy as
function of the magnetic field for the coupling constant
$\alpha = 1$ and $\alpha = 4$. We have listed the results of Larsen
$(\alpha=1)E_{LM}$, Kartheuser and Negrete $(\alpha=4)E_{KN}$, Lepine and
Matz E_{LM}, the present result E and $\Delta = (E-E_{LM}) \cdot 100/E_{LM}$.
The cyclotron resonance ω_c is given in units of the
phonon frequency ω, the energies are given in units of
the phonon energy $\hbar\omega$.

$\alpha = 1$

ω_c	E_{LA}	E_{LM}	E	Δ
0.0	-1.	-1.	-1.01303	1.30
0.2	-0.9151	-0.91806	-0.92921	1.21
0.5	-0.78668	-0.79387	-0.80260	1.10
1.0	-0.56954	-0.58317	-0.58914	1.02
2.0	-0.12527	-0.15003	-0.15442	2.93

$\alpha = 4$

ω_c	E_{KN}	E_{LM}	E	Δ
0.0	-2.158	-4.072	-4.256	4.52
0.1	-2.157	-4.071	-4.237	4.08
0.5	-2.143	-4.040	-4.155	2.82
1.0	-2.101	-3.964	-4.043	1.99
5.0	-1.181	-2.866	-3.005	4.85
10.0	+0.540	-1.098	-1.417	29.05

3. CONCLUSION

The present approach leads to a ground state energy
which is lower than that obtained by other theories, with the
exception of the zero magnetic field result of Adamowski et al[6].
As mentioned earlier, we believe that the present results are an
upper bound to the exact ground state energy. The difference
with the result of Lepine and Matz is largest for small and
strong magnetic field. At small magnetic field the present

approach reduces to the one of Marshall and Chawla[9] which
gives a lower result than that reported in Ref.10. At strong
magnetic field the anisotropy in the effective interaction
(which have not been taken into account by Lepine and Matz)
becomes important.

REFERENCES

* Work supported by E.S.I.S. and a CDC-grant.
° Aspirant of the National Foundation of Scientific Research
  (N.F.W.O.)
+ Also at R.U.C.A.

1. H. Fröhlich, Adv.Phys. $\underline{3}$, 325 (1954).

2. R.P. Feynman, Phys.Rev. $\underline{97}$, 660 (1955).

3. Y. Osaka, Progr.Theor.Phys. $\underline{22}$, 437 (1957).

4. T.D. Schultz, in Polarons and Excitons, edited by C.E. Kuper
   and G.P. Whitfield (Oliver and Boyd, Edinburgh and London
   1963), p.71.

5. R.W. Hellwarth and P.M. Platzmann, Phys.Rev. $\underline{128}$, 1599 (1962).

6. J. Adamowski, B. Gerlach and H. Leschke, in Proceedings
   of the Workshop on Functional Integration, Theory and
   Applications, Louvain-la-Neuve, November 6-9, 1979.

7. D.M. Larsen, in Polarons in Ionic Crystals and Polar
   Semiconductors, edited by J.T. Devreese (North-Holland,
   Amsterdam and London 1972) p.237.

8. R. Evrard, E.P. Kartheuser and J.T. Devreese, Phys.Status
   Solidi $\underline{4}$, 431 (1970).

9. J.T. Marshall and S.M. Chawla, Phys.Rev. $\underline{B10}$, 4283 (1970).

10. Y. Lepine and D. Matz, Can. J. Phys. $\underline{54}$, 1979 (1976).

11. E.P. Kartheuser and P. Negrete, Phys. Status Solidi B $\underline{57}$,
    77 (1973).

ELECTRON-PHONON INTERACTION $g(\vec{q})$ IN POLAR MATERIALS

R. Salchow[*], R. Liebmann[+] and J. Appel

I. Institut für Theoretische Physik
Universität Hamburg
2 Hamburg 36, Jungiusstr. 9
Federal Republic of Germany

The interaction of electrons and lattice vibrations in polar
materials is investigated taking into account their dielectric
properties by using the microscopic dielectric function, $\varepsilon_{\vec{G}\vec{G}'}(\vec{q})$.
Terms nondiagonal in the reciprocal lattice vectors, $\vec{G} \neq \vec{G}'$,
account for the local field effects.

We present our results for the screening of the electron-phonon
interaction $g(\vec{q})$ in polar insulators and in polar metals. We
apply the dipole model to the screening by the valence electrons.
In addition to this deformation polarization we consider for the
polar metals the screening by free conduction electrons. In the
case of the sodium-tungsten bronzes, the $\vec{q}$-dependence of the
interaction is discussed in some detail. In our model this
dependence is caused to some extent by the phonons and up to
20 % by the screening. For the metallic bronzes, the screening
by conduction electrons dominates that by core electrons. We
discuss our principal approximation, namely neglect of the $\vec{q}$-
dependence of the overlap matrix elements of the tightly bound
valence electrons.

We have investigated quantitatively the electron-phonon
interaction (epi) between electrons and lattice vibrations in
polar solids taking account of their microscopic dielectric
properties.

The lattices of polar solids have bases with differently
charged ions. To optical lattice vibrations belong a displacement
and a deformation polarization. The total polarization causes by

virtue of its local electric field a strong coupling between
electrons and optical phonons.

A conduction electron in such a substance polarizes its
surrounding; the polarization acts back on the electron. Moving
through the crystal the electron carries along this local
distortion of the lattice and forms with it a quasi-particle,
the polaron. Its effective mass is larger and, therefore, the
mobility is smaller than that of the Bloch electron.

The matrix element for the scattering of a Bloch electron
from state $\vec{k}$ to $\vec{k}'$ due to a change in the potential,
$\delta V(\vec{r}) = \Sigma\ \delta\vec{R}.\nabla U(\vec{r}-\vec{R})$, is given by :

$$g = <\psi_{\vec{k}'}(\vec{r})\,|\,\delta V(\vec{r})\,|\,\psi_{\vec{k}}(\vec{r})> \ . \tag{1}$$

We take into account that the electrons of the ion core are not
carried along rigidly with the displacement $\delta\vec{R}$ and that a
screening of the potential change $\delta V(\vec{r})$ occurs due to the
deformation of the outer electrons. Thus, we have an effective
change in the potential,

$$\delta V^{eff}(\vec{r}) = \int d^3r'\ \delta V(\vec{r}')\varepsilon^{-1}(\vec{r},\vec{r}'). \tag{2}$$

Here $\varepsilon^{-1}(\vec{r},\vec{r}')$ is the inverse dielectric function. Now, the
matrix element for the effective electron-phonon interaction
is given by

$$\bar{g} = <\psi_{\vec{k}'}(\vec{r})\,|\,\delta V^{eff}(\vec{r})\,|\,\psi_{\vec{k}}(\vec{r})> \ . \tag{3}$$

This can be fouriertransformed into

$$\bar{g}(\vec{q},\vec{k}\vec{k}') = \sum_{\vec{G}\vec{G}'}\ <\psi_{\vec{k}'}(\vec{r})\,|\,e^{-i(\vec{q}+\vec{G})\vec{r}}\,|\,\psi_{\vec{k}}(\vec{r})>$$

$$\times\ \varepsilon^{-1}(\vec{q}+\vec{G},\vec{q}+\vec{G}').\Sigma\ \delta\vec{R}.(\vec{q}+\vec{G}')\ U(\vec{q}+\vec{G}'). \tag{4}$$

In polar metals the conduction electrons contribute in
addition to the screening given by the inverse dielectric
function, ε^{-1}.

A phenomenological theory for the electron-phonon interaction $\bar{g}$ in semiconductors has been developped by Fröhlich[1]. The
interaction of a free electron with the continuum polarization
is investigated in the long wavelength limit, $\vec{k}' - \vec{k} = \vec{q} \to 0$. The
screening is expressed through the macroscopic dielectric constants
ε_∞ and ε_o for optical and infrared frequencies :
$$\bar{g}^{-2} \propto \alpha \propto (1/\varepsilon_\infty - 1/\varepsilon_o), \text{ independent of } q.$$

In the present paper we calculate the screening for the whole range of values for the phonon wavevector $\vec{q}$ by means of the microscopic dielectric function, $\varepsilon_{GG'}(\vec{q})$. This function is non-diagonal in the reciprocal latticevectors $\vec{G}$ and $\vec{G}'$ and thereby takes into account the local field effects[2]. These effects are of special interest in polar materials with strongly localized core electrons because the microscopic polarization may vary appreciably within an atomic distance. Furtheron we consider discrete ions instead of a continuum.

In the random phase approximation (RPA) we extract a microscopic expression for the epi g from the functional derivative method. $\bar{g}$ depends on $\varepsilon_{GG'}^{-1}$, the electron-ion potential U, the phonons and the Bloch functions.

The dielectric function $\varepsilon_{GG'}(\vec{q})$ is given by

$$\varepsilon_{GG'}(\vec{q}) = \delta_{GG'} - v(\vec{q}+\vec{G})\chi_{GG'}^{o} , \qquad (5)$$

χ^{o} being the irreducible polarizability. In the RPA χ^{o} is represented through the bubble, $\chi^{o} = \bigcirc$. The quantity $v(q+G)$ is the Coulomb potential. The inversion of the matrix ε yields

$$\varepsilon_{GG'}^{-1}(\vec{q}) = \delta_{GG'} + v(\vec{q}+\vec{G})\chi_{GG'}(\vec{q}) . \qquad (6)$$

The density correlation function χ and the function χ^{o} are connected through Dyson's equation.

The inversion of the infinite matrix $\varepsilon_{GG'}$ is possible without great numerical effort in two special cases. For free electrons, ε is given by the Lindhard function and the inversion is trivial. The Lindhard function which applies to the nearly-free conduction electrons in metals has a singularity at $q = 2k_F$ with k_F being the Fermi vector. For tightly bound electrons either of the cores or in narrow bands, the LCAO-Ansatz leads to a separable form of χ^{o} and ε :[3]

$$\chi_{GG'}^{o}(\vec{q}) = \sum_{ss'} A_{s}(\vec{q}+\vec{G})N_{ss'}(\vec{q})A_{s'}^{*}(\vec{q}+\vec{G}') . \qquad (7)$$

Here A_s may be interpreted as a form factor for a generalized "charge density wave", and $N_{ss'}$ as the quantum-mechanical probability of the charge-density wave s being coupled to the wave s'. The subscript s stands for a set of indices indicating site and symmetries of the orbitals.

Because of the separable form of χ^{o}, eq.(6), $\varepsilon_{GG'}$ can be inverted. The result is given by :[3]

$$\varepsilon_{\vec{G}\vec{G}'}^{-1}(\vec{q}) = \delta_{\vec{G}\vec{G}'} + v(\vec{q}+\vec{G})\chi_{\vec{G}\vec{G}'}(\vec{q}) , \tag{8}$$

$$\chi_{\vec{G}\vec{G}'}(\vec{q}) = \sum_{ss'} A_s(\vec{q}+\vec{G})S_{ss'}^{-1}(\vec{q})A_s^*(\vec{q}+\vec{G}) , \tag{9}$$

$$S_{ss'}(\vec{q}) = N_{ss'}^{-1}(\vec{q}) - \sum_{\vec{G}} A_s^*(\vec{q}+\vec{G})A_{s'}(\vec{q}+\vec{G})v(q+G). \tag{10}$$

Thus, the inversion problem for $\varepsilon_{\vec{G}\vec{G}'}$ is reduced to that one for the much smaller matrix $S_{ss'}$. The exchange interaction in Hartree-Fock approximation can be included in the matrix S by replacing $\chi^0 = \bigcirc$ by a series of ladder diagrams leading to an additional term on the right hand side of eq.(10) : $v_{ss'}^{ex}$. As is often done for strongly localized electrons, we adopt the contact approximation $v_{ss'}^{ex} \sim \delta_{ss'}$; thereby we consider only those contributions to $v_{ss'}^{ex}$ for which the orbitals are at the same lattice-site.

In metals we have to take into account besides the localized electrons also the conduction electrons in the partially filled bands. We adopt the method of ref.3 and split up $\varepsilon_{\vec{G}\vec{G}'}$ as follows :

$$\varepsilon_{\vec{G}\vec{G}'}(\vec{q}) = \varepsilon_o(\vec{q}+\vec{G}) \cdot \delta_{\vec{G}\vec{G}'} - v(\vec{q}+\vec{G})\chi_{\vec{G}\vec{G}'}^o , \tag{11}$$

so

$$\varepsilon_{\vec{G}\vec{G}'}^{-1}(\vec{q}) = (1/\varepsilon_o(\vec{q}+\vec{G})).$$
$$\cdot (\delta_{\vec{G}\vec{G}'} + (v(\vec{q}+\vec{G})/\varepsilon_o(\vec{q}+\vec{G})).$$
$$\cdot \sum_{ss'} A_s(\vec{q}+\vec{G})\tilde{S}_{ss'}^{-1}(q)A_s^*(\vec{q}+\vec{G}')). \tag{12}$$

with

$$\tilde{S}_{ss'}(\vec{q}) = N_{ss'}^{-1}(\vec{q}) - \sum_{\vec{G}''} A_s^*(\vec{q}+\vec{G}'')A_{s'}(\vec{q}+\vec{G}'').$$
$$\cdot (v(\vec{q}+\vec{G}'')/\varepsilon_o(\vec{q}+\vec{G}'')). \tag{13}$$

Here the function ε_o describes the transitions within the conduction band for which we use the Lindhard function. Transitions between valence and conduction band are neglected.

For polar insulators the authors have shown elsewhere[4] how the results of earlier, phenomenological theories can be obtained from the RPA dielectric function, eqs.(8)-(10), namely those of a) Born and Huang for local electric fields[5], b) Clausius and Mosotti for the macroscopic screening and c) Fröhlich for the electron-phonon interaction[1]. The principal approximation is the neglect of the overlap of the electron orbitals centered on different sites together with the dipole approximation for the matrix elements of the dielectric function. The error introduced by the neglect of the overlap tends to be compensated by using the effective ionic charges (Szigeti charges)[6].

Here we study microscopically the screening of the electron-phonon interaction $g(\vec{q})$ in polar metals. As an application we consider g and $\bar{g}$ in the sodium-tungsten bronzes, Na_xWO_3. These non-stochiometric compounds have some properties causing our interest in calculating the epi :
1) For x < 0.2 they are insulators, for x > 0.2 polar metals.
2) For different x they exhibit different crystal structures. For x > 0.48 the Na-W bronzes crystallize in the cubic perovskit structure as do $SrTiO_3$ and $BaTiO_3$. Accordingly a strong epi is expected. By the technique of diffusion-dilution the cubic perovskit phase can be kept down to x = 0.22.[7] For x < 0.2 the WO_6 octahedrons have tretragonal distortions and in the range 0.2 < x < 0.48 there is another tetragonal phase (T 1) with groups of four octahedrons rotated by 45°.
3) In some other phases the bronzes have an antiferro - or ferroelectric structure.
4) In the T 1-phase the bronzes have been found supraconducting with a transition temperature between 1 and 3 K, depending on x.[8] This is an experimental indication for a strong epi.
5) The Na-atoms act as electron donators giving their electrons to the conduction bands without changing the band-structure.[9]

For the cubic perovskit phase the band-structure is known.[9] The conduction bands predominantly originate from the tungsten 5d orbitals which hybridize with the oxygene 2p orbitals. It has been shown that the conduction bands depend on only two of the components of the wavevector $\vec{k}$.[10] For the dielectric function of the conduction electrons, ε_o in eq.(12), we, therefore, use the twodimensional Lindhard function with the Fermi wavevector k_F depending on the Na-concentration x. In fig.1 the Lindhard function is shown as a function of q for the twodimensional case;[11] the singularity in the derivative will manifest itself in the screened epi. As for the valence electrons, we assume that they are strongly localized and hence replace the valence bands by atomic levels; in other words, the valence electrons are identified with the core electrons. The deformation of the cores which is expressed as $AS^{-1}A^*$ in eq.(12) is described by atomic dipole transitions. The atomic polarizabilities are taken from other investigations on pervoskites.[12]

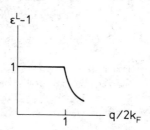

<u>Fig.1</u> : The Lindhard function ε^L-1 (two dimensions); note the
singularity in the derivative at $q = 2 \; k_F$.

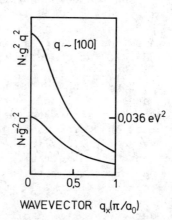

<u>Fig.2</u> : Electron-phonon interaction excluding (upper curve) and
including (lower curve) the electronic polarization for
a longitudinal optical phonon, as a function of the x-
component of the phonon wavevector $\vec{q}$.

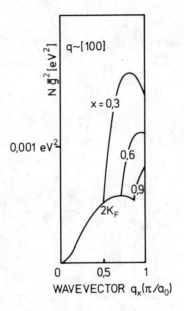

<u>Fig.3</u> : Electron-phonon interaction screened by the tightly bound core electrons as well as by the conduction electrons.

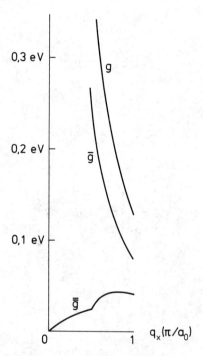

Fig.4 : Comparison of the electron-phonon interaction unscreened
        (g), screened by core electrons ($\bar{g}$) and screened by core
        as well as conduction electrons ($\bar{\bar{g}}$).x = 0.3.

Furtheron, we choose a model pseudo-potential in order to calculate the epi, eq.(4).[13] The phonons are obtained from a simple force-constant model. The parameters, namely the ionic polarizabilities, the wave functions of the outer core electrons and the pseudo-potential have to be fitted together in such a way that the acoustic sum rule is fulfilled.[14] This rule is a test of the consistency of the parameters and guarantees that the acoustic phonons have zero frequency for $q \to 0$.

The result of our calculation for the epi in an insulator is shown in fig.2 for the case of an optical phonon in [100]-direction. The upper curve corresponds to the unscreened interaction (rigid ion model), the lower to the screened epi taking account of the deformation of the ion core. In fig. 2 $g^2.q^2$ resp. $\bar{g}^2.q^2$ is shown for in an insulator g^2 and $\bar{g}^2$ increases as $1/q^2$ for small q. Fig.2 shows that the greater part of the q-dependence of $\bar{g}^2.q^2$ is caused by the phonon system, only 10-20 % emerges from the q-dependence of ε^{-1}. We expect this effect to be larger if the q-dependence of the overlap is taken into account. In the present calculation the neglect of overlap is compensated by a q-independent effective Szigeti-charge.[6] The maximum of $g^2.q^2$ at $q \to 0$ has its origin in the maximum dipole moment caused by optical phonons for $q \to 0$. From fig.2 we can conclude for $q \to 0$ that in WO_3 Fröhlich's constant α has the value 6.2 corresponding to a strong epi.[15]

In the metallic bronzes the epi is screened additionally by the conduction electrons. For x = 0.3, 0.6 and 0.9 $\bar{g}^2$ is shown for optical phonons in Fig.3. The singularities at $q = 2k_F$ originate in the $2k_F$-anomalies of the two-dimensional Lindhard function, Fig.1. For $q \to 0$ $\bar{g}^2$ tends to zero because ε_L^{-1} increases with q^2. The jumps in the derivatives at $q = 2k_F$ occur at different values of q for k_F depends on x which is determining the number of conduction electrons.

Finally we demonstrate in fig.4 that the screening by the conduction electrons dominates the screening by the tightly bound core electrons even at small concentrations x of the conduction electrons.

REFERENCES

* Present address: Hochschule der Bundeswehr Hamburg,
FB Maschinenbau, Holstenhofweg 85, 2000 Hamburg 70.

+ Present address: Universität Frankfurt, Institut für
Theoretische Physik, Robert-Meyer-Str. 8, 6 Frankfurt/Main.

1. H. Fröhlich, Adv.Phys. 3, 325 (1954).

2. R. Liebmann, R. Salchow and J. Appel, Proc. Meeting on One-
dimensional Conductors, in Lecture Notes in Physics, Vol.34,
Springer (1974).

3. W. Hanke, Physical Review B8, 4585 and 4591 (1973).

4. R. Salchow, R. Liebmann and J. Appel, to be published.

5. M. Born and K. Huang, Dynamical theory of crystal lattices,
Clarendon Press, Oxford 1954.

6. B. Szigeti, Proceedings of the Royal Society A 204, 51 (1950).

7. P.A. Lightsey, thesis, Cornell University, Ithaca N.Y. (1972).

8. H.R. Shanks, Solid State Communications, 15, 753 (1974).

9. L. Kopp, B.N. Harmon and S.H. Liu, Solid State Communications
22, 677 (1977).

10. W.A. Kamitakahara, B.N. Harmon, J.G. Taylor, L. Kopp,
H.R. Shanks and J. Rath, Phys.Rev.Lett. 36, 1393 (1976).
T. Wolfram, Phys.Rev.Lett. 29, 1383 (1972).

11. A.M. Afanas'ev and Yu. Kagan, JETP 16, 1030 (1963).

12. R. Migoni, H. Bilz and D. Bäuerle, Stuttgart 1977.

13. We apply a model pseudo-potential which is similar to that
one in transition metals: Th. Starkloff and J.D. Joannopoulos,
Physical Review B16, 5212 (1977).

14. R.M. Pick, M.H. Cohen and R.M. Martin, Physical Review B1,
910 (1970).

15. J. Appel, in Solid State Physics 21, Ed. Seitz/Turnbull,
Academic Press, N.Y. and London 1968.

EXPERIMENTAL EVIDENCE OF THE EXISTENCE OF AN

INTERMEDIATE EXCITONIC POLARITON IN CuCl

E. Ostertag and Vu Duy Phach*

Laboratoire de Spectroscopie et d'Optique du Corps
Solide, Associé au C.N.R.S. n° 232
Université Louis Pasteur
5, rue de l'Université - 67000 Strasbourg - France

When two laser beams propagate in opposite directions in thin
samples of CuCl at 4.2 K, with photon energies adjusted so that
their sum amounts exactly to the energy of a biexciton in this
material, biexcitons of very small wave vectors are created.
Their radiative recombination gives rise to a new narrow
emission line, which has been attributed to the process in
which the observed photon belongs to the lower branch and the
remaining particle to the "intermediate" branch of the polariton
dispersion curve, as given by a two-oscillator model.

INTRODUCTION

Biexcitons can be created directly, in CuCl, by two-
photon absorption. In most of the experiments which have been
performed until now[1-8], the two photons originated from the
same laser beam. When the energy of these photons is adjusted
to 3.186 eV, which amounts to exactly one half of the biexciton
fundamental energy in CuCl, this results in the creation of a cold
gas of biexcitons[2] initially at $\underset{\sim}{K}_o = 2\underset{\sim}{g}_i$, where $\underset{\sim}{g}_i$ is the wave

vector of the lower polariton having that energy. The radiative
recombination of such biexcitons with either a longitudinal
exciton or a lower polariton left in the crystal has been observed,
yielding two narrow emission lines called respectively N_L and N_T

in ref.2. When the laser is detuned slightly from that value,
resonant Raman scattering processes via the biexciton intermediate
state to the same final states are observed.

389

Experiments using two counterpropagating beams of equal[9] or nearly equal[10-12] energies have recently shown the possibility of creating biexcitons at values of $\underset{\sim}{K}$ respectively equal to zero or finite but smaller than $\underset{\sim}{K}_o$. In some of these experiments[9,11,12] circularly polarized light was used in order to eliminate the formation of biexcitons at $\underset{\sim}{K}_o = 2\underset{\sim}{q}_i$ from a single beam, at resonance. The two-photon transition is indeed forbidden for two photons of same circular polarization, but allowed if they propagate in opposite directions[13].

We report here a similar experiment, performed in CuCl with two linearly polarized laser beams of opposite directions, the energies of which are adjusted far off resonance so as to create biexcitons at finite but largely varying wave vectors.

EXPERIMENTAL SET-UP

The experiment was performed with two tunable dye lasers[14], pumped by the same nitrogen laser, as shown in fig.1. The active medium of both cavities was a half saturated solution of αNND in ethanol. The two beams, propagating in opposite directions, were then focused into thin platelet samples of CuCl, immersed in liquid helium. Careful adjustments were made, by means of a photocell associated with a fast oscilloscope and of a telescope, to ensure both the temporal and spatial coincidence of the two laser pulses in the crystal. The direction of one of the lasers was fixed, at normal incidence ($i_2 = 0$), but the angle of incidence i_1 of the other beam could be varied from 0° to about -30°. The emission coming out of one of the illuminated faces of the sample was focused onto the entrance slit of a 0.75 m spectrograph. The angle r of the direction of observation with the normal to the crystal surfaces was fixed at 30°. The detection of the spectra was made by means of an optical multichannel analyzer, associated with a microcomputer[15,16].

EXPERIMENTAL RESULTS AND DISCUSSION

The two laser beams were adjusted at normal incidence ($i_1 = i_2 = 0$). Their energies $\hbar\omega_1$ and $\hbar\omega_2$ were set at different values, chosen each time so that their sum $\hbar\omega_1 + \hbar\omega_2$ amounts exactly to the energy $E_M = 6.372$ eV of the biexciton in its ground state. As can be seen in figure 2, they will yield polaritons having wave vectors q_1 and q_2 of different magnitudes, and thus create biexcitons of finite wave vector $K = q_1 + q_2$. By simply changing the couple of values (ω_1, ω_2), it should then be possible to vary K between 0 and a finite amplitude which, due to the strong curvature of the lower polariton dispersion curve, can be rather large.

Figure 3 shows the emission spectra obtained for different couples (ω_1, ω_2) satisfying the previous condition of resonance. They exhibit three lines. Two of them are located at the energies of 3.1638 eV and 3.1699 eV, which are known[8] to be the positions of the luminescence lines N_L and N_T. A new line appears on the low energy side of the N_L line, and its position varies when the pair (ω_1, ω_2) is varied. It should be noted first that all three lines disappear as soon as one of the lasers is detuned from its value corresponding to the resonance by a few tenths of a meV. This shows that the observed emissions are due to really created biexcitons.

Due to the small wave vector of these "ultra-cold" biexcitons, their dissociation into two polaritons belonging respectively to the lower and the upper branch is now possible. Itoh and Suzuki[10], who have shown the possibility of this process, have observed the Raman line associated with the upper polariton leaving the crystal. We believe that the new line observed in our present experiment is the complementary manifestation of this process, being a lower polariton which comes out of the crystal, whereas a transverse polariton of the "upper" branch remains in it. The denomination "upper" polariton should however be revised, as will be discussed next.

In order to corroborate that interpretation we calculated the energy of this line as a function of (ω_1, ω_2) by the same self-consistent computer alogirthm as described in references 4 and 6. The polariton dispersion relation which was used for the initial and final polariton states involved was derived from a two-oscillator model[17] taking into account the effects of the excitons Z_3 and Z_{12} of CuCl. The resulting dispersion curve has thus three branches (Fig.2) : a lower branch located below the Z_3 exciton, a true upper branch above the Z_{12} exciton, and a branch situated between these two excitons which we suggest to call "intermediate polariton" branch in order to avoid any confusion. The curves shown in figure 4 indicate the calculated energy $\hbar\omega$ of the photon emitted in the direction of observation as a function

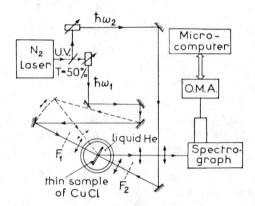

Fig.1 : Block diagram of the experimental set-up.

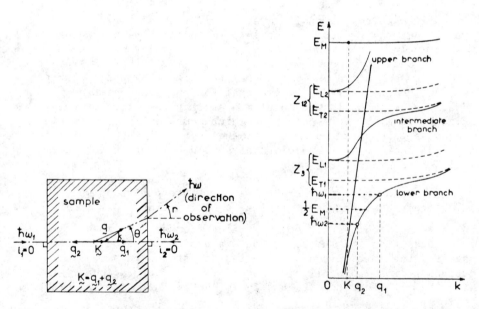

Fig.2 : Construction of the wave vector $\underset{\sim}{K}$ at which the biexcitons
        are created.

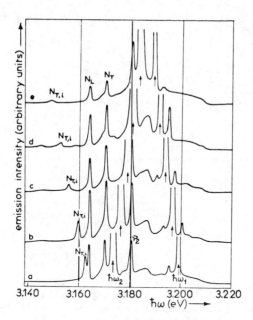

Fig. 3: Luminescence spectra of CuCl, at 4.2 K, due to biexcitons
created at a small wave vector $\underline{K}$. The energies $\hbar\omega_1$ and $\hbar\omega_2$

of the two counterpropagating laser beams, and the
corresponding calculated value of $|\underline{K}|$ are, respectively:
a) 3.1989 eV, 3.1731 eV, 1.91×10^5 cm^{-1}
b) 3.1963 eV, 3.1758 eV, 1.11×10^5 cm^{-1}
c) 3.1936 eV, 3.1786 eV, 6.73×10^4 cm^{-1}
d) 3.1913 eV, 3.1808 eV, 4.25×10^4 cm^{-1}
e) 3.1888 eV, 3.1832 eV, 2.12×10^4 cm^{-1}

of the energy $\hbar\omega_1$ of one of the two impinging photons for the
following processes:
(I)   lower polariton detected, longitudinal exciton left in the
      crystal,
(II) lower polariton detected, lower polariton left,
(III) lower polariton detected, intermediate polariton left.
The experimentally observed positions of the new line, which we
have tentatively attributed to process (III) and denoted for that
reason by N_{Ti}, are also indicated on this figure. The very good

agreement between experiments and calculated positions at all
pair values (ω_1, ω_2) is a strong argument in favour of this
interpretation.

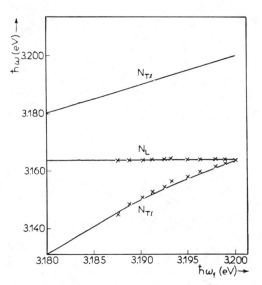

<u>Fig. 4:</u>   Experimental (×) and calculated positions of the emission
          lines, as a function of the energy $\hbar\omega_1$ of one of the
          incident photons, the other one being adjusted to the
          resonance condition $\hbar\omega_2 = E_M - \hbar\omega_1$.

        As can be expected for small values of the angle θ
between $\underset{\sim}{K}$ and the direction of observation inside the crystal,
the $N_{T\ell}$ line corresponding to process (II) is located at almost
the energy of the incident photons $\hbar\omega_1$. Due to the strong Rayleigh
diffusion of the incident laser lines, this recombination cannot
be seen in our spectra.

        The positions of the N_L line experimentally observed
are also indicated in figure 4. We believe that this line is due
as well to the process (I) described above as to the recombination
of biexcitons created at larger wave vectors, as will be discussed
hereafter.

        Due to the small absorption coefficient at the excitation
energies used and to the small thickness of the sample, the
configuration adopted ($i_1 = i_2 = 0$) leads to an overlapping of
each laser beam with the part of the other beam which is reflected
in the direction of the first one.   As can be seen from figure 2,
these superpositions result in the creation of biexcitons at
$\underset{\sim}{K} \simeq 2q_i$, thus leading to the formation of a cold gas of biexcitons,
similar to the case in which a single laser is used.   The radiative
recombination of such cold biexcitons is then responsible for the
appearance of the N_T line in our spectra and, partially, for that
of the N_L line.

To summarize, we have observed the recombination of the "ultra-cold" biexcitons towards the "intermediate" and lower polariton branches calculated by using a two-oscillator model for CuCl. The experimental results show also that the radiative recombination occurs with the wave vector at which the biexcitons are created, which suggests that almost no relaxation takes place.

REFERENCES

*Present address: Max Planck Institut für Festkörperforschung Stuttgart, Federal Republic of Germany.

1. N. Nagasawa, N. Nakata, Y. Doi and M. Ueta, Journal of the Physical Society of Japan, 39, 987 (1975).

2. R. Levy, C. Klingshirn, E. Ostertag, Vu Duy Phach and J.B. Grun, Physica Status Solidi, (b) 77, 381 (1976).
    B. Hönerlage, Vu Duy Phach and J.B. Grun, Physica Status Solidi, (b) 88, 545 (1978).

3. N. Nagasawa, T. Mita and M. Ueta, Journal of the Physical Society of Japan, 41, 929 (1976).

4. E. Ostertag, thesis, University of Strasbourg (1977).

5. N. Nagasawa, T. Mita, T. Itoh and M. Ueta, Journal of the Physical Society of Japan, 43, 1295 (1977).

6. E. Ostertag, A. Bičas and J.B. Grun, Physica Status Solidi, (b) 84, 673 (1977).

7. M. Ojima, T. Kushida, Y. Tanaka and S. Shionoya, Solid State Communications, 24, 841 (1977).

8. Vu Duy Phach, A. Bivas, B. Hönerlage and J.B. Grun, Physica Status Solidi, (b) 86, 159 (1978).

9. L.L. Chase, N. Peyghambarian, G. Grynberg and A. Mysyrowicz, Optics Communications, 28, 189 (1979).

10. T. Itoh and T. Suzuki, Journal of the Physical Society of Japan, 45, 1939 (1978).

11. N. Nagasawa, T. Mita and M. Ueta, Journal of the Physical Society of Japan, 45, 713 (1978).

12. N. Nagasawa, T. Mita and M. Ueta, Journal of the Physical Society of Japan, 47, 909 (1979).

13. E. Hanamura and H. Haug, Physics Reports, 33c, 209 (1977).

14. Vu Duy Phach, thesis, University of Strasbourg (1980).

15. E. Ostertag, Review of Scientific Instruments, 50, 1344 (1979).

16. E. Ostertag, Euromicro Journal, 5, 309 (1979).

17. E. Ostertag, to be published.

SOME PECULARITIES OF THE PIEZO- AND MAGNETO-OPTIC

EFFECTS ON EXCITONIC POLARITONS IN CuBr

A. Daunois, J.L. Deiss, and J.C. Merle

Laboratoire de Spectroscopie et d'Optique du
Corps Solide, Associé au C.N.R.S. n° 232
Université Louis Pasteur
5, rue de l'Université - 67000 Strasbourg - France

The stress-induced and magneto-optic effects on the Z_{12} exciton in CuBr investigated at 1.8 K in different geometries cannot be explained satisfactorily by the simple models customarily used. The analysis of the experimental data has been achieved successfully by the polariton formalism.

Les effets dus à la pression et magnéto-optiques sur l'exciton Z_{12} dans CuBr, étudiés à 1,8 K dans différentes géométries, ne peuvent être expliqués de façon satisfaisante par les modèles simples habituellement utilisés. L'analyse des résultats expérimentaux a été effectuée avec succès par le formalisme du polariton.

INTRODUCTION

The simple model of hydrogenic excitons is inadequate to explain correctly the experimental features related to highly degenerate excitons. It has been shown recently[1,2] that, in such a case, it is necessary to take into account the exchange interaction and the finiteness of the exciton wavevector, in addition to this degeneracy.

The application of an external perturbation, which can lift some degeneracies and/or mix different exciton states, can give rise to splittings and/or to a change of the selection rules. Consequently, a detailed analysis of these effects will help to understand the internal structure and the interaction mechanisms of such degenerate excitons.

In CuBr, a zinc-blende crystal, the 1S (Z_{12}) exciton associated with the Γ_8 valence and Γ_6 conduction bands has a total representation $\Gamma_3 + \Gamma_4 + \Gamma_5$ which gives a degeneracy of 8. Due to the absence of inversion symmetry in the zinc-blende lattice, k-linear terms are present in the expression giving the energy of the Γ_8 degenerate valence band and mix at $\vec{k} \neq 0$ the dipole forbidden triplet states $\Gamma_3 + \Gamma_4$ with the dipole allowed singlet state Γ_5. Thus, the triplet states become weakly allowed and give rise to a weak and sharp line in the intense reflectivity anomaly due to the singlet Γ_5. The exciton radius being small in CuBr, the exchange interaction is important and the singlet-triplet Δ_1 and longitudinal-transverse Δ_{TL} separations are particularly large ($\Delta_1 \simeq 1.1$ meV, $\Delta_{TL} \simeq 12.3$ meV)[3].

Using a purely excitonic model, we applied previously the matrix formulation elaborated by Cho[1] to uniaxial stress measurements in the geometry $\vec{P} // 001$[4]. The splittings, polarization dependence and the variations with the stress of the 1S exciton are well explained by the excitonic model. But, it is still inappropriate to explain all the details of magnetic field experiments or uniaxial stress measurements, for instance for $\vec{P} // 1\bar{1}0$.

We will point out in this paper that some features can be satisfactorily explained only by the polariton formalism, in particular, the appearance of unexpected components and the non-convergence of the $\Gamma_3 + \Gamma_4$ components at zero field.

POLARITON SCHEME

It is well known that the best results in the quantitative analysis of a reflectivity spectrum are given by the polariton formalism[5]. However, it was suitable only in relatively simple cases. The recently elaborated models for degenerate excitons[6,7] allow now such an analysis.

With a set of n exciton states, coupled with a photon in a definite polarization, 2(n+1) polariton branches are obtained in the E-k space. For a given photon energy E, n+1 branches only are excited in a reflectivity experiment on a half-infinite crystal. The Maxwell equations and the additional boundary conditions (ABC) allow the calculation of the normal reflectivity. Following Pekar[5] and Cho[6], we suppose that the amplitudes of the n excitons are zero at the surface. Moreover, it is not necessary to introduce

any dead layer in the calculation because of the small exciton
radius.

Figure 1 shows an example of the calculated and
experimental reflectivity spectra of the 1S exciton for $\vec{k}$ // 110,
$\vec{P}$ // 001 and $\vec{\varepsilon}$ // 1$\bar{1}$0. The agreement between the two curves is very
good and the model is able to reproduce correctly the lineshape
as well as its variations with the applied stress. In this case,
the selection rules and the shifts of the different components
have been also explained successfully with the excitonic model[4].
But the quantitative analysis of the reflectivity spectrum is
achievable only with the polariton scheme. Now, the excitons which
are considered here have a complex dispersion. Their lifetimes
must be defined for each exciton branch and are at most k-dependent.
Such a complicated behavior has not been taken into account by the
model. This explains that the calculation reproduces generally
quite well the experimental positions but not always the exact
shape of the structures.

INTERPRETATION OF UNUSUAL EFFECTS

Some experimental features we have obtained cannot be
explained by the simple excitonic model and must be interpreted
by the polariton theory. To understand at least roughly the
reflectivity spectra, it is necessary to consider the behavior
of the excitons and polaritons in the E-k space. The existence of
polariton branches with complex k-values, in addition to the usual
real and purely imaginary ones, complicates the problem even without any broadening effect. The complex k-value branches, existing
only in presence of the k-linear terms, and the imaginary ones are
considered in the calculation but can be neglected for a
qualitative analysis because they correspond to evanescent waves[5,7].
Figure 2 shows the polariton exciton dispersions for the branches
with real k-values without broadening. The calculated and
experimental reflectivity spectra in the vicinity of the triplet
states Γ_3, Γ_4 are also given in this figure for $\vec{k}$ // 110 and
$\vec{\varepsilon}$ // 001. In this geometry, two triplet states become dipole allowed
for $\vec{k} \neq 0$, but one structure only is calculated and observed. It
corresponds well to the high density of states associated with the
point noted 2 on the figure. For the point noted 3 at small $\vec{k}$, the
polariton branches have also a high density of states. But, the
triplet states being forbidden at $\vec{k} = 0$, the polariton states have
a fully excitonic character and cannot give any contribution to
the dielectric constant $\varepsilon(\omega,\vec{k})$. From this figure, the following
conclusion can be drawn : the number of observed structures is
not automatically connected with the number of active exciton
states.

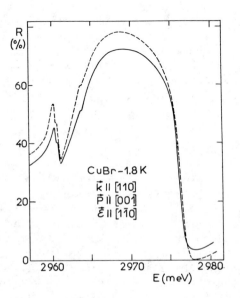

Fig.1 : Calculated (---) and experimental (———) reflectivity spectra
of CuBr at 1.8 K for P = 4.5 kg/mm^2. The parameters used in
the calculation here and in the following figures are given
in references 4 and 7. The excitonic dampings are .1, .1
and .2 meV for the Γ_3, Γ_4 states and .2 meV for the Γ_5 state.

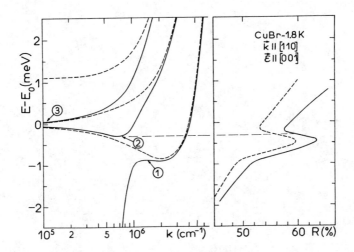

Fig. 2:   Left.   Exciton (---) and polariton (——) dispersion for the free crystal of CuBr.   The $\Gamma_3 + \Gamma_4$ triplet states are

associated to the two lower energy exciton branches which converge at $\vec{k} = 0$.   The higher exciton branch is associated to the Γ_5 singlet state.

Right.   Corresponding experimental (---) and calculated (——) reflectivity spectra.   The dampings used in the calculation are .05 meV for the Γ_3, Γ_4 excitons and .2 meV for the Γ_5 exciton.

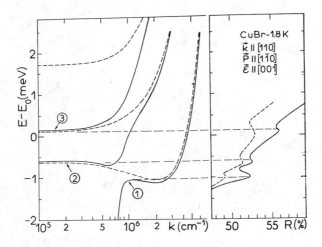

Fig.3 : Same as figure 2, but with P = 2 kg/mm^2.

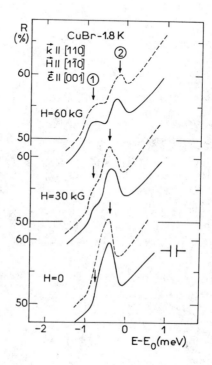

<u>Fig.4</u> : Calculated (---) and experimental (——) reflectivity
spectra of CuBr at 1.8 K for different values of the
magnetic field.

An unexpected situation has been observed on the triplet exciton Γ_3, Γ_4 in a crystal of CuBr stressed along the $|1\bar{1}0|$ axis: the normal reflectivity in the $|110|$ direction exhibits a total of six structures while the degeneracy of these states is only five.

For $\vec{\varepsilon}$ // $1\bar{1}0$, we observe three singularities, as expected by the exciton model where three triplet exciton states are active in this polarization. In the other polarization $\vec{\varepsilon}$ // 001, the two other triplet exciton branches are involved but three structures are observed. This supplementary structure, unpredictable by the group theory, cannot be explained by the excitonic model. The exciton and polariton dispersions in this polarizations and the corresponding calculated and experimental reflectivities are reported in figure 3. The geometries of figures 2 and 3 being the same ($\vec{k}$ // 110, $\vec{\varepsilon}$ // 001), the exciton states for P = 0·and P ≠ 0 are identical since the stress does not presently mix the states active in two different polarizations. Thus the dispersion curves of figure 3 are the deformed curves of figure 2. The two polariton singularities noted points 1 and 2 are associated to a single exciton branch and give rise to two different reflectivity structures. It can be noticed that the point 3 gives also a reflectivity structure, which was not the case in figure 2. The explanation is that, now, the triplet states are dipole allowed even at $\vec{k}$ = 0, because of the stress. The polariton states, for small $\vec{k}$, are no longer purely excitonic and contribute to the energy transport in the crystal. As noticed previously, the agreement between the calculated and experimental reflectivity curves is not very good, but the main result is that the calculation is able to account for the observed appearance of the supplementary structure and to explain the puzzling existence of three reflectivity structures corresponding to only two excitonic branches.

Another unusual feature we observed is the following : if we plot the positions of the three experimental reflectivity singularities of figure 3 as a function of the applied stress, we do not obtain a convergence of the levels towards an unique point for P = 0 but rather towards three points corresponding to the energies of the points 1, 2 and 3 of figure 2.

An example of this non-convergence can be seen in fig.4 where the experimental and calculated reflectivity curves are shown for different values of the magnetic field $\vec{H}$ // $1\bar{1}0$, in the polarization $\vec{\varepsilon}$ // 001. Here again, the exciton and polariton branches, perturbed by the magnetic field, are the same than in figures 2 and 3. The arrows in figure 4 show the positions of the two main structures associated to the points 1 and 2 for the different field values. The extrapolated position for H = 0 of the lower energy component corresponds also to the point 1 of fig.2.

CONCLUSION

        The polariton formalism allows a quantitative analysis
of the behavior of highly degenerate excitons in CuBr under the
influence of external perturbations. This analysis takes into
account, in addition to the degeneracy of the excitonic states,
the exchange interaction, the k-linear terms as well as the terms
related to the external perturbations.

        This analysis allows to point out the following main
facts :
        - the overall quantitative agreement between the
calculated and the experimental results shows that the polariton
model we have used is well adapted to the study of degenerate
excitons and leads to a better comprehension of the observed
phenomena.
        - the experimental features of the zero-field or
perturbed reflectivity spectra principally explained by the
presence of the k-linear terms in the Γ_8 valence band confirm
the importance of these terms[3,7].
        - the influence of the Luttinger parameters γ_2, γ_3
related to the light and heavy holes and to the warping of
this valence band has also been considered but they cannot be
responsible for the experimentally observed features, similar
conclusions have also been obtained in the case of the Hyper-
Raman scattering[3,8].

REFERENCES

1. K. Cho, Physical Review, B14, 4463 (1976).

2. E.O. Kane, Physical Review, B11, 3850 (1975).

3. E. Suga, K. Cho and M. Bettini, Physical Review, B13, 943 (1976).
   A. Bivas, Vu Duy Phach, B. Hönerlage, U. Rössler and J.B. Grun,
   Physical Review, B20, 3442 (1979).

4. C. Wecker, A. Daunois, J.L. Deiss, P. Fiorini and J.C. Merle,
   Solid State Communications, 31, 649 (1979).

5. J.J. Hopfield, Physical Review, 112, 1555 (1958).
   S.I. Pekar, Soviet Physics Solid State, 6, 785 (1958);
   Journal of Physics and Chemistry of Solids, 5, 11 (1958).

6. K. Cho, Solid State Communications, 27, 305 (1978).

7. P. Fiorini, J.C. Merle and M. Simon, to be published.

8. B. Hönerlage, U. Rössler, Vu Duy Phach, A. Bivas and J.B. Grun,
   to be published.

THEORY OF OPTICAL BISTABILITY OF DIELECTRIC CRYSTALS

J. Goll and H. Haken

Institut für Theoretische Physik der
Universität Stuttgart

Pfaffenwaldring 57, 7000 Stuttgart 80, Germany

We calculate the non-linear transmissivity of a dielectric crystal
for an impinging electromagnetic wave. We investigate the
phenomenon of optical bistability and discuss the influence of
the detuning.

1. INTRODUCTION

The phenomenon of optical bistability[1-5] of saturable
absorbers inside a Fabry-Perot was suggested in 1969 by Szoeke,
Daneu, Goldhar and Kurnit[6]. They showed by a simple model that
two stable values of the transmitted light field intensity would
exist in a certain range of input intensities provided the
transmissivity of the mirrors, T, and the absorption in the
absorptive medium, αL (α absorption coefficient, L length of
the cavity), were small enough with $\alpha L/T \gg 1$.

In this paper we want to discuss whether the phenomenon
of optical bistability can even be expected for dielectric crystals
without mirrors. We discuss an experimental situation, where an
electromagnetic plane wave $E^{(I)}$ is impinging on the crystal (see
fig.1). $E^{(R)}$ is the reflected, $E^{(T)}$ the transmitted light wave.
$E^{(F)}$ is the forward field and $E^{(B)}$ the backward field in the
crystal. Our aim is to find the transmissivity $|E^{(T)}|^2/|E^{(I)}|^2$
in dependence on the detuning of the light field frequency with
respect to the resonance frequency of the atoms. Furthermore we

405

shall discuss the model of a crystal with a mirror at its right
end face.

2. EQUATIONS OF MOTION

In our calculations we investigate the interaction of
the light field with noninteracting excitons. In our simple model
we do not include the effect of spatial dispersion, i.e. we discuss
the limiting case that the total mass of an exciton tends to
infinity. In reality these excitons of infinite mass can be either
impurity atoms in a host crystal or excitons in a semiconductor or
molecular crystal. For small exciton masses and low damping of the
excitons the occurence of spatial dispersion would lead to the
appearance of additional "anomalous" waves at high enough
frequencies. However, our model of infinite exciton mass is
qualitatively adequate for the case of small exciton masses, too,
provided the damping is strong enough, or the frequency is low
enough, so that an "anomalous" wave does not appear for the applied
frequency.

Starting point of our calculations are the Heisenberg
equations of motion for the exciton amplitude, the inversion and
the light field amplitude[7] :

$$\frac{d}{dt} B_1^+ = i\nu B_1^+ - gA_1^- D_1 \tag{1}$$

$$\frac{d}{dt} D_1 = 2g(A_1^+ B_1^+ + A_1^- B_1) \tag{2}$$

$$\frac{d}{dt} A_1^- = \frac{d}{dt} A_1^- \Big|_{H_{light}} - GB_1^+ \tag{3}$$

In the case of excitons B_1^+ is the creation operator of an exciton
at a lattice site 1. In the case of impurity atoms B_1^+ annihilates
an electron in the ground state and creates an electron in the
excited state at lattice site 1. For simplicity we shall always
designate B_1^+ as the exciton amplitude. D_1 is the inversion. The
coupling constant G is proportional to N/V. In the case of excitons
N/V is the density of elementary cells. For impurity atoms embedded
in a host crystal, N/V is the density of impurity atoms. c is the
light velocity in the crystal for frequencies far beyond the
resonance frequency ν of the excitons (impurity atoms). A_1^- and A_1^+
are the negative and positive frequency part of the vector potential
of the light field, respectively.

Fig.1 : Experimental situation.

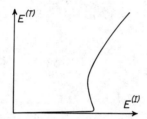

Fig.2 : For impurity atoms in a host crystal with a very high
refractive index (n=4 in the figure) the effect of
optical bistability is possible without a mirror.

By averaging the quantum-mechanical equations of motion and factorizing the averages of products of operators we proceed to semiclassical equations. Introducing damping terms in a phenomenological way we obtain the following equations of motion :

$$\frac{d}{dt} B_1^* = i\nu B_1^* - \gamma B_1^* - g A_1^- D_1 \tag{4}$$

$$\frac{d}{dt} D_1 = 2g(A_1^+ B_1^* + A_1^- B_1) - \gamma_{\parallel}(D_1 + \psi(0)) \tag{5}$$

$$\frac{d}{dt} A_1^- = \frac{d}{dt} A_1^-\Big|_{H_{light}} - G B_1^* \tag{6}$$

where

$$\gamma = 1/T_2, \quad \gamma_{\parallel} = 1/T_1 \tag{7}$$

T_1 and T_2 are the homogeneous longitudinal and transverse relaxation times, respectively. $\psi(0)$ is the wave function of the relative motion of electron and hole for electron-hole separation $x = 0$. In the case of impurity atoms, $\psi(0) = 1$.

3. STATIONARY SOLUTIONS

In the stationary state we decompose the exciton amplitude and the light field amplitude into a complex stationary amplitude and into a rapidly oscillating phase factor :

$$B_1^*(t) = \tilde{B}_1^* \exp(i\Omega t), \quad A_1^-(t) = \tilde{A}_1^- \exp(i\Omega t) \tag{8}$$

We now split the vector potential $\tilde{A}_1^-$ into a forward and a backward wave :

$$\tilde{A}_1^- = \tilde{A}_1^{-(F)} e^{-iKl} + \tilde{A}_1^{-(B)} e^{iKl} \tag{9}$$

Using eq. (9) and the following equation which can easily be derived :

$$\frac{d}{dt} A_1^-\Big|_{H_{light}} = -c \frac{d}{dl}\{\tilde{A}_1^{-(F)} e^{-iKL}\} + c \frac{d}{dl}\{\tilde{A}_1^{-(B)} e^{iKL}\} \tag{10}$$

we obtain from equation (6) after elimination of the exciton amplitude and the inversion :

$$0 = i(cK-\Omega)\tilde{A}_1^{-(F)}e^{-iKl} + i(cK-\Omega)\tilde{A}_1^{-(B)}e^{iKl}$$

$$- c(\frac{d}{dl}\tilde{A}_1^{-(F)})e^{-iKl} + c(\frac{d}{dl}\tilde{A}^{-(B)})e^{iKl}$$

$$+ \frac{Gg\psi(0)}{i(\nu-\Omega)-\gamma}(\tilde{A}_1^{-(F)}e^{-iKl} + \tilde{A}_1^{-(B)}e^{iKl})$$

$$\{1 + A_S^{-2}(|\tilde{A}_1^{-(F)}|^2 + |\tilde{A}_1^{-(B)}|^2 \tag{11}$$

$$+ \tilde{A}_1^{-(F)}\tilde{A}_1^{+(B)}e^{-2iKl} + \tilde{A}_1^{+(F)}\tilde{A}_1^{-(B)}e^{2iKl})\}^{-1}$$

where

$$A_S^{-2} = \frac{4g^2\gamma/\gamma_\parallel}{(\nu-\Omega)^2 + \gamma^2} \tag{12}$$

is the inverse of the saturation intensity.

We now decompose the complex amplitudes $\tilde{A}^{-(F)}$ and $\tilde{A}^{-(B)}$, respectively, into the modulus and the phase factor :

$$\tilde{A}_1^{-(F)} = A_1^{(F)}e^{-i\Phi_1^{(F)}} \quad , \quad \tilde{A}_1^{-(B)} = A_1^{(B)}e^{i\Phi_1^{(B)}} \tag{13}$$

Multiplying eq.(11) by the denominator of the last term in eq.(11) we obtain an equation of the following structure

$$e^{-iKl}(...)_1 + e^{iKl}(...)_2 + e^{i3Kl}(...)_3 + e^{-i3Kl}(...)_4 = 0 \tag{14}$$

where the expressions in brackets are independent on $e^{\pm inKl}$. Neglecting the third and fourth terms in eq.(14) as rapidly oscillating terms and putting $(...)_1$ as well as $(...)_2$ equal to zero :

$$(...)_1 = 0, \quad (...)_2 = 0 \tag{15a},(15b)$$

we obtain a unique decomposition of the vector potential $\overline{A}_1$ into a backward and a forward wave. From the real and imaginary parts of eqs.(15a), (15b) we obtain the following differential equations

$$\frac{d}{dl}a_1^{(F)} = -\alpha_{1,NL}^{(F)}a_1^{(F)} \tag{16}$$

$$\frac{d}{dl}\Phi_1^{(F)} = -(K\frac{\Omega}{c}) + \delta\alpha_{1,NL}^{(F)} \tag{17}$$

where

$$a_1^{(F)} = A_S^{-1}A_1^{(F)}, \quad a_1^{(B)} = A_S^{-1}A_1^{(B)} \tag{18}$$

are the normalized amplitudes of the forward and the backward
wave, respectively, and

$$\delta = \frac{\nu - \Omega}{\gamma} \tag{19}$$

is the normalized detuning. The abbreviation $\alpha_{1,NL}^{(F)}$ is given
by :

$$\alpha_{1,NL}^{(F)} = \alpha \frac{1 + a_1^{(F)^2}}{(1 + a_1^{(F)^2} + a_1^{(B)^2})^2 - a_1^{(F)^2} a_1^{(B)^2}} \tag{20}$$

with

$$\alpha = \frac{\gamma G g \psi(0)/c}{(\nu-\Omega)^2 + \gamma^2} \tag{21}$$

The quantity 2α is the classical absorption coefficient in the
case of homogeneous line broadening. Analogous results can be
derived for the backward wave, where only the superscripts (F)
and (B) have to be interchanged. The coefficient on the right
hand side of eq.(20) which depends non-linearly on the intensities
of the forward and the backward wave describes the saturation
behaviour of the excitons (impurity atoms).
From eq.(16) and the corresponding equation for the backward wave
the following conservation law can be derived :

$$\frac{a_1^{(F)^2} a_1^{(B)^2}}{(1 + a_1^{(F)^2})(1 + a_1^{(B)^2})} = const = C \tag{22}$$

Using this conservation law, eq.(16) can be solved

$$z_1 \exp(z_1 - \frac{C}{z_1}) = z_o \exp(z_o - \frac{C}{z_o}) \exp(-2\alpha)(1-C)l) \tag{23}$$

where

$$z_1 = a_1^{(F)^2}(1-C) - C \tag{24}$$

We determine the wave vector K from the condition that
the linear part of $d\Phi_1^{(F)}/dl$ or $d\Phi_1^{(B)}/dl$ is equal to zero :

$$K - \frac{\Omega}{c} - \delta\alpha = 0 \tag{25}$$

or, explicitly,

$$cK - \Omega = \frac{(\nu-\Omega) G g \psi(0)}{(\nu-\Omega)^2 + \gamma^2} \tag{25b}$$

Eq.(25) describes the dispersion law at low intensities. If $Gg\psi(0)$ is big enough and γ is small enough, a forbidden energy gap appears at the resonance frequency (polariton gap). If the density of impurity atoms is small enough, no forbidden energy gap occurs, since $G \sim N/V$. For low densities of impurity atoms the dispersion law (25) is just slightly different from the dispersion of the free light field in the crystal. Using eq.(25) we obtain :

$$\frac{d}{dl} \phi_1^{(F)} = \delta\{\alpha_{1,NL}^{(F)} - \alpha\} \tag{26}$$

The total, non-linear wave vector of the forward wave is given by

$$K_{1,NL}^{(F)} = K + \frac{d}{dl} \phi_1^{(F)} = \frac{\Omega}{c} + \delta\alpha_{1,NL}^{(F)} \tag{27a}$$

Thus we obtain the following result for the index of refraction

$$n_1^{(F)} = \frac{c_o}{c} + \frac{c_o}{\Omega} \delta\alpha_{1,NL}^{(F)} \tag{27b}$$

where c_o is the light velocity in vacuum. An analogous expression can be derived for the backward wave. Due to saturation the dispersion law tends to the dispersion of the free light field in the crystal, when the intensity is high enough.

Using eqs.(16), (22) and (26) we find the solution

$$\phi_1^{(F)} = -\frac{\delta}{2}\{1na_1^{(F)^2} - a_1^{(F)^2} - \frac{1}{1-C} 1n(a_1^{(F)^2}(1-C) - C)$$

$$+ \frac{C}{1-C} \frac{1}{a_1^{(F)^2}(1-C) - C}\} + const^{(F)} \tag{28}$$

where $const^{(F)}$ is a constant of integration. An analogous expression holds for $\phi_1^{(B)}$.

4. BOUNDARY CONDITIONS. OPTICAL TRANSMISSIVITY

The tangential components of the electric field strength E and of the magnetic field strength H are continuous across a dielectric-vacuum surface. Using the fact that

$$H^{-(I)} = E^{-(I)}, \quad H^{-(R)} = -E^{-(R)}, \quad H^{-(T)} = E^{-(T)} \tag{29}$$

and

$$E_1^- = -\frac{1}{c_o}\frac{\partial}{\partial t}A_1^-, \qquad H_1^- = \frac{\partial}{\partial l}A_1^- \tag{30}$$

the following relations can be derived at the end faces $l = L$
and $l = 0$ of the crystal :

$$A_S e^{(T)} = \tilde{A}_L^-, \qquad A_S e^{(T)} = i\frac{c_o}{\Omega}\frac{d}{dl}\tilde{A}_1^-\Big|_{l=L} \tag{31a},(31b)$$

$$A_S(e^{(I)} + e^{(R)}) = \tilde{A}_o^-, \quad A_S(e^{(I)} - e^{(R)}) = i\frac{c_o}{\Omega}\frac{d}{dl}\tilde{A}_1^-\Big|_{l=0}$$
$$\tag{32a},(32b)$$

where

$$E^{-(j)} = -A_S\, i\,\frac{\Omega}{c_o}\, e^{(j)}\, \exp(i\Omega t) \qquad j = I,R,T \tag{33}$$

Subtracting (31b) from (31a) yields

$$0 = a_L^{(F)}e^{-i\phi_L^{(F)} - iKL}\frac{c_o}{c}(\frac{c_o}{c} - 1 + \frac{c_o}{\Omega}(\delta-i)\alpha_{L,NL}^{(F)})$$
$$- a_L^{(B)}e^{i\phi_L^{(B)} + iKL}\frac{c_o}{c}(\frac{c_o}{c} + 1 + \frac{c_o}{\Omega}(\delta-i)\alpha_{L,NL}^{(B)}) \tag{34}$$

Thus the transmitted light field $e^{(T)}$ is given by :

$$|e^{(T)}|^2 = a_L^{(F)^2}\left|1 + \frac{\frac{c_o}{c} - 1 + \frac{c_o}{\Omega}(\delta-i)\alpha_{L,NL}^{(F)}}{\frac{c_o}{c} + 1 + \frac{c_o}{\Omega}(\delta-i)\alpha_{L,NL}^{(B)}}\right|^2 \tag{35}$$

Eq.(34) yields two relations, one between $a_L^{(B)}$ and $a_L^{(F)}$, and one
between the phases $\phi_L^{(B)}$ and $\phi_L^{(F)}$. Since all the phases are
determined to within a constant (in the framework of the rotating
wave approximation) we put the total phase of the forward wave
at $l = L$ equal to zero :

$$\phi_L^{(F)} + KL = 0 \tag{36}$$

With the aid of eq.(36) we can determine the integration constant
$\text{const}^{(F)}$. Using this result and eq.(23), eq.(28) yields the
following result for $\phi_o^{(F)}$:

$$\phi_o^{(F)} = -\frac{\Omega}{c}L + \delta\ln(a_L^{(F)}/a_o^{(F)}) \tag{37}$$

For $\phi_o^{(B)}$ we obtain :

$$\phi_o^{(B)} = -\frac{\Omega}{c} L + \delta \ln(a_L^{(B)}/a_o^{(B)}) - 2\delta \alpha L$$

$$+ \, \mathrm{arctg}(\frac{c_o}{\Omega} \alpha_{L,NL}^{(B)})/(1 + \frac{c_o}{c} + \frac{c_o}{\Omega} \delta \alpha_{L,NL}^{(B)})) \tag{38}$$

$$- \, \mathrm{arctg}((\frac{c_o}{\Omega} \alpha_{L,NL}^{(F)})/(-1 + \frac{c_o}{c} \delta \alpha_{L,NL}^{(F)}))$$

Using eqs. (32a) and (32b) the incident light field $e^{(I)}$ is given by

$$2e^{(I)} = \exp(-i\phi_o^{(F)})a_o^{(F)}(\frac{c_o}{c} + 1 + \frac{c_o}{\Omega}(\delta - i)\alpha_{o,NL}^{(F)})$$

$$\tag{39}$$

$$- \exp(i\phi_o^{(B)}) \, a_o^{(B)} (\frac{c_o}{c} - 1 + \frac{c_o}{\Omega} (\delta - i)\alpha_{o,NL}^{(B)})$$

5. MIRROR AT THE RIGHT END FACE

We describe a mirror at the right end face by a dielectric plate characterized by a refractive index n_m and an absorption constant κ which are taken as intensity independent. For simplicity, we assume that the backward wave in the mirror can be neglected at the interface mirror-crystal. We calculate the transmissivity using the boundary conditions that the tangential components of H and E are steady.

6. DISCUSSION

As a simple example, certainly not realistic, but illustrative, we discuss the limiting case

$$n \gg 1, \ \alpha L \ll 1, \ n\alpha L \gg 1 \tag{40}$$

In this limit

$$a_o^{(F)} \approx a_L^{(F)} (1 + \alpha_{L,NL}^{(F)} L) \tag{41a}$$

$$a_o^{(B)} \approx a_L^{(B)} (1 - \alpha_{L,NL}^{(B)} L) \tag{41b}$$

$$a_L^{(F)} \approx |e^{(T)}| \, \frac{(n+1)}{2n}, \tag{42a}$$

$$a_L^{(B)} \approx |e^{(T)}| \, \frac{(n-1)}{2n} \tag{42b}$$

$$2\left|e^{(I)}\right| \approx \frac{1}{2n}\left|\,\left|e^{(T)}\right|\{(n+1)^2 - (n-1)^2 e^{i(\Phi_o^{(B)} + \Phi_o^{(F)})}\}\right.$$

$$+ \frac{\alpha L\left|e^{(T)}\right|}{1 + \frac{3}{4}\left|e^{(T)}\right|^2}\{(n+1)^2 + (n-1)^2 e^{i(\Phi_o^{(B)} + \Phi_o^{(F)})}\}\left.\right| \quad (43)$$

In order to get an optical bistability the non-linear part $(\sim\left|e^{(T)}\right|/(1 + \frac{3}{4}\left|e^{(T)}\right|^2))$ must be sufficiently large in comparison with the linear part $(\sim\left|e^{(T)}\right|)$, so that the function $\left|e^{(I)}\right| = \left|e^{(I)}\right|(\left|e^{(T)}\right|)$ has a maximum and a subsequent minimum. Eq.(43) shows the important influence of the phase $\Phi_o^{(F)} + \Phi_o^{(B)}$. For $\Phi_o^{(F)} + \Phi_o^{(B)} = 2m\pi$ (m integer), the nonlinearity becomes most effective. The total phase $\Phi_o^{(F)} + \Phi_o^{(B)}$ is given by

$$\Phi_o^{(F)} + \Phi_o^{(B)} = -\frac{2\Omega}{c}L - 2\delta\alpha L - \delta\alpha L \frac{a_L^{(F)^2} - a_L^{(B)^2}}{(1 + a_L^{(F)^2} + a_L^{(B)^2})^2 - a_L^{(F)^2} a_L^{(B)^2}}$$

$$(44)$$

For exact resonance ($\delta = 0$, $\Omega = \nu$) the optimal situation is achieved, if $(\nu L/c) = m\pi$. Otherwise the phase may be optimized by detuning. However, with increasing detuning the absorption coefficient decreases

$$\alpha = \frac{\alpha_{res.}}{1 + \delta^2} \quad (45)$$

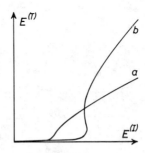

<u>Fig.3</u> : When the refractive index is low, a mirror is necessary
to obtain a bistability. For a given crystal length L the
phase may be optimized by detuning.
a) $\delta = 0$, b) $\delta = 0.04$ (with mirror).

Thus it depends on the parameters $(\alpha_{res.}, L, \gamma/\nu)$, whether one can switch on the bistability by detuning or not.

For strong detuning and large values of L a multi-stability can occur due to the intensity dependence of the phase $\Phi_o^{(F)} + \Phi_o^{(B)}$.

Acknowledgement - The author wish to thank Dipl. Phys. H. Ohno for fruitful discussions and his aid with the numerical calculations. One of the authors (JG) wants to thank the Deutsche Forschungsgemeinschaft for financial support.

REFERENCES

1. S.L. McCall, Physical Review A 9, 1515 (1974).

2. R. Bonifacio, L.A. Lugiato, Optics Communications 19, 172 (1976).

3. R. Bonifacio, L.A. Lugiato, Lettere al nuovo cimento 21, 505 (1978).

4. T.N.C. Venkatesan, S.L. McCall, Applied Physics Letters 30, 282 (1977).

5. H.M. Gibbs, S.L. McCall, T.N.C. Venkatesan, Physical Review Letters 36, 1135 (1976).

6. A. Szöke, V. Daneu, J. Goldhar, N.A. Kurnit, Applied Physics Letters 15, 376 (1969).

7. J. Goll, H. Haken, Physical Review A 18, 2241 (1978).

AUTHOR INDEX

Aikala, O., 319
Alper, T., 43
Alsem, W.H.M., 113
Amzallag, E., 59
Appel, J., 379
Ashkenazi, J., 255
Ausloos, M., 347

Babcenco, A., 15
Baldereschi, A., 281
Balkanski, M., 59, 205
Balzer, R., 89
Barfuss, H., 189
Batifol, E., 225
Bénoit à la Guillaume, C., 243
Berggren, K.-F., 183
Bianconi, A., 271
Bijvank, E.W., 95
Biskupski, G., 29
Bliek, L., 333
Bohnlein, G., 189
Bonnet, A., 37
Bonneville, R., 217
Bourgoin, J.C., 83

Calais, J.L., 305
Campagna, M., 271
Carabatos, C., 295
Celik, H., 43
Chang, H., 205
Chemla, D.S., 225
Conan, A., 37
Copland, G.M., 265, 313
Costa-Quintana, J., 275
Coussement, R., 165
Cox, R.T., 355
Czaja, W., 121

Daunois, A., 397
De Bruyn, J., 165
De Hosson, J.Th.M., 113
Deiss, J.L., 343, 397
Demangeat, C., 151
Den Hartog, H.W., 75, 95, 105
De Potter, M., 165
Devreese, J.T., 373
Dubois, H., 29

Evrard, R., 15

Fischer, K., 271
Fishman, G., 67, 217
Freunek, P., 189

Gerardy, J.M., 347
Gerlach, B., 365
Giersberg, E.J., 89
Gogolin, O., 343
Goll, J., 405
Goltzene, A., 127
Graf, C.J.F., 313

Hackeloer, H.J., 113
Haken, H., 405
Hansen, O.P., 1, 7
Hautojarvi, P., 143
Heremans, J., 1
Héritier, M., 235
Hess, F., 75
Hirlimann, C., 205
Hofmann, R., 189
Hohenstein, H., 189
Hugel, J., 295

Isomaki, H., 287

417

SUBJECT INDEX

α-Sn, 21
AgBr, 121
Ag, 135
Al_2O_3, 355
Al, 135
Alkaline earth fluoride crys-
     tals, 95
Aluminium, 323
Amorphous diamond, 165
Antimony, 43
As, 189

Biexcitons, 225
Bismuth, 1, 7
Brillouin scattering, 67

Carbon-doped iron, 143
CdS, 225
Compton profile, 305
Correlation energy, 275
CsAu, 281
CuBr, 343, 397
CuCl, 225, 389
Cu, 135
Cyclotron masses, 43

DC conductivity, 37
Density of states, 287
Dielectric crystals, 405
Directed Compton profiles, 319
Dislocations, 113

Electron-donor interaction,
     183
Electron-electron interact-
     ion, 183

Electron-hole drops, 243
Electron-hole pairs, 249
Electron-hole plasma, 235
Electron-phonon interaction, 53,
     379
Electron density deformation
     maps, 255
Electronic energy levels, 151
Electron transport, 7
Electron velocity runaway, 53
Energy gap, 287
EPR, 105, 127
Exchange energy, 217
Exciton, 59, 205, 235
Excitonic polaritons, 397

Fluorescence, 339
Force constants, 157
4-f insulators, 275
Free energy, 365
Free excitons, 243

GaP, 15, 197, 205
$Gd^{3+}-M^+$ centers, 95
Ground state energy, 373

Hall effect, 1, 197
Hydrogenic exciton, 231

Impurity, 75
Impurity vibrational amplitudes,
     157
Infrared absorption, 197, 347
Intrinsic birefringence, 343
Ion implanted diamond, 165
Ion implanted graphite, 165